Bioremediation of Recalcitrant Organics

BIOREMEDIATION

The *Bioremediation* series contains collections of articles derived from many of the presentations made at the First, Second, and Third International In Situ and On-Site Bioreclamation Symposia, which were held in 1991, 1993, and 1995 in San Diego, California.

First International In Situ and On-Site Bioreclamation Symposium

1(1)   *On-Site Bioreclamation: Processes for Xenobiotic and Hydrocarbon Treatment*

1(2)   *In Situ Bioreclamation: Applications and Investigations for Hydrocarbon and Contaminated Site Remediation*

Second International In Situ and On-Site Bioreclamation Symposium

2(1)   *Bioremediation of Chlorinated and Polycyclic Aromatic Hydrocarbon Compounds*

2(2)   *Hydrocarbon Bioremediation*

2(3)   *Applied Biotechnology for Site Remediation*

2(4)   *Emerging Technology for Bioremediation of Metals*

2(5)   *Air Sparging for Site Bioremediation*

Third International In Situ and On-Site Bioreclamation Symposium

3(1)   *Intrinsic Bioremediation*

3(2)   *In Situ Aeration: Air Sparging, Bioventing, and Related Remediation Processes*

3(3)   *Bioaugmentation for Site Remediation*

3(4)   *Bioremediation of Chlorinated Solvents*

3(5)   *Monitoring and Verification of Bioremediation*

3(6)   *Applied Bioremediation of Petroleum Hydrocarbons*

3(7)   *Bioremediation of Recalcitrant Organics*

3(8)   *Microbial Processes for Bioremediation*

3(9)   *Biological Unit Processes for Hazardous Waste Treatment*

3(10) *Bioremediation of Inorganics*

Bioremediation Series Cumulative Indices: 1991-1995

For information about ordering books in the Bioremediation series, contact Battelle Press. Telephone: 800-451-3543 or 614-424-6393. Fax: 614-424-3819. Internet: sheldric@battelle.org.

Bioremediation of Recalcitrant Organics

Edited by

Robert E. Hinchee and Daniel B. Anderson
Battelle Memorial Institute

Ronald E. Hoeppel
U.S. Naval Facilities Engineering Services Center

BATTELLE PRESS
Columbus • Richland

Library of Congress Cataloging-in-Publication Data

Hinchee, Robert E.
    Bioremediation of recalcitrant organics / edited by Robert E. Hinchee,
    Daniel B. Anderson, Ronald E. Hoeppel.
        p.   cm.
    Includes bibliographical references and index.
    ISBN 1-57477-008-X (hc : acid-free paper)
    1. Organic compounds—Environmental aspects—Congresses. 2.
    Bioremediation—Congresses. I. Hinchee, Robert E. II. Anderson,
    Daniel B. III. Hoeppel, Ronald E.
    TD196.073B56 1995
    628.5′2—dc20   95-32254
                                                                          CIP

Printed in the United States of America

Additional copies may be ordered through:
Battelle Press
505 King Avenue
Columbus, Ohio 43201, USA
1-614-424-6393 or 1-800-451-3543
Fax: 1-614-424-3819
Internet: sheldric@battelle.org

CONTENTS

FOREWORD

This book and its companion volumes (see overleaf) comprise a collection of papers derived from the Third International In Situ and On-Site Bioreclamation Symposium, held in San Diego, California, in April 1995. The 375 papers that appear in these volumes are those that were accepted after peer review. The editors believe that this collection is the most comprehensive and up-to-date work available in the field of bioremediation.

Significant advances have been made in bioremediation since the First and Second Symposia were held in 1991 and 1993. Bioremediation as a whole remains a rapidly advancing field, and new technologies continue to emerge. As the industry matures, the emphasis for some technologies shifts to application and refinement of proven methods, whereas the emphasis for emerging technologies moves from the laboratory to the field. For example, many technologies that can be applied to sites contaminated with petroleum hydrocarbons are now commercially available and have been applied to thousands of sites. In contrast, there are as yet no commercial technologies commonly used to remediate most recalcitrant compounds. The articles in these volumes report on field and laboratory research conducted both to develop promising new technologies and to improve existing technologies for remediation of a wide spectrum of compounds.

The editors would like to recognize the substantial contribution of the peer reviewers who read and provided written comments to the authors of the draft articles that were considered for this volume. Thoughtful, insightful review is crucial for the production of a high-quality technical publication. The peer reviewers for this volume were:

D. A. Abramowicz, *GE Corporate R&D Center*
Peter Adriaens, *University of Michigan*
Robert Ahlert, *Rutgers University*
Todd A. Anderson, *Iowa State University*
Jean-Pierre Arcangeli, *Technical University of Denmark*
Mick Arthur, *Battelle Columbus*
Robin L. Autenrieth, *Texas A&M University*
Bum-Han Bae, *Texas A&M University*
Daniel Ballerini, *Institut Français du Pétrole*
Pamela E. Bell, *Environmental Protection Systems*
Marlene L. Bennett, *BHP Research* (Australia)
Poul L. Bjerg, *Technical University of Denmark*
David Boone, *Oregon Graduate Center*
Gene Bowlen, *Accutech Remedial Systems*
G. Briseld, *University of Oslo*
Barbara J. Butler, *University of Waterloo* (Canada)
Peter R. Cali, *U.S. Army Corps of Engineers*
Douglas C. Cameron, *University of Wisconsin*

B.A. Campbell, *Roy F. Weston, Inc.*
Lisa M. Carmichael, *University of North Carolina, Chapel Hill*
John H. Carson, *OHM Remediation Services Corporation*
Daniel P. Cassidy, *University of Notre Dame*
Frank J. Castaldi, *Radian Corporation*
Keh-Ping Chao, *Polytechnic University*
Thomas G. Chasteen, *Sam Houston State University*
Jian-Shin Chen, *University of Minnesota*
Dennis Chilcote, *BioTrol, Inc.*
David Cosgriff, *Champion International Corporation*
Scott D. Cunningham, *DuPont Co.*
Durell Dobbins, *BioTrol, Inc.*
I.J. Dortch, *Shell Development Co.*
Bobby F. Dowden, *The Natural Solution, Inc.*
Eric Drescher, *Battelle Columbus*
Dick van Elsas, *IPO-DLO* (The Netherlands)
Mark Emptage, *DuPont Co.*
Françoise Fayolle, *Institut Français du Pétrole*
Eric A. Foote, *Battelle Columbus*
Herb Fredrickson, *U.S. Army Corps of Engineers*
David L. Freedman, *University of Illinois at Urbana-Champaign*
Arun R. Gavaskar, *Battelle Columbus*
Edwin Gelderich, *U.S. Environmental Protection Agency*
Domenic Grasso, *University of Connecticut*
A. Greene, *University of Canberra*
Douglas Gunnison, *U.S. Army Corps of Engineers*
Francis Hsu, *DuPont Co.*
Wendy Huang, *Battelle Columbus*
Michael H. Huesemann, *Battelle Pacific Northwest*
Peter J. Hutchinson, *The Hutchinson Group, Ltd.*
Danny Jackson, *Lockheed ESAT*
Donald Johnstone, *Washington State University*
A. M. Jones, *National Research Council of Canada*
Don Kampbell, *U.S. Environmental Protection Agency*
David Kaplan, *U.S. Army*
John Lawrence (Canada)
Richard Madura, *Tulane University*
B. Mahro, *Tech University of Hamburg-Harburg*
Minna Männistö, *Helsinki University of Technology*
Eric H. Marsman, *TAUW Milieu b.v.* (The Netherlands)
F.C. Michel, *Michigan State University*
Ralph Moon, *HSA Environmental, Inc.*
Dave Nagels, *Carnegie-Mellon University*
Joseph Odencrantz, *Environmental Systems & Technologies, Inc.*
John Patterson, *Continental Recovery Systems*
Judy Pennington, *U.S. Army Corp of Engineers*

Brent M. Peyton, *Battelle Pacific Northwest*
Frederic K. Pfaender, *University of North Carolina*
Susan M. Pfiffner, *University of Tennessee*
Jaakko Puhakka, *Tampere University of Technology* (Finland)
John F. Quensen, *Michigan State University*
Roger Reeves, *Massey University* (New Zealand)
Alan J. Sheehy, *University of Canberra*
Roald Sorheim, *SINTEF* (Norway)
Steve J. Vesper, *University of Cincinnati*
Timothy Vogel, *Rhône-Poulenc Industrialisation*
John Waid, *La Trobe University* (Australia)
Mary Watwood, *Idaho State University*
N. Lee Wolfe, *U.S. Environmental Protection Agency*
William Woods, *Sybron Chemicals, Inc.*
Robert Wyza, *Battelle Columbus*

The figure that appears on the cover of this volume was adapted from the article by Omori (see page 214).

Finally, I want to recognize the key members of the production staff, who put forth significant effort in assembling this book and its companion volumes. Carol Young, the Symposium Administrator, was responsible for the administrative effort necessary to produce the ten volumes. She was assisted by Gina Melaragno, who tracked draft manuscripts through the review process and generated much of the correspondence with the authors, co-editors, and peer reviewers. Lynn Copley-Graves oversaw text editing and directed the layout of the book, compilation of the keyword indices, and production of the camera-ready copy. She was assisted by technical editors Bea Weaver and Ann Elliot. Loretta Bahn was responsible for text processing and worked many long hours incorporating editors' revisions, laying out the camera-ready pages and figures, and maintaining the keyword list. She was assisted by Sherry Galford and Cleta Richey; additional support was provided by Susan Vianna and her staff at Fishergate, Inc. Darlene Whyte and Mike Steve proofread the final copy. Judy Ward, Gina Melaragno, Bonnie Snodgrass, and Carol Young carried out final production tasks. Karl Nehring, who served as Symposium Administrator in 1991 and 1993, provided valuable insight and advice.

The symposium was sponsored by Battelle Memorial Institute with support from many organizations. The following organizations cosponsored or otherwise supported the Third Symposium.

Ajou University–College of Engineering (Korea)
American Petroleum Institute
Asian Institute of Technology (Thailand)
Biotreatment News
Castalia
ENEA (Italy)
Environment Canada

Environmental Protection
Gas Research Institute
Groundwater Technology, Inc.
Institut Français du Pétrole
Mitsubishi Corporation
OHM Remediation Services Corporation
Parsons Engineering Science, Inc.
RIVM–National Institute of Public Health and the Environment
  (The Netherlands)
The Japan Research Institute, Limited
Umweltbundesamt (Germany)
U.S. Air Force Armstrong Laboratory–Environics Directorate
U.S. Air Force Center for Environmental Excellence
U.S. Department of Energy Office of Technology Development
  (OTD)
U.S. Environmental Protection Agency
U.S. Naval Facilities Engineering Services Center
Western Region Hazardous Substance Research Center–
  Stanford and Oregon State Universities

Neither Battelle nor the cosponsoring or supporting organizations reviewed this book, and their support for the Symposium should not be construed as an endorsement of the book's content. I conducted the final review and selection of all papers published in this volume, making use of the essential input provided by the peer reviewers and other editors. I take responsibility for any errors or omissions in the final publication.

Rob Hinchee
June 1995

Use of Alternative Growth Substrates to Enhance PAH Degradation

Piper C. Tittle, Yu-Ting Liu, Stuart E. Strand, and H. David Stensel

ABSTRACT

Freshwater and saltwater polycyclic aromatic hydrocarbons (PAH)-degrading enrichments were developed from seed from a manufactured gas plant site and contaminated marine sediment, respectively. Both enrichments were able to maintain specific degradation rates of 3- and 4-ring PAHs after growth with salicylate or phthalate, which increased their biomass concentrations by a factor of 9 to 10. Phthalate was a more effective alternative substrate than was salicylate. Specific degradation rates of phenanthrene and anthracene by the freshwater enrichment were increased after growth with phthalate. Growth with phthalate increased the specific degradation rates of phenanthrene and pyrene by the saltwater enrichment.

INTRODUCTION

PAHs are compounds that contain varying numbers of conjugated benzene rings. Most PAHs found in the environment are produced by the pyrolysis of organic material (Gibson and Subramanian 1984). PAH contamination of soils, groundwater, sludges, and marine sediments has resulted from many industrial operations, including manufactured gas plants, coal coking, wood treating, and petroleum-related operations. Some PAHs are known carcinogens and mutagens and are cited on the U.S. Environmental Protection Agency (U.S. EPA) priority pollutant list (Keith and Telliard 1979).

PAH degradation is achieved under aerobic conditions by many bacteria. Observed laboratory degradation rates in soils and sediments have been very low, with losses ranging from 6.0 to 7.7%, 3.0 to 12.0%, 0.4 to 6.4%, and 0.2 to 4.5% per day for anthracene, phenanthrene, fluoranthene, and pyrene, respectively (Bauer and Capone 1985; Heitkamp and Cerniglia 1988; Park et al. 1990). Treatment of a contaminated soil in a bioslurry reactor required 30 to 60 days to meet U.S. EPA PAH final limits for landfilling (U.S. EPA 1993). The low biodegradation rate of PAHs may be related to their low bioavailability because they have high solids-partitioning coefficients and low solubilities. For example, freshwater aqueous solubilities for 3- and 4-ring PAHs range from 1.29 mg/L for phenanthrene to 0.135 mg/L for pyrene. The low-soluble PAH concentrations may limit PAH degradation rates and thus the amount of PAH-degrading biomass produced.

The basic hypothesis of this research was that alternative, more readily available substrates could be used to grow PAH-degrading bacteria. Higher concentrations of these bacteria in soils, sludges, or sediments should increase PAH degradation. Compounds that are less likely to be consumed by competing non-PAH-degrading bacteria were selected as alternative growth substrates. Salicylate and phthalate were chosen as both are intermediates of known bacterial PAH oxidation pathways. Salicylate has been reported as an intermediate of naphthalene, phenanthrene, and anthracene degradation; and phthalate is a known intermediate of phenanthrene and pyrene degradation (Cerniglia 1992). Cardinal and Stenstrom (1991) showed that a mixed culture grown mainly with salicylate could enhance naphthalene degradation but not phenanthrene degradation. No rate data were reported, and the degradation of higher-ring PAHs was not studied. In this research, the degradation rates of 3- to 4-ring PAH compounds by freshwater and marine PAH-degrading enrichments were determined before and after growth on the alternative substrates.

EXPERIMENTAL PROCEDURES

Freshwater PAH-degrading enrichments were developed from a sludge sample from an oil recovery operation at a former manufactured gas plant site. Marine enrichments were seeded from contaminated sediments found in Puget Sound, Washington. These enrichments were maintained in aerobic stirred 2-L Erlenmeyer flasks and were fed a PAH mixture of phenanthrene, anthracene, fluoranthene, and pyrene. The PAHs were introduced by placing a nylon bag containing crystals of the respective PAHs in the reactor solution. A mean cell retention time (MCRT) of 60 days was maintained for the freshwater enrichments by weekly wasting of the reactor liquid and replacement with a nutrient medium. The saltwater enrichment was similarly maintained at a 30-day MCRT. The nutrient medium consisted of a phosphate buffer, trace minerals and nitrogen supplied as sodium nitrate (Tittle 1994). In addition, the saltwater medium had 24.1 g/L of sodium chloride and 10.0 g/L of sodium sulfate.

Short-Term Growth with Alternative Substrates

Samples of the PAH-degrading enrichment were placed in separate reactors and fed a solution of the alternative substrates and nutrient medium over a period of 7 to 10 days. For some growth conditions, a small amount of the PAH mixture was added. The final biomass VSS concentration due to growth on either salicylate or phthalate was 9 to 10 times greater than the initial biomass VSS concentration. The following short-term growth enrichments, indicated by their feed composition, were developed from the freshwater PAH enrichment: 100% salicylate, 96% salicylate and 4% PAH mixture; 100% phthalate, 96% phthalate and 4% PAH mixture; and low-feed PAH mixture. The percentages of the salicylate and phthalate plus PAH mixtures are based on their fraction of chemical oxygen demand (COD) in the feed. The low-feed PAH mixture was fed with the same amount of COD as that from PAHs in the alternative substrate and 4% PAH mixture. The following short-term growth enrichments were developed from the saltwater PAH enrichment: 100% salicylate, 99% salicylate and 1% PAH mixture and 100% phthalate, 99% phthalate and 1% PAH mixture. Prior to batch PAH degradation tests, all the short-term growth enrichments were rinsed twice with nutrient medium.

Batch PAH Degradation Tests

Degradation rates of phenanthrene, anthracene, and pyrene for the freshwater enrichments and phenanthrene, fluoranthene, and pyrene for the saltwater enrichments were determined as individual PAH compounds in batch tests. Multiple microcosms containing the PAH compound, and a 4-mL solution of nutrient medium and enrichment cultures, were prepared in 8-mL glass vials with Teflon™-lined screw caps. Test biomass concentrations ranged from 20 to 200 mg/L VSS. The PAH compound was added to the empty glass vial in pentane, which was evaporated to leave a crystallized PAH coating on the glass. The total PAH concentration was generally in the range of 5 mg/L, based on the 4-mL liquid volume, so the soluble PAH concentration should have been at maximum solubility. An abiotic control with formaldehyde (5% v/v) was prepared for each experiment. The vials were incubated in a shaker at 20°C, and duplicate vials were sacrificed for PAH analysis at each time point. Five to seven time points were used for each experiment. Three days were sufficient to follow phenanthrene degradation, and 18 to 20 days were needed to follow anthracene, fluoranthene, and pyrene degradation.

Analytical Methods

The PAH compounds were extracted into cyclohexane and analyzed with high-performance liquid chromatography (HPLC); 4 mL of cyclohexane was added to each vial, and the vials were shaken for 1 h at 250 rpm. The vials were centrifuged, cyclohexane was removed, and 2-methylnaphthalene was added as an internal standard prior to HPLC analysis. A linear water-acetonitrile solvent gradient was used with a Millipore-Waters HPLC and Supelcosil LC-PAH 25-cm × 4.6-mm column. An ultraviolet (UV) detector at a wavelength of 254 nm was used. External standards were prepared and run for each set of experiments.

Biomass was quantified by total suspended solids (TSS) and volatile suspended solids (VSS), according to Standard Methods #2540D and E (American Public Health Association 1985).

RESULTS

PAH concentrations generally decreased in a linear fashion in the batch vial tests until the total PAH concentrations were below the maximum solubility concentrations. The linear portion of the data was used with the initial VSS concentrations to calculate specific PAH degradation rates for each of the PAH compounds degraded by the PAH enrichments and alternative substrate enrichments.

Figure 1 summarizes the specific PAH degradation rates for the freshwater PAH enrichments grown only on the PAH mixture and after growth on the

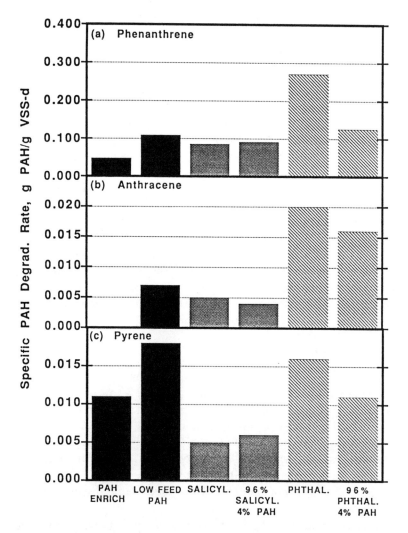

FIGURE 1. Specific PAH degradation rates by freshwater enrichment in batch tests following different growth conditions.

salicylate and phthalate; and salicylate and phthalate with a small amount of the PAH mixture. The phenanthrene-specific degradation rates are 1 order of magnitude greater than that for anthracene and pyrene. The anthracene was not degraded by the PAH enrichment, but was degraded after growth under the low-PAH feed condition. Pyrene or phenanthrene may have been present after growth of the enrichment under the low-PAH feed condition and may have supported cometabolic degradation of the anthracene. Degradation of the three PAHs tested was maintained after growth with either salicylate or phthalate. Growth with phthalate resulted in much higher specific degradation rates for phenanthrene and anthracene and the same degradation rate for pyrene as the PAH-fed enrichment. Salicylate-fed cultures had phenanthrene- and anthracene-specific degradation rates similar to those for the PAH-fed enrichments, but salicylate was not effective in maintaining the specific degradation rates for pyrene.

The specific degradation rates for the saltwater enrichments grown on the PAH mixture and alternative substrates are compared in Figure 2 for the degradation of phenanthrene, fluoranthene, and pyrene. Both salicylate and phthalate were shown to be effective alternative substrates, producing enrichments that maintained specific PAH degradation rates similar to the rates of the PAH-fed enrichment. Specific degradation rates for phenanthrene and pyrene were much higher for enrichments grown with phthalate.

Specific phenanthrene degradation rates were lower for the saltwater enrichments than for the freshwater enrichments by a factor of 2.5 or greater, but pyrene degradation rates were of the same magnitude. Salicylate was more effective in maintaining specific pyrene degradation rates for the saltwater enrichments than it was for the freshwater enrichments.

The addition of a small amount of the PAH mixture during growth with alternative substrates improved the PAH degradation rates for the salicylate-grown saltwater enrichments, but had little effect on the freshwater enrichments grown with salicylate. The addition of a small amount of the PAH mixture did not improve specific PAH degradation rates for either type of enrichment grown with phthalate.

SUMMARY

The results of this research showed that freshwater or saltwater PAH-degrading enrichments could be grown rapidly with salicylate or phthalate and still maintain PAH degradation abilities. In some cases the PAH degradation rate was increased. Phthalate was found to be a more effective alternative substrate than salicylate, based on observations of higher specific PAH degradation rates. These findings suggest an opportunity to increase bioremediation rates of contaminated soils, sludges, or groundwaters because the population of PAH-degrading bacteria can be quickly increased. PAH degradation rates can be increased provided that the PAH removal rate is not limited instead by desorption rates from the solid phase. Further research is needed to test this

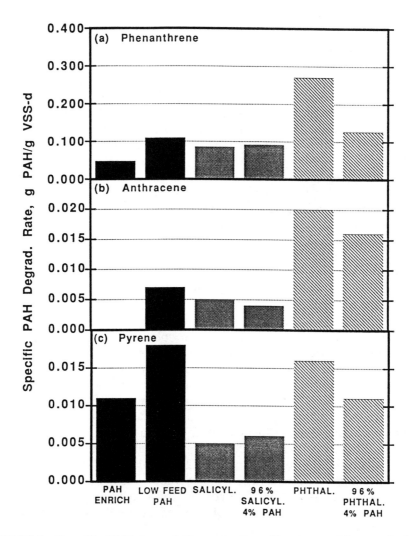

FIGURE 2. Specific PAH degradation rates by saltwater enrichment in batch
tests following different growth conditions.

potential in a solid matrix and compare potential biodegradation rates to
desorption rates.

REFERENCES

American Public Health Association. 1985. *Standard Methods for Examination of Water and
Wastewater, 17th ed.*, Washington DC.
Bauer, J. E., and D. G. Capone. 1985. "Degradation and Mineralization of the Polycyclic
Aromatic Hydrocarbons Anthracene and Naphthalene in Intertidal Marine Sediments."
Applied and Environmental Microbiology. 50 (1): 81-90.

Cardinal, L. J., and M. K. Stenstrom. 1991. "Enhanced Biodegradation of Polyaromatic Hydrocarbons in the Activated Sludge Process." *Research Journal Water Pollution Control Federation.* 63(7): 950-957.

Cerniglia, C. E. 1992. "Biodegradation of Polycyclic Aromatic Hydrocarbons." *Biodegradation.* 3: 351-368.

Gibson, D. T., and V. Subramanian. 1984. "Microbial Degradation of Aromatic Hydrocarbons." In D. Gibson (Ed.), *Microbial Degradation of Organic Compounds,* pp. 181-250, New York, NY.

Heitkamp, F., and C. E. Cerniglia. 1988. "Mineralization of Polycyclic Aromatic Hydrocarbons" by a Bacterium Isolated from Sediment Below an Oil Field." *Applied and Environmental Microbiology.* 47(6): 1612-1614.

Keith, L. H., and W. A. Telliard. 1979. "Priority Pollutants—a Perspective View." *Environmental Science and Technology.* 13(4): 416-423.

Park, K. S., Sims, R. C., and R. R. Dupont. 1990. "Transformation of PAHs in Soil Systems." *Journal of Environmental Engineering* 116(3): 632-640.

Tittle, P. C. 1994. "Degradation of Polyaromatic Hydrocarbon Compounds after Growth on Alternate Substrates." M. S. Thesis. University of Washington. Seattle, WA.

U. S. Environmental Protection Agency. 1993. *Pilot-Scale Demonstration of a Slurry-Phase Biological Reactor for Creosote-Contaminated Soil.* EPA/540/A5-91/009. Office of Research and Development, Washington DC.

Polycyclic Aromatic Hydrocarbon Biodegradation by a Mixed Bacterial Culture

Günter Dreyer, Joachim König, and Manfred Ringpfeil

ABSTRACT

Biodegradation of polycyclic aromatic hydrocarbons (PAHs) (according to U.S. Environmental Protection Agency standards), which are a complex mixture of organic compounds, was demonstrated using a bacterial mixed culture selected from a contaminated site by the BIO-PRACT GmbH. The investigations were carried out in a laboratory fermenter using emulsified tar oil as the substrate to determine the following: (1) concentration of the single PAH and of the sum of PAHs relative to fermentation time, (2) carbon dioxide (CO_2) and oxygen (O_2) content in the outflowing air during fermentation, (3) chemical oxygen demand (COD) of the broth, and (4) toxicity of the broth before and after fermentation according to the bioluminescence test (DIN 38412, part 34/1). The results of this model experiment indicated that the investigated mixed culture is able to effectively metabolize the PAHs contained in tar oil, including the higher condensed compounds such as benzo(a)pyrene. In the first 8 days of fermentation, the PAH sum decreased to below 5% of the starting concentration connected with a five-fold reduction of the toxic effect on *Vibrio fischeri*. The PAH degradation rate correlated with the rate of COD decrease, the rate of evolving CO_2, and the consumption of O_2.

INTRODUCTION

Bioremediation can be performed using externally bred microbial cultures that are adapted to the degradation of specific organic pollutants. It is widely accepted that microorganisms can degrade lower PAHs with 2 to 4 rings (Cerniglia et al. 1989, Weißenfels et al. 1991, Bryniok 1994). Higher condensed five- and six-ring systems are assessed to be more resistant to microbial attack. Little is known about the biodegradation of complex mixtures of PAHs (Cerniglia 1992), especially in the presence of additional organic compounds.

Biodegradation of the 16 priority U.S. Environmental Protection Agency PAHs from actual pollutants, particularly from tar oil, was investigated using a bacterial mixed culture isolated from soil samples from an industrial site. In cooperation with G. Maue and W. Dott from Technische Universität Berlin, Germany, the cultivable microorganisms were identified within the mixed culture and their relationship to each other was determined during fermentation (Maue and Dott 1995).

EXPERIMENTAL PROCEDURES AND MATERIALS

The experimental arrangement consisted of a computer-controlled, laboratory-scale bioreactor (3 L; 30°C; 1,200 rpm; Biostat B, Braun Biotech Internat., Melsungen, Germany) connected to an outflowing air analyzer registering CO_2 and O_2 (70 L/h; EGAS-2, Braun Biotech Internat., Melsungen, Germany). To ensure a constantly high grade of emulsion of the tar oil, the content of the bioreactor was pumped by a peristaltic pump with two heads (240 mL/h; Masterflex, Cole Parmer Instrument Co., Chicago, USA), through an ultrasonic generating device (amplitude: 45 μm, 7 s pulsed; U.S. \$30, Emich Ultraschall GmbH, Berlin, Germany). The setup is shown in Fig. 1.

The mixed culture was precultivated in a 300-mL shaking flask. After 3 days, the titer of the inoculum was about 10^7 colony-forming units (CFUs), per mL.

Fermentation was carried out in the first 24 h without ultrasonic treatment. Samples for the high-performance liquid chromatographic (HPLC) analysis

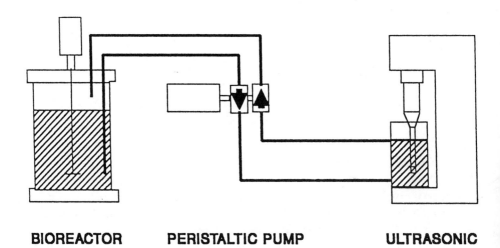

BIOREACTOR   **PERISTALTIC PUMP**   **ULTRASONIC**

FIGURE 1. Schematic setup of equipment for the investigations of PAH biodegradation.

and for the COD determination were taken at the start of the fermentation and then after 1, 2, 4, 6, and 8 days. Analysis of the samples was carried out in a double set. CO_2 and O_2 concentrations were continuously recorded.

Growth Conditions

Mineral salts medium (per liter deionized water): 2.8 g ammonium chloride, 1.36 g potassium dihydrogen phosphate, 0.32 g magnesium sulfate 7-hydrate, 0.04 g copper(II) sulfate 5-hydrate, 0.036 g manganese(II) sulfate 1-hydrate, 0.018 g zinc chloride, 0.005 g cobalt(II) sulfate 7-hydrate, 0.0005 g ferric chloride; adjusted to pH 6.5 by the addition of dipotassium hydrogen phosphate. The mineral salts medium was supplemented with 0.25 wt. % tar oil (TEWE GmbH, Erkner, Germany) as a carbon source and 0.025 wt. % emulsifying agent (AUF GmbH, Berlin, Germany).

Chemical Analyses

The PAHs were quantitatively extracted from the total culture fluids using solid-phase extraction columns (Baker separation column Octadecyl 7020-07; Baker Chemikalien, Groß-Gerau, Germany; Kicinski et al. 1989) and were analyzed by HPLC (Shimadzu, Japan) using a Bakerbond PAH 16-plus column (Baker Chemikalien, Groß-Gerau, Germany) at the following parameters: elution: binary gradient (acetonitrile/water); flowrate: 0.6 mL/min; temperature: 35°C; detection: (1) diode array detector (SPD-M10A; Shimadzu, Japan), (2) programmable fluorometer (RF-551; Shimadzu, Japan); standard: SRM 1647 b (Promochem GmbH, Wesel, Germany). The COD value was determined according to DIN 38409, part 43.

RESULTS AND DISCUSSION

Corresponding to the maximum of CO_2 production after a period of approximately 30 h (Fig. 2), the greatest rate of PAH degradation was observed (Fig. 3). Simultaneously, the COD value of the fermenter samples decreased from an initial value of 4,781 mg/L to the end value of 982 mg/L, including biomass (Fig. 4). Disregarding possible stripping effects and chemical transformations, these observations indicate that a significant part of the PAHs was mineralized during fermentation.

The HPLC records demonstrate the degradation of the main components of the tar-oil PAHs (Fig. 5) using ultraviolet detection, as well as the degradation of higher condensed PAHs (Fig. 6) using fluorescence detection. Quantitative values are given in Table 1. It seems obvious that the investigated mixed culture is able to metabolize the PAHs present in the tar-oil sample, including the higher condensed compounds such as benzo(a)pyrene.

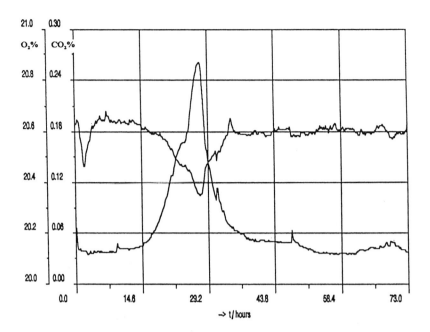

FIGURE 2. Carbon dioxide (CO_2) and oxygen (O_2) content in the outflowing air during the fermentation.

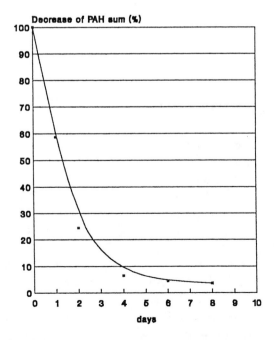

FIGURE 3. PAH sum in relation to fermentation time.

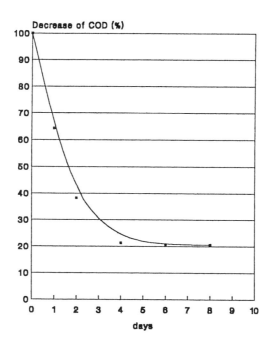

FIGURE 4. Chemical oxygen demand of samples in relation to fermentation time.

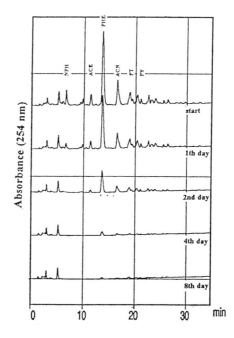

FIGURE 5. HPLC records of the main components of tar-oil PAHs.

TABLE 1. Decrease of PAH compounds during the fermentation.

PAH (mg/L)	start	1st day	2nd day	4th day	6th day	8th day
Naphthalene	161.5	174.9	65.0	61.2	—	—	—	—	—
Acenaphthylene	—	—	—	—	—	—	—	—	—
Acenaphthene	31.6	61.5	43.5	42.3	22.6	24.7	1.8	1.9	1.0	—	0.4	0.4
Fluorene	35.3	37.8	24.8	23.8	10.1	11.8	2.3	2.4	1.5	1.4	0.8	—
Phenanthrene	98.8	105.9	71.4	68.0	30.1	36.5	8.6	8.5	6.6	6.2	3.6	1.3
Anthracene	17.2	18.3	11.5	10.9	5.8	6.8	3.1	3.1	1.0	0.9	0.4	0.2
Fluoranthene	49.9	56.4	39.1	37.9	19.5	21.8	9.7	8.6	7.7	5.1	9.2	6.5
Pyrene	49.1	52.2	38.0	36.1	20.2	22.3	4.5	4.2	3.5	3.4	4.3	2.5
Benzo[a]anthracene	7.8	8.5	5.5	5.1	2.9	3.2	1.2	1.2	1.0	1.0	1.2	0.9
Chrysene	12.4	13.6	7.3	7.7	4.4	4.7	1.8	1.7	1.5	1.6	1.6	1.3
Benzo[b]fluoranthene	2.1	2.3	1.7	1.6	1.0	1.1	0.6	0.5	0.5	0.5	0.6	0.5
Benzo[k]fluoranthene	1.1	1.2	0.7	0.7	0.2	0.3	—	—	—	—	—	—
Benzo[a]pyrene	2.1	2.2	1.6	1.6	1.0	1.1	0.6	0.6	0.5	0.5	0.6	0.5
Dibenzo[a,h]anthracene	0.5	0.5	—	—	—	—	—	—	—	—	—	—
Benzo[g,h,i]perylene	—	—	—	—	—	—	—	—	—	—	—	—
Indeno[1,2,3,-cd]pyrene	—	—	—	—	—	—	—	—	—	—	—	—
PAH sum	468.9	534.8	310.1	296.9	117.8	134.3	34.2	32.7	24.8	21.5	22.7	14.1

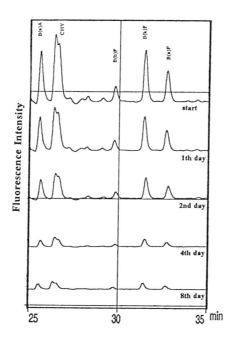

FIGURE 6. HPLC records of the higher condensed PAHs.

This conclusion is supported by the result of the toxicity studies. The toxic effect on *Vibrio fischeri*, expressed in GL_{20}-values of the fermenter samples, was decreased from 256 at fermentation start to 48 at the end of fermentation.

REFERENCES

Bryniok, D. 1994. "PAK-Abbau in Mehrphasensystemen." In B. Weigert (Ed.), *Biologischer Abbau von polycyclischen aromatischen Kohlenwasserstoffen*, pp. 91-107. Kolloquium, TU Berlin/Germany, 18-19 November.

Cerniglia, C. E., and M. A. Heitkamp. 1989. "Microbial Degradation of Polycyclic Aromatic Hydrocarbons (PAH) in the Aquatic Environment." In U. Varanasi (Ed.), *Metabolism of Polycyclic Aromatic Hydrocarbons in the Aquatic Environment*, pp. 41-68. CRC Press, Boca Raton, FL.

Cerniglia, C. E. 1992. "Degradation of PAH in Soil—State of the Art of U.S. Research Work." In DECHEMA, *Soil Decontamination Using Biological Processes*, pp. 10-11. Preprints International Symposium, Karlsruhe/Germany, 6-9 December.

Kicinski, H. G., S. Adamek, and A. Kettrup. 1989. "Festphasenextraktion und HPLC-Bestimmung von PAHs aus Trinkwasser-, Boden- und Altölproben." *Chemie für Labor und Betriebe* 40: 537-541.

Maue, G., and W. Dott. 1995. "Activity Tests with PAH-Metabolizing Soil Bacteria for In Situ Bioremediation." In R. E. Hinchee, G. S. Douglas, and S. K. Ong (Eds.), *Monitoring and Verification of Bioremediation*, pp. 127-133. Battelle Press, Columbus, OH.

Weißenfels, W. D., M. Beyer, J. Klein, and H. J. Rehm. 1991. "Microbial Metabolism of Fluoranthene: Isolation and Identification of Ring Fission Products." *Appl. Microbiol. Biotechnol.* 34: 528-535.

Biological PAH Degradation in Dredged Sludges

Michel G. A. Huis in 't Veld, Joan Werners,
Johan J. van Veen, and Hans J. Doddema

ABSTRACT

Sediment-associated PAHs are biodegraded by fungi under low pH conditions. The 16 polycyclic aromatic hydrocarbons (PAHs) targeted by the U.S. Environmental Protection Agency (U.S. EPA) are converted by more than 95% within a period of 20 weeks. The process appears to be a metabolic conversion, because PAH removal occurred in absence of additional growth substrates.

INTRODUCTION

In the Netherlands 10 to 20 million tons of dredging sediment are contaminated with heavy metals and PAHs. The combination of PAHs and heavy metals makes the cleanup of the (mostly freshwater) sediment difficult. A simple and cost-effective process is not yet available, and as a consequence dumping in storage sites is the current practice. However, storage capacity is limited, and a cost-effective sediment cleanup technology has to be developed. A promising technique may be biotechnological heavy metal removal, combined with PAH biodegradation. On a laboratory scale, it has been demonstrated that heavy metals can be leached to below legal requirements by creating a pH of 1 to 3 by adding biologically produced sulfuric acid (Holmes 1991, Joziasse et al. 1990). This study investigates PAH biodegradation under these bioleaching conditions.

At neutral pH, PAHs can be degraded by bacteria and yeasts, as well as by fungi (Cerniglia 1981, Leahy & Colwell 1990). Bacteria probably do not play a significant role in PAH bioconversion under bioleaching conditions; they cannot survive a pH as low as pH 3. Fungi are better equipped; their pH optimum is in the lower range. Moreover, these organisms have several enzymatic systems and extracellular ligninases that mediate PAH degradation (Sanglard et al. 1986).

However, the applicability of fungal PAH bioconversion under bioleaching conditions still is unclear for the following reasons:

1. So far, all research has been done at pH higher than 5. Under neutral pH conditions, biological PAH degradation with many organisms under varying conditions has been described.
2. Most of the PAH-degrading fungi only partially convert the PAH molecules via cometabolic mechanisms. Addition of growth substrates (e.g., glucose) and special conditions (i.e., nitrogen limitations) are required, which may make treatment very expensive.
3. Besides microbial limitations, physicochemical limitations also exist, such as low bioavailability. These are known to restrict the rate and extent of the PAH bioconversion (Afferden et al. 1992).

The objective of this study is to investigate the possibility of degrading PAHs with fungi under low pH conditions. For this purpose, fungi that use PAHs as a sole carbon and energy source under low pH conditions were isolated. Moreover, the capabilities of these fungi to convert sediment-associated PAHs, under bioleaching conditions, were tested with batch experiments.

MATERIALS AND METHODS

All fungi used were isolated from soil samples by enrichment in a mineral medium, with hexadecane as the sole carbon and energy source. Subsequently, the isolates were tested for growth on a mixture of PAHs (anthracene, pyrene, benzo[a]pyrene, fluoranthene, and perylene) as the only carbon and energy source. These tests were performed in the dark at a temperature of 30°C, and in a liquid mineral medium, with a pH of 3, that contained 30 mM NH_4NO_3 and 10 mM KH_2PO_4 at pH 3.

The medium used in the sediment batch experiments was made with demineralized water plus the following mineral concentrations (μM): NH_4NO_3 30,000; KH_2PO_4, 10,000; $FeCl_3$, 1.6; H_3BO_3, 1.6; $CuSO_4*5H_2O$, 0.80; KI, 1.20; $MnCl_2*4H_2O$, 6.00; $Na_2MoO_4*2H_2O$, 1.60; $ZnSO_4*7H_2O$, 4.20; $CoCl_2$, 5.20; $AlK(SO_4)_2*12H_2O$, 0.80; $CaCl_2*2H_2O$, 1,800; and NaCl, 3,400. A pH of 3 was achieved by adding concentrated sulfuric acid.

Two sediments, obtained from locations in the Netherlands, were used: Malburger Haven (MH) and 1^e Rijks Binnenheven (RB). The sediments were air-dried, ground, sieved (over a Retsch SK1 strainer with a grit size of 1 mm), distributed (by a Retsch PT distributing apparatus) into samples of 6 g dry weight each, and put in 300-mL conical flasks. Medium was added so that the dry weight was 5% (wt/vol). Fungi were added to the slurry, and the pH was adjusted to pH 3. The flasks were placed on a shaker at 30°C. The pH of the slurry was measured every week and adjusted to pH 3 with sulfuric acid when necessary (the first 3 weeks).

Every 4 weeks samples were taken from duplicate flasks and filtered over 8-μm paper filters. The amounts of PAHs in the sediment and water phase were quantified, using the external standards of the 16 EPA PAHs and dibenzo[a,h]anthracene as an internal standard.

The PAHs were measured according to standard NEN-5771, using a Waters™ high-performance liquid chromatograph (HPLC). Measurement requires an rp Chrompack 100*4.6 mm Chromspher 3 PAH column (catalog no. 29324), with an acetonitrile/water mixture (80/20% v/v) as the elutant, applied at a flowrate of 1 mL/min. PAH detection was performed with an ultraviolet (UV)-Var detector (Chrompack UV-Var spectrophotometer), with the wavelength set at 254 nm, and a fluorescence spectrophotometer (Millipore), with 254 nm and 339 nm as the excitation and emission wavelengths, respectively. PAH recoveries were between 78 to 113%, except that from naphthalene (60 to 90%). This difference is due to the method of analysis; during the analysis the extraction solvent (acetone) is evaporated, and part of the volatile PAH (i.e., naphthalene) also evaporates.

RESULTS AND DISCUSSION

Figure 1 illustrates the percentage of PAH degraded in the sediments for 14 PAHs, for a period of 5 months. The values shown are averages of two

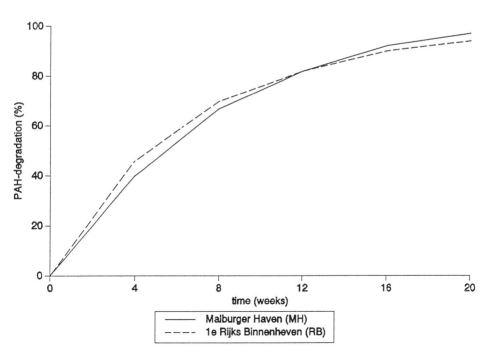

FIGURE 1. PAH degradation in two sediments.

duplicates. The degradation appears to follow an exponential curve and in the first month is very high—about 40% of the initial amount of PAHs are degraded. All the PAHs tested were degraded, including the largest, which have a molecular weight of 276 g/m [benzo(ghi)perylene and indeno(1,2,3-cd)pyrene, see Tables 1 and 2].

At the end of the experiment, the total amount of PAHs is still higher than the set limits for reuse of the soil. The reuse limits for soil are 40 μg PAH per g soil for RB and 30 μg/g for MH. The total amount of PAHs is 102 μg/g for MH and 209 μg/g for RB after 20 weeks of incubation. When the first-order degradation constant k remains constant, the set limits for reuse of the soil for PAHs should be reached in approximately 8 weeks for MH and 12 weeks for RB.

The degradation of the smaller PAHs is faster than for the higher-molecular-weight PAHs. The initial concentration is also higher. The initial degradation rates ($r_{initial}$) of 13 EPA PAHs are listed in Tables 1 and 2. The degradation data fit a first-order reaction model ($C_t = C_0*e^{-k*t}$), where C_t = PAH concentration at time t, C_0 = initial concentration, k = first-order reaction constant, and t is the time. The average degradation constant k is 0.020 day^{-1} for both MH and RB sediments.

CONCLUSIONS

The results show that PAHs are degradable by fungi under acidic conditions. At a pH of 3, 95% of the initial amount of PAHs in dredged sediments is

TABLE 1. PAH degradation in dredged sediment from Malburger Haven.

PAH	$C_{initial}$[a] (μg/g)	C_{end}[b] (μg/g)	$r_{initial}$[c] (μg/g.d)	k[d] (d^{-1})
naphthalene	602	41	7.4	0.018
fluorene	310	27	4.1	0.017
phenanthrene	162	2	1.6	0.019
anthracene	306	5	5.7	0.027
fluoranthene	227	11	2.0	0.016
pyrene	249	10	2.8	0.019
benzo(a)anthracene	117	3	1.0	0.021
chrysene	131	ND[e]	2.9	0.027
benzo(b)fluoranthene	58	ND	1.0	0.022
benzo(k)fluoranthene	27	ND	0.5	0.030
benzo(a)pyrene	39	ND	0.3	0.018
benzo(ghi)perylene	81	2	1.2	0.028
indeno(1,2,3-cd)pyrene	44	ND	1.2	0.035

(a) $C_{initial}$: initial PAH concentration.
(b) C_{end}: concentration after 5 months of incubation.
(c) $r_{initial}$: initial degradation rate.
(d) k: first-order degradation constant.
(e) ND: not detectable.

TABLE 2. PAH degradation in dredged sediment from 1^e Rijks Binnenheven.

PAH	$C_{initial}$[a] (μg/g)	C_{end}[b] (μg/g)	$r_{initial}$[c] (μg/g.d)	k[d] (d^{-1})
naphthalene	497	24	8.7	0.024
fluorene	321	12	6.1	0.024
phenanthrene	318	16	5.1	0.021
anthracene	778	38	13.0	0.023
fluoranthene	512	26	10.6	0.024
pyrene	237	24	3.6	0.017
benzo(a)anthracene	180	15	1.6	0.015
chrýsene	67	5	1.4	0.021
benzo(b)fluoranthene	104	7	1.3	0.018
benzo(k)fluoranthene	101	10	1.2	0.015
benzo(a)pyrene	78	6	1.1	0.018
benzo(ghi)perylene	102	13	0.9	0.014
indeno(1,2,3-cd)pyrene	142	13	2.0	0.017

(a) $C_{initial}$: initial PAH concentration.
(b) C_{end}: concentration after 5 months of incubation.
(c) $r_{initial}$: initial degradation rate.
(d) k: first-order degradation constant.

removed within 20 weeks. The average degradation rate over the first 4 weeks is 4.5 μg PAHs per g dry weight of sediment per day. The average first-order reaction constant (k) is 0.02 day^{-1}, for both sediments, during the 20 weeks of incubation.

Biological cleanup of PAH-contaminated sediments appears to be possible in a treatment time of less than 6 months. No organic substrates have to be added to the sediment, and thus there are no additional costs. The total amount of PAH in the sediment after 20 weeks of degradation is still too high for reusing the sediment, but the cleanup requirements for reusable soil probably can be reached within 8 to 12 weeks of additional incubation. Current work includes additional time for incubation.

These results have shown that, in principle, PAH-contaminated sediments can be cleaned up. Further research has to be done to develop this process for in situ application.

REFERENCES

Afferden, M. van, M. Beyer, and J. Klein. 1992. "Significance of bioavailability for the microbial remediation of PAH contaminated soils." *DECHEMA-Biotechnol. Conf.*; 5, Part B. pp. 1009-1012.
Cerniglia, C.E. 1981. "Aromatic Hydrocarbons: Metabolism by bacteria, fungi and algae." *Rev. Biochem. Toxicol.* 3:321-361.

Holmes, D.S. 1991. "Biorecovery of metals from mining, industrial and urban wastes." *Bioconversion of Waste Materials to Industrial Products.* Elsevier Applied Science, London/New York.

Joziasse, J., W.G. van Marwijk, and J.G.H. Brouwer. 1990. *Extractie van metalen uit waterbodems.* POSW, executed by TNO Apeldoorn, the Netherlands, fase I, 1989-1990. RIZA, Report no. 91071.

Leahy, J.G. and R.R. Colwell. 1990. "Microbial degradation of hydrocarbons in the environment." *Microb. Rev. 54* :305-315.

Sanglard, D., M.S.A. Leisola, and A. Fiechter. 1986. "Role of extracellular ligninases in biodegradation of benzo[a]pyrene by *Phanerochaete chrysosporium.*" *Enzym. Microb. Techn. 8* :209-212.

Biodegradation of Polycyclic Aromatic Hydrocarbons in Rhizosphere Soil

*A. Paul Schwab, M. Katherine Banks,
and M. Arunachalam*

ABSTRACT

Increased contaminant biodegradation in soil in the presence of plants has been demonstrated for several classes of organic compounds. Although enhanced dissipation of polycyclic aromatic hydrocarbons (PAHs) was observed previously in the rhizosphere of several plant species, the mechanism of this effect has not been assessed. A laboratory experiment was conducted to test the importance of cometabolism and the presence of common rhizosphere organic acids on the loss of PAHs (pyrene and phenanthrene) from soil. The role of cometabolism in the mineralization of pyrene was tested by observing the impact of adding phenanthrene to soil containing ^{14}C-pyrene and observing the effects on $^{14}CO_2$ generation. Adding phenanthrene apparently induced cometabolism of pyrene, particularly in the presence of organic acids. In a subsequent experiment, mineralization of pyrene to $^{14}CO_2$ was significantly greater in soil from the rhizospheres of warm-season grasses, sorghum (*Sorghum bicolor* L.) and bermuda grass (*Cynodon dactylon* L.), compared to soil from alfalfa (*Medicago sativa* L.), which did not differ from sterilized control soil. A highly branched, fine root system appears to be more effective in enhancing biodegradation than taproots, and the presence of organic acids increases rates of PAH mineralization.

INTRODUCTION

PAHs have long been the subject of environmental study because they are recalcitrant, persistent, and suspected to be toxic, mutagenic, and carcinogenic (Southworth 1977). PAH compounds are generally hydrophobic and nonvolatile, and soil organic matter acts as an adsorbent to immobilize them. Once incorporated into the soil, biodegradation seems to be the most significant means of PAH decomposition (Cerniglia et al. 1986).

Vegetation has been known to play an important role in the biodegradation of organic compounds in soil (Cunningham and Berti 1993). The beneficial effects of vegetation on the biodegradation of hazardous organics may be due to volatilization, plant uptake or accumulation, or accelerated biodegradation of the contaminant due to the rhizosphere microflora (Banks and Schwab 1993). The proliferation of plant roots serves as a means to stimulate growth of soil microorganisms and to increase their probability of contact with the contaminant (Aprill and Sims 1990). Fibrous root systems, being finer and more profuse, increase the total rhizoplane area and establish a more active microbial population. Other studies have employed different plant species to investigate the relative degradation rates for different contaminants in soil (Hsu and Bartha 1979; Walton and Anderson 1990). Legumes also have exhibited the capacity to accelerate the degradation of recalcitrant compounds (Hsu and Bartha 1979; Banks and Schwab 1993).

Biodegradation of phenanthrene by soil bacteria is well documented (Cerniglia et al. 1979). However, only a few studies have been conducted on the biodegradation of higher (4 or more rings) PAHs (Weisenfels et al. 1990). Cometabolism is the oxidation of nongrowth substrates during the growth of an organism on another carbon and energy source. McKenna and Heath (1976) studied the cometabolism of PAHs and found that certain species such as *Flavobacterium* sp. could degrade pyrene in the presence of phenanthrene as the growth substrate. Cometabolism of other multiringed PAHs also was demonstrated. These studies strongly suggest that higher-ringed PAHs that cannot act as a sole source of carbon to the microorganisms can be biodegraded when lower-ringed PAHs (3 rings or less) are present to serve as cosubstrates.

In a previous study (Schwab and Banks 1994), enhanced dissipation of PAHs was demonstrated in the presence of plants compared to the absence of plants, but the mechanism of enhancement was not clear. This study was conducted to help elucidate two possible reasons for increased PAH dissipation in the root zone, cometabolism and biodegradation enhancement by root exudation of low-molecular-weight organic acids. The objectives of this laboratory study were to (1) quantify the mineralization of pyrene in rhizosphere soils using phenanthrene as a cometabolic substrate, (2) evaluate the mineralization of phenanthrene in rhizosphere soils collected from the natural habitat of different legumes and grasses, and (3) identify the most effective plant species.

EXPERIMENTAL PROCEDURES AND MATERIALS

Cometabolism of Pyrene

Uniformly labeled ^{14}C-pyrene was mixed with unlabeled pyrene in acetone to make a solution of 185 MBq/L and 100 mg pyrene/L. Ten grams of soil (dry weight) were placed in microcosms and amended to result in 10 mg pyrene/kg and a total of 185 kBq ^{14}C. Evolved CO_2 was trapped in Carbosorb II (Packard

Instrument Company, Downer Grove, Illinois), and the ^{14}C was determined daily by liquid scintillation.

Four treatments were used in the experiment: (1) rhizosphere soil (from alfalfa plants) irrigated with an organic acid solution; (2) rhizosphere soil irrigated with distilled/deionized water; (3) nonrhizosphere soil irrigated with distilled/deionized water; and (4) autoclaved, nonrhizosphere soil irrigated with distilled/deionized water. The organic acid solution was designed to mimic the composition of major organic acids in the soil solution from an active rhizosphere (Jones 1991) and contained 15 μM succinic acid and 10 μM formic acid. For all treatments, irrigation was applied to maintain 80% moisture-holding capacity for maximum microbial activity; microcosms were weighed every second day and brought to proper moisture content by applying the appropriate treatment solution. For the sterile control, the soil was autoclaved twice to help ensure sterility; however, microbial regrowth over the 100-day trial period is possible. After 60 days, phenanthrene was added to achieve a concentration of 100 mg/kg. Evolved [14]CO_2 was monitored daily for another 40 days.

Dissipation of Phenanthrene in Rhizosphere Soil

In this microcosm study, the volatilization and mineralization of ^{14}C-phenanthrene was examined in soil from the rhizosphere of three different plant species: alfalfa (*Medicago sativa* L.), sorghum (*Sorghum bicolor* L.), and bermuda grass (*Cynodon dactylon* L.). The same soil series was used for each plant species. A sterile control was prepared by autoclaving the soil. To each microcosm, 10 g of soil was added, and each soil was amended with 185 MBq/kg and 100 mg phenanthrene/kg. All treatments were prepared in quadruplicate. Carbon dioxide ([14]CO_2) was trapped in Carbosorb™, and volatile ^{14}C-organic compounds were captured in Scintilene™; all solutions were analyzed daily by liquid scintillation. Total residual ^{14}C in the soil was determined by wet combustion, trapping CO_2, followed by liquid scintillation counting.

RESULTS AND DISCUSSION

Cometabolism of Pyrene

During the first 60 d of the experiment, the production of [14]CO_2 from ^{14}C-pyrene was highest in the rhizosphere soil irrigated with organic acids (Fig. 1). This treatment was used to simulate rhizosphere conditions with root organic exudation, and it is apparent that the amendment stimulated biodegradation of pyrene. The increased mineralization of pyrene apparently was the result of enhanced microbial populations supported by the organic acid carbon sources. The autoclaved soil irrigated only with water served as the control for this experiment. The daily [14]CO_2 measurements for the control microcosms were barely, but significantly, above background counts. At any given time, the cumulative [14]CO_2 from the control soils were 5 to 10 times smaller than the

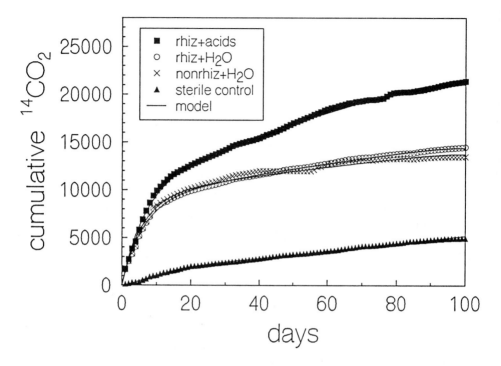

FIGURE 1. Mineralization of ^{14}C-pyrene in the absence of phenanthrene (days 1 to 60) or in its presence (days 61 to 100).

nonsterile soils. The slight amount of mineralization is unlikely the result of abiotic mineralization. Perhaps the control soil was not completely sterilized by autoclaving but contained a small population of microorganisms capable of mineralizing the pyrene at a slow rate.

Phenanthrene was added to all soils on day 61. Because there was no obvious change in the rates of mineralization, a model was used to assess possible cometabolism. The $^{14}CO_2$ evolution data were fitted to a deterministic three-half order model (Brunner and Focht 1984) to quantify the differences between the treatments:

$$CO_2 = S_0 (1 - e^{-k_1 t - k_2 t^2/2}) = k_0 t$$

where CO_2 is the cumulative evolved $^{14}CO_2$, t is time, S_0 is the mineralizable pool of ^{14}C-labeled pyrene, k_1 is the first-order constant for degradation per unit time, k_2 is a constant representing growth of the biomass, and k_0 is a zero-order rate constant for the mineralization of ^{14}C that has been incorporated into the soil humus. The parameters for the model for the organic treatments before and after phenanthrene addition are given in Table 1.

TABLE 1. Values of parameters for the three-half model for CO_2 accumulation for soils with various treatments.

Treatment	S_0		k_1		k_2		k_0	
	1-60[a]	1-100[b]	1-60	1-100	1-60	1-100	1-60	60-100
rhiz. + acids	9835	11498	0.117	0.141	0.014	−0.002	142	113
rhiz. + water	8817	9580	0.150	0.155	0.005	−0.003	67	54
nonrhiz + water	8275	9820	0.157	0.170	0.004	−0.003	87	44
sterile control	1170	1277	0.014	0.024	0.013	0.010	41	39
lsd (P<0.05)	535		0.015		0.002		14	

(a) Parameters fit to the model for $^{14}CO_2$ accumulation prior to phenanthrene addition (days 1 to 60).
(b) Parameters fit to the model for $^{14}CO_2$ accumulation for the entire experiment (days 1 to 100).

For the first 60 days, the more rapid mineralization in the presence of organic acids is reflected in the significantly higher parameters of the modeling equation, particularly for the total mineralizable ^{14}C (S_0) and the zero-order constant for the degradation of organic matter (k_0). When all data (including after phenanthrene was added) are fitted to the equation, the impact of the cometabolite is suggested. The rate constants (k_0, k_1, k_2) did not change significantly for any of the treatments, but the total mineralizable pyrene (S_0) increased significantly in all treatments, particularly the treatment with the organic acids in which S_0 increased 17%. Consequently, one could conclude that pyrene, which is not readily biodegradable, may be cometabolized in the presence of phenanthrene. However, the observed changes in S_0 simply may be the result of being able to fit more data to the model; extending the time period from 60 to 100 days enabled the fitting routine to more clearly define the modeling parameters.

Dissipation of Phenanthrene in Rhizosphere Soils

Previous greenhouse studies (Schwab and Banks 1994) demonstrated that pyrene and anthracene degrade readily in rhizosphere soils, with greater than 95% dissipation within 28 days. Although the parent compound had been altered, the ultimate fate of the phenanthrene was not determined. Therefore, a short-term dissipation study using ^{14}C-phenanthrene was initiated to account for mineralization, volatilization, and residual (soil bound) ^{14}C. The PAH was incubated in microcosms containing soil that was collected from the root zone of various plants.

A small percentage (0.11%) of the phenanthrene was mineralized after 14 d in sterile soil (Fig. 2). Mineralization in the alfalfa rhizosphere soil was not significantly different from the sterile control, but $^{14}CO_2$ generation in the bermuda grass and sorghum soils was 3 to 4 times greater than in the sterile control. Although the rhizosphere effect is statistically significant, mineralization cannot account for previously observed dissipation of phenanthrene. Similarly, volatilization was ≤0.11% in all treatments (Table 2) and could not account for the dissipation of phenanthrene.

In contrast to the small recoveries in the mineralized and volatilized fractions, the residual ^{14}C was ≥88% in all cases. Apparently, a systematic negative error kept our recoveries below 100%; nevertheless, it is obvious that the vast

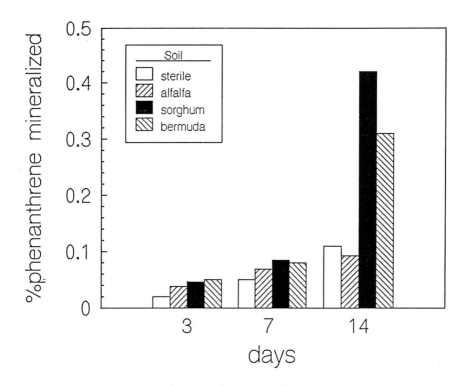

FIGURE 2. Mineralization of ^{14}C-phenanthrene in rhizosphere soil from several species of plants. Sterilized soil served as a control.

TABLE 2. Recovery of ^{14}C during the 14-day ^{14}C-phenanthrene study.

Soil	% ^{14}C recovered		
	Mineralization	Volatilization	Residual
sorghum	0.46	0.11	88
bermuda grass	0.31	0.06	91
alfalfa	0.09	0.09	89
sterile	0.11	0.04	92

majority of the biodegraded phenanthrene remained in the soil. The phenanthrene was most likely converted to microbial biomass or degradation products and incorporated into soil organic matter. Future investigations will attempt to identify and quantify these products.

REFERENCES

Aprill, W., and R.C. Sims. 1990. "Evaluation of the use of prairie grasses for stimulating polycyclic aromatic hydrocarbon treatment in soil." *Chemosphere 20*:253-265.

Banks, M.K., and A.P. Schwab. 1993. "The benefits of vegetation in the bioremediation of hydrocarbons: A review." *Air and Waste Management 93*(89.05):1-10.

Brunner, W., and D.D. Focht. 1984. "Deterministic three-half-order kinetic model for microbial degradation of added carbon substrates in soil." *Appl. Environ. Microbiol. 47*:167-172.

Cerniglia, C.E., R.L. Herbert, R.H. Dodge, P.J. Szaniszlo, and D.T. Gibson. 1979. "Some approaches to studies on the degradation of aromatic hydrocarbons by fungi." In A.L. Bourquin and H. Pritshard (Eds.), *Microbial Degradation of Pollutants in Marine Environments*. EPA-600/9-79-012, pp. 360-369.

Cerniglia, C.E., D.W. Kelly, J.P. Freeman, and D.W. Miller. 1986. "Microbial metabolism of pyrene." *Chem. Biol. Interact. 57*:203-216.

Cunningham, S.D., and W.R. Berti. 1993. "Remediation of contaminated soils with green plants: An overview." *In Vitro Cell Devel. Bio. Plant. 29*:207-212.

Hsu, T.S., and R. Bartha. 1979. "Accelerated mineralization of two organophosphate insecticides in the rhizosphere." *Appl. Environ. Microbiol. 37*:36-41.

Jones, R.D. 1991. "Soil chemistry in a brome meadow under long-term fertilization." Ph.D. Dissertation, Dept. of Agronomy, Kansas State Univ., Manhattan, KS.

McKenna, E.J., and R.D. Heath. 1976. *Biodegradation of polynuclear aromatic hydrocarbon pollutants by soil and water microorganisms*. WRC Research Report No. 113, University of Illinois, Water Resources Cntr., Urbana, IL. pp. 1-25.

Schwab, A.P., and M.K. Banks. 1994. "Biologically mediated dissipation of polyaromatic hydrocarbons in the root zone." In Anderson, T.A. and J.R. Coats. (Eds.) *Bioremediation through Rhizosphere Technology*. Amer. Chem. Soc. Symp. Ser. 563, pp. 132-141.

Southworth, G.R. 1977. *Transport and Transportation of Anthracene in Natural Waters: Process Rate Studies*. U.S. Department of Energy, Oak Ridge TN.

Walton, B.T., and T.A. Anderson. 1990. "Microbial degradation of trichloroethylene in the rhizosphere: Potential application to biological remediation of waste sites." *Appl. Environ. Microbiol. 56*:1012-1016.

Weisenfels, W.D., M. Beyer, and J. Klein. 1990. "Degradation of phenanthrene, fluorene, and fluoranthene by pure bacterial cultures." *Appl. Microbiol. Biotechnol. 32*:479-484.

Ultrasonic Degradation of Polycyclic Aromatic Hydrocarbons

Joon Kyu Park and Teh-Fu Yen

ABSTRACT

Trace amounts of the polycyclic aromatic hydrocarbons (PAHs) present in the environment are of concern due to their hazardous nature. Conventional processes for degrading PAH are time consuming and inefficient, and they require costly equipment; furthermore, they contribute to other problems. Ultrasound can provide the highly energetic hydrogen atom and hydroxyl radicals, whose reactivity accounts for the destruction of organic solutes, in a very short time. The degradation mechanisms and the degradation efficiency for 1-methylnaphthalene, anthracene, phenanthrene, pyrene, 1,12-benzoperylene, and coronene, in dilute organic solvent-water solutions, were studied under ultrasound on a laboratory scale. The extent of degradation is followed by ultraviolet (UV)-visible spectrometry. The effects of additional agents, such as hydrogen peroxide, sodium borohydride, and dissolved gas, were investigated. The catalytic effects of ferrous chloride and nickel chloride were investigated. A total of 70 to 83% of PAH was degraded as a result of the additional agents and 2 hours of ultrasound. The higher percentage of degradation was caused by the greater abundance of free radicals, especially hydroxyl, for cleavage. However, with the catalysts, the added PAH degradation efficiency was not significant.

INTRODUCTION

Bioremediation has been a powerful method for degrading a wide variety of chemical compounds, especially organic pollutants. However, it is usually time consuming, costly, and inefficient. Therefore, as an alternative or for supporting bioremediation, chemical and physical processes have been used. For this reason, ultrasound can be employed in such a manner due to its high efficiency and flexible applications.

Ultrasonic radiation is known to decompose water molecules in the bubbles into free radicals such as hydroxyl ($\cdot OH$), hydrogen ($H\cdot$), and hydroperoxyl ($HO_2\cdot$). Evidence for the formation of free radicals by ultrasound in an aqueous solution has been demonstrated recently by Makino et (1983). The hydroxyl radical is particularly reactive with carbon-chlorine and carbon-carbon double bonds and is capable of aromatic ring cleavage. Dechlorination of chloroform in water with ultrasound in combination with hydrogen peroxide has reached 94% efficiency in 2 h for the removal of chloroform (Chen et al. 1990). The ultrasonic irradiation of hydrogen sulfide in water results in its rapid oxidation (Kotronarou et al. 1992). The upgrading of petroleum and the conversion of asphaltenes into gas oil and resins at room temperature and atmospheric pressure with ultrasound has been achieved in a very short time (Lin and Yen 1993). The recovery and separation of petroleum from tar/oil sands has been accomplished successfully in situ with multiple ultrasound transducers mounted on drill bits (Kalbfleisch 1991). The recent application of ultrasound to environmental fields has shown that ultrasound is a powerful tool for environmental use in bench-scale as well as field applications.

PAH compounds make up a class of concern due to their carcinogenic, mutagenic, teratogenic, and chronic low-level hazard potential (Neff 1979). Therefore, the U.S. Environmental Protection Agency (U.S. EPA) has identified 16 unsubstituted PAHs as priority pollutants. Because PAHs are known to be stored in plant and animal tissues, increasing concentrations of these organics may be established through trophic-level magnification. Therefore, wastewater resulting from industrial activities, especially those of the coal and petroleum industries, represents a threat to human health and the environment as a whole (Neff 1979).

Several methods are known to degrade PAHs. However, significant degradation in a short period of time has not been achieved by those methods. For example, degradation of photosensitive PAHs by photooxidation under UV light resulted in at best a 73% efficiency in 5 h, while a 25% degradation of only low-molecular-weight PAHs was achieved using white-rot fungi in a 3-week period (Park 1995). The use of ultrasound to degrade PAHs is little known. D'Silva et al. (1990) in a preliminary study stated that a cavitation process resulting from ultrasonic irradiation is considered to be the cause of the destruction of the aromatic rings in PAH compounds because, when the cavities implode, high energy occurs. In aquatic systems, sonication produces hydrogen and hydroxyl radicals, which consequently produce hydrogen peroxide and molecular hydrogen, and their reactivity destroys the PAH compounds.

In this study, we investigate ultrasound as an alternative method by which to destroy the PAHs. To increase the degradation of PAHs, some agents (hydrogen peroxide, sodium borohydride) as well as dissolved gas (argon) are used. To accelerate the cavitation reaction, catalytic effects are considered.

EXPERIMENTAL

The experimental investigation was divided into three parts: (1) sample preparation, (2) sonication process, and (3) quantification of degradation efficiency by UV-visible spectrophotometer.

Sample Preparation

Standard samples were prepared to (1) compare the degradation efficiencies in different solvent contents; (2) determine the degradation efficiencies of PAHs; (3) determine the effects of some agents, such as hydrogen peroxide and sodium borohydride, to extend the degradation efficiency; (4) determine the effect of the dissolved gas; (5) determine the effect of different pHs of solution, and (6) determine the catalytic effect.

PAHs used in this investigation were 1-methylnaphthalene (Aldrich Chemical Company, Inc., 98%), anthracene (Lancaster Synthesis, 99%), phenanthrene (Matheson Coleman & Bell Manufacturing Chemists, 99%), pyrene (Aldrich Chemical Company, Inc., 99%), 1,12-benzoperylene (Aldrich Chemical Company, Inc., 98%), and coronene (Aldrich Chemical Company, Inc., 99%). Among these PAHs, we focus on the results of 1-methylnaphthalene.

Several standard solutions of each PAH in different concentrations (0, 10, 20, and 30 (μg/g as different ratios of organic solvent and water mixture) were prepared for calibration, and 30 (μg/g of each solution was used for the sonication process. The sample solution was wrapped with aluminum foil to prevent degrading by photooxidation and was used in 48 h to make sure that all experiments were done under the same concentration conditions.

To compare the degradation efficiencies of PAHs in different solvent contents, each PAH was dissolved in 100% of organic solvent or different ratios of organic solvents with water (by weight). To dissolve PAHs in the appropriate organic solvents, ethanol was used for 1-methylnaphthalene, anthracene, phenanthrene, and pyrene. Benzene was used for 1,12-benzoperylene and coronene. PAHs were dissolved in organic solvents first, then mixed with water.

To determine how much other agents affect the degradation efficiencies, the following procedures were followed: a solution of 1% (by wt.) of hydrogen peroxide (Mallinckrodt Chemical Co.) was applied to each before the sonication process, and one of 0.5% of sodium borohydride (Wilshire Chemical Co.) was applied every 30 min during the sonication process.

To determine the pH effect, the solution was adjusted to pH 2.0 with HCl and to pH 12.0 with NaOH, whereas the pH of the original solution was around 6.8. To maintain the pH during the experiment, the pH of the solution was monitored and adjusted by HCl or NaOH.

To determine the catalyst effect, ferrous chloride (J. T. Baker Chemical Co.), sodium borohydride, nickel chloride (Mallinckrodt Chemical Co.), sodium borohydride, ferrous chloride-hydrogen peroxide, and nickel chloride-hydrogen peroxide were used (Lian 1993).

Sonication Process

The instrument used in this analysis was manufactured by Sonics & Materials, Inc. Sonication in this experiment was at 20 kHz, and the intensity at the titanium tip of the 0.25-in (0.6-cm) sonication probe was ~35 W/cm^2 (Suslick 1988). In each experiment, 16 mL of PAH solution was taken from the stock solution and put into the 30-mL vial. The sonication reaction was performed in this solution with a constant temperature 22°C. During the sonication process, 5 m³/min of 99% compressed argon was applied to achieve a higher degradation efficiency (Suslick 1988). To determine the effect of argon, every solution was sonicated twice, once with argon and once without it.

Quantification Process

After the sonication process, solutions of each PAH were scanned and the specific wavelengths where a peak absorbance was found were determined using a Beckman 25 UV-visible spectrophotometer. A calibration curve was made with the absorbance reading of standard solutions at those specific wavelengths. Then, the absorbance reading of the sonicated solution at a specific wavelength was determined and applied to the calibration curve to determine the degradation efficiency.

RESULTS

Basically, ultrasonic irradiation of PAHs in a solution of organic solvent with water resulted in a decrease of PAH with time. Figure 1 shows that the degradation efficiency of PAHs in 100% organic solvent remained less than 10%. Figure 2 shows the degradation trend that the more benzene rings a PAH

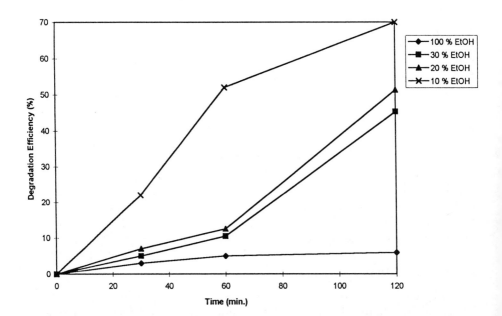

FIGURE 1.   Sonication of 1-methylnaphthalene in different ethanol:water ratios.

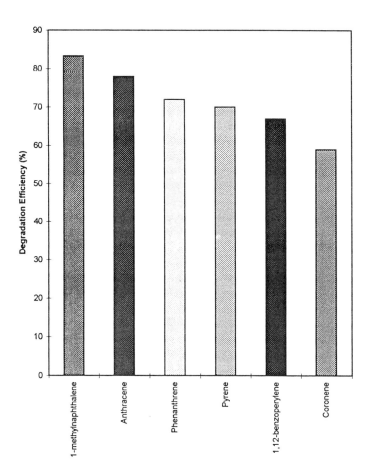

FIGURE 2. Comparison of degradation efficiencies of different PAHs.

has, the more difficult it is to degrade. In aqueous solution, the electron potential of free radicals produced by the ultrasonic process is different depending
on the pH of the solution. Figure 3 shows how the pH of the solution affects
the degradation of PAHs. The effects of various agents and catalysts on increasing the degradation of 1-methylnaphthalene are shown in Table 1.

DISCUSSION

The primary chemical effect of ultrasound is cavitation, which produces
transient high pressures and temperatures. When water is used as a solvent,
the water present in the cavities undergoes thermal dissociation to give hydrogen atoms (H•) and hydroxyl radicals (•OH) (Makino et al. 1983). PAHs present near the bubble/water interface can also undergo thermal decomposition

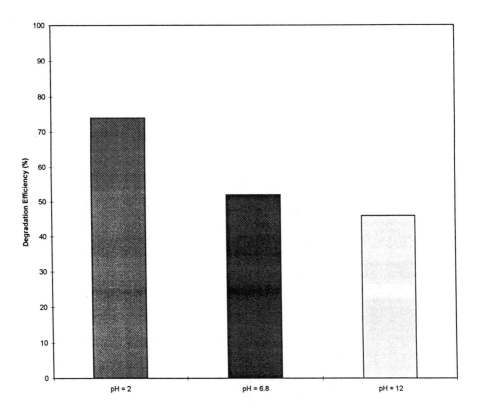

FIGURE 3. Sonication of 1-methylnaphthalene with pH effect.

(Kotronarou 1992). Secondary reactions take place in the liquid phase between PAH molecules and excited species, mainly •OH in the case of aqueous solutions. The •OH is a powerful and efficient chemical oxidant; it has a one-electron oxidation potential of $E° = +1.8$ V in neutral solutions and 2.7 V in acidic solutions (Klaning et al. 1985). For this reason, oxidation is the dominant reaction in this experiment. Therefore, producing more •OH during the reaction and making the solution more acidic can improve the degradation efficiency. Hydrogen peroxide was an efficient oxidizing agent in this experiment because it produced more •OH during sonication. Also, the rate of degradation of PAH increased with a decrease in pH in more acidic solutions because the •OH produced by the sonication process has a greater electron potential. Depending on the ratio of organic solvent and water, various degradation efficiencies are achieved. In more aqueous solutions, there is more degradation of PAH because more •OH is produced. Water produces more •OH than do organic solvents, which have a higher vapor pressure and a lower tensile strength (D'Silva et al. 1990), due to the hydrophobicity effect as indicated by Figure 1 (Suslick 1988). To study the effects of dissolved gas during sonication, argon was chosen because it is the best dissolved gas for cavitation (Suslick 1988).

TABLE 1. Degradation efficiencies (%) for 1-methylnaphthalene under ultrasound with various agents.

Conditions	60 min	120 min
Ultrasound only	10.5	45.3
A[a]	37.5	75.4
H_2O_2[b]	28.4	57.2
H_2O_2 - A	52.0	83.3
$NaBH_4$[c]	12.5	44.2
$NaBH_4$ - A	48.9	80.1
$FeCl_2$ - $NaBH_4$ -A[d]		
0.1%	44.0	—
0.05%	48.5	78.8
0.025%	51.3	—
$NiCl_2$ - $NaBH_4$ - A		
0.1%	51.1	—
0.05%	31.4	80.8
0.025 %	28.8	—
$FeCl_2$ - H_2O_2 - A[e]		
0.1%	53.1	—
0.05%	49.2	79.2
0.025%	43.2	—
$NiCl_2$ - H_2O_2 - A		
0.1%	31.6	—
0.05%	39.3	81.2
0.025%	50.4	—

(a) Flowrate of A in all experiments is 5 m^3/min.
(b) 1% (by wt) of H_2O_2 is applied before sonication.
(c) 0.5% (by wt.) of $NaBH_4$ is applied every 30 min during sonication.
(d) Different amounts of $FeCl_2$ are applied in each experiment.
(e) Different amounts of $NiCl_2$ are applied in each experiment.

With argon, 20 to 30% more degradation is achieved. The geometry of the chemical structure of PAH affects degradation efficiency. The more benzene rings the PAH has, the less degradation efficiency is achieved. Although PAHs have the same number of benzene rings, the PAH that has a straight structure has a better degradation efficiency than the PAH that has a bent or branched structure. It is expected that catalysts form the catalytic complexes; however, the increased degradation efficiency with catalysts is not significant (Roth et al. 1994). We concluded that PAH compounds can be degraded very efficiently in a short period of time. Oxidizing environments created a more vigorous cavitation process, which can degrade PAH compounds faster and more efficiently.

REFERENCES

Chen, J. R., X. W. Xu, A. S. Lee, and T. F. Yen. 1990. "A Feasibility Study of Dechlorination of Chloroform in Water by Ultrasound in the Presence of Hydrogen Peroxide." *Environ. Technol.*, *11*: 829-836.

D'Silva, A. P., S. K. Laughlin, S. J. Weeks, and W. H. Buttermore. 1990. "Destruction of Polycyclic Aromatic Hydrocarbons with Ultrasound." *Polycyclic Aromatic Compounds, Vol. 1*(3): 125-135.

Kalbfleisch, H.L. 1991. "Major New Developments in Ultrasonics Provide Functional Systems for Recovery of Petroleum from Tar/Oil Sands 'In Situ'." *5th Unitar International Conference on Heavy Crude & Tar Sands*, Caracas, Venezuela. p. 140.

Klaning, U. K., K. Sehested, and J. Holcman. 1985. "Standard Gibbs Energy of Formation of the Hydroxyl Radical in Aqueous Solution. Rate Constant for the Reaction $ClO_2^- + O_3 \leftrightarrow O_3^- + ClO_2$." *Journal of Physical Chemistry, 89*: 760.

Kotronarou, A., G. Mills, and M.R. Hoffman. 1992. "Oxidation of Hydrogen Sulfide in Aqueous Solution by Ultrasonic Irradiation." *Environmental Science & Technology, 26*: 2420-2428.

Lian, H. 1993. "Environmental Coating and Encapsulating Techniques: Application of Asphalt System." Ph.D Thesis, University of Southern California, Los Angeles, CA.

Lin, J. R., and T. F. Yen. 1993. "An Upgrading Process Through Cavitation and Surfactants." *Energy and Fuel, 7*: 111-118.

Makino, K., M. M. Massoba, and P. J. Riesz. 1983. "Chemical Effects of Ultrasound on Aqueous Solutions. Formation of Hydroxyl Radicals and Hydrogen Atoms." *Journal of Physical Chemistry, 87*: 1369.

Neff, J. M. 1979. *Polycyclic Aromatic Hydrocarbons in the Aquatic Environment*, Applied Science Publisher Ltd., London.

Park, J. K. 1995. unpublished.

Roth, J. A., S. R. Dakoji, R. C. Hughes, and R. E. Carmody. 1994. "Hydrogenolysis of Polychlorinated Biphenyls by Sodium Borohydride with Hydrogeneous and Heterogeneous Nickel Catalysts." *Environ. Sci. Technol., 28*: 80-87.

Suslick, K. S. 1988. In K.S.Suslick (Eds.), *Ultrasound: Its Chemical, Physical, and Biological Effects*, VCH, New York, NY.

Ex Situ Bioremediation of Polycyclic Aromatic Hydrocarbons in Laboratory Systems

Britt-Marie Pott and Tomas Henrysson

ABSTRACT

Biological remediation of soil contaminated with polycyclic aromatic hydrocarbons (PAHs) is a promising innovative technology. This project was started to get a better understanding of the parameters that influence the degradation of PAHs of different molecular weights. The degradability of those substances that are the most recalcitrant will be crucial for the success of a biological treatment system. Four substances, naphthalene, anthracene, benzo(a)anthracene, and dibenzo-(a,h)anthracene (2, 3, 4, and 5 aromatic rings respectively), were chosen as model contaminants. Water content and addition of nutrients were varied to determine the effects of these parameters on the degradation. The biological degradation of the four PAHs was followed. Removal of the substances with 2, 3, and 4 rings, respectively, was successful. No difference in the rate of removal of the five-ringed member (dibenzo-(a,h)anthracene) compared to the control was detected. Neither naphthalene nor anthracene was affected by water addition, but small effects were detected when nutrients were added. Degradation of benzo(a)anthracene (4 aromatic rings) was clearly stimulated by addition of water, but the effects of nutrient addition were inconclusive and need further study.

INTRODUCTION

Soil polluted with PAHs is a common problem worldwide. The PAHs comes from different sources such as coal gasification, wood impregnation, and petroleum products. Several methods to remediate PAH-contaminated soil can be used (U.S. EPA 1993). Biological methods seem to have the potential to be the least expensive alternative (Maier 1992). It has been shown that indigenous microbes can be stimulated to degrade PAH both in situ and ex situ (Mueller et al. 1989a). At many sites, bioremediation in situ is not feasible due to the low permeability of the soil or a too small microbial community. In such cases ex situ treatment might be a better alternative (Morgan & Watkinson 1990).

The ex situ treatment of PAH-polluted soil can be performed in different ways. Composting is one method where the soil is mixed with organic compounds such as straw or wood chips. Solid-phase treatment with tilling and irrigation is another method that can be used (Harmsen 1991). In most of these methods, indigenous microorganisms are stimulated to degrade the PAHs. However, it is also possible to add microorganisms to the treatment.

It is well known that PAHs become more recalcitrant with increasing numbers of rings (Morgan et al. 1991). This is important because many remediation goals are based not only on the total PAH value, but also on individual compounds. Thus, it is very important to optimize degradation of those compounds that are most recalcitrant, i.e., those compounds that take the longest time to reach the remediation goals. It has been shown that PAHs with at least 4 rings can be used as the sole carbon and energy source for bacteria (Mueller et al. 1989b). However, compounds with 5 or 6 rings can be degraded in a biological treatment (Morgan et al. 1991).

This project is aimed at a better understanding of the degradation of PAHs by indigenous microbes. Therefore, parameters that influence degradation were screened. The tested parameters were the water content of the soil, temperature, and nutrient addition. To study how PAHs with a different number of rings are influenced by these parameters, a soil was spiked with four PAHs with 2, 3, 4, and 5 rings, respectively. Bioremediation was followed by analysis on a gas chromatograph coupled to a mass spectrometer (GC/MS).

EXPERIMENTAL PROCEDURES AND MATERIALS

The reactors used for biodegradation were closed amber glass bottles (1 L) containing 100 g of soil. The soil was contaminated in the lab and the fate of the added PAHs was monitored. Fresh air entered the reactors when they were sampled.

Contaminated Soil

Clay and sand were mixed and sieved through a net (14 mesh) to remove stones and large particles. The mixed soil contained 1.9% organic matter and had a water-retaining capacity of 22%. The PAHs used to contaminate the soil were naphthalene, anthracene, benzo(a)anthracene, and dibenzo(a,h)anthracene. All four PAHs were dissolved in a mixture of hexane and acetone (3+1) and then added to the soil in small portions. Between each addition of contaminants, the soil was thoroughly mixed and the solvents were allowed to evaporate.

Sampling and Sample Treatment

Samples of 1 g were collected in triplicate, and 2.5 mL of hexane was added. The samples were placed in boiling water for 6 h. After filtration, the solvent was analyzed on a GC/MS (GC: Varian Star 3400, MS: Varian Saturn II). The GC column was a 30-m DB 5, ID 0.25 mm, film thickness 0.25 μm programmed as follows: 60°C for 2.7 min, raised to 220°C (30°C/min), raised to 275°C (15°C/min), raised to 320°C (10°C/min), and held at 320°C for 4.8 min. The injector was started at 50°C, raised to 320°C (100°C/min), and held at 320°C for 2 min.

Parameters Tested

Three reactors were run as controls and kept cold during the test. Addition of nutrients and variation of the water content were two parameters tested. Table 1 is an outline of the variation in the parameters.

RESULTS

The biological degradation of naphthalene, anthracene, and benzo(a)anthracene was significant in all reactors stored at room temperature. In Figure 1, differences in degradation rates between the four PAHs tested are clearly visible. Three groups of analyzed samples failed due to problems with the analysis; one of these groups was the sample at time zero. All samples taken at day 6 and day 28 gave results that were only 50% of the expected value. These five groups of samples were omitted in the results. The amount of naphthalene at the start of the test should be the same as for the three other components, but actually is much lower. One possible reason for the loss of naphthalene could be that it evaporates during the contamination of the soil. All reactors stored at room temperature during the experiment followed the same pattern of degra-

TABLE 1. Parameters tested.

Reactor	Temperature, °C	Water content[a]	Nutrients
1	+6	11%	—
2	−18	11%	—
3	+25	11%	—
4	+25	14%	—
5	+25	14%	N+P[b]
6	+25	14%	N+P[c]
7	+25	14%	Plant fertilizer[b]
8	+25	14%	Plant fertilizer[c]
9	+6	14%	—

(a) 11% and 14% water content corresponds to 50% and 64%, respectively, of the water-retaining capacity.

(b) The theoretical COD value of the added PAHs was calculated. Nitrogen and phosphorus were then added in the following proportions: COD:N:P at 100:5:1.

(c) As above, but the added amount of nitrogen and phosphorus was tenfold.

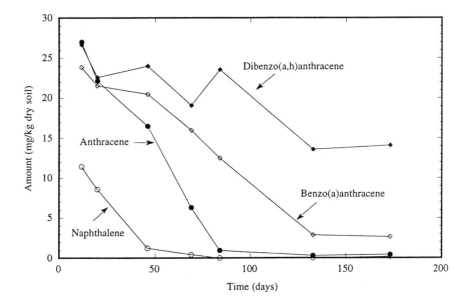

FIGURE 1. Degradation of the four PAHs used at room temperature and with 11% water content (Reactor 3 in Table 1).

dation for the four PAHs. Storage at +6°C or at –18°C resulted in a much slower removal of these PAHs (Figure 2) than storage at room temperature (+25°C). The loss of PAHs at –18°C was the same as at +6°C.

Removal of naphthalene was very fast in all of the reactors stored at room temperature. Within 50 days, the naphthalene had disappeared from these reactors. Anthracene was degraded at rates varying between 0.36 and 0.57 mg anthracene per kg soil per day. Water addition did not affect the rate to a significant extent. The results indicate that addition of nutrients stimulates the degradation of anthracene, but further tests have to be carried out to verify this. Benzo(a)anthracene also was degraded in all reactors stored at room temperature. A higher content of water clearly stimulated the degradation compared to the control (Figure 3). The rate of removal in the reactor with the higher water content was 0.55 mg benzo(a)anthracene per kg soil per day, compared to 0.22 mg per kg per day in the reactor with less water. Nutrients did not seem to have any great effect on the degradation rate. So far the degradation of dibenzo(a,h)anthracene is inconclusive. The removal is about the same in all reactors, including those stored at lower temperatures. The study will continue to establish if the degradation of this 5-ring compound is significant or not.

DISCUSSION

The results clearly show the difference in degradation rates between PAHs with a varying number of aromatic rings. This points out the importance of

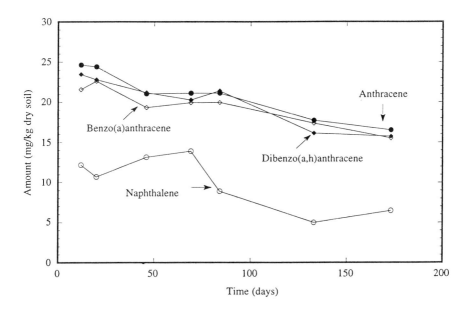

FIGURE 2. Disappearance of the four PAHs in a control reactor, stored at +6°C, with 11% water content (Reactor 1 in Table 1).

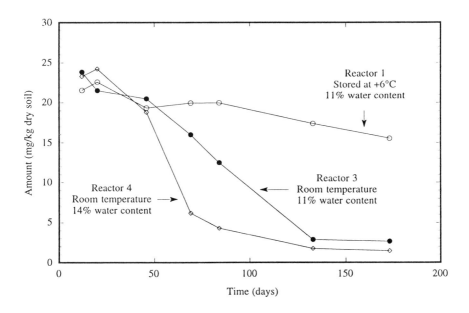

FIGURE 3. Degradation of benzo(a)anthracene in three different reactors (the reactor numbers refer to Table 1).

optimizing bioremediation methods based on the degradation rates of high-molecular-weight PAHs instead of the total concentration of PAHs.

The water content of the soil also seems to influence the degradation rates of at least some compounds at the small variations used in this experiment. Therefore, it is important to carefully measure and control the water content of the soil when bioremediation is used.

Nutrient addition to the soil did not make a conclusive difference in degradation rates in these experiments. This can possibly be explained by the fact that there already are enough nutrients for efficient degradation in the soil used in these experiments. These results must be further examined and verified.

REFERENCES

Harmsen, J. 1991. "Possibilities and limitations of landfarming for cleaning contaminated soils." In R. E. Hinchee and F. Olfenbuttel (Eds.), *On-Site Bioreclamation*, pp. 255-272. Butterworth-Heinemann, Stoneham, MA.

Maier, M. 1992. *Den naturliga lösningen. Biologisk efterbehandling av förorenad mark.* Utlandsrapport, USA 9205, Sveriges Tekniska Attachéer.

Morgan, D. J., A. Battaglia, J. R. Smith, A. C. Middleton, D. V. Nakles, and D. G. Linz. 1991. "Evaluation of a biodegradation screening protocol for contaminated soil from manufactured gas plant sites." In C. Akin and J. Smith (Eds.), Proceedings from *Gas, Oil, Coal and Environmental Biotechnology III*, Chicago, pp. 55-74.

Morgan, P., and R. J. Watkinson. 1990. "Assessment of the potential for *in situ* biotreatment of hydrocarbon-contaminated soils." *Wat. Sci. Tech.* 22(6): 63-68.

Mueller, J. G., P. J. Chapman, and P. H. Pritchard. 1989a. "Creosote-contaminated sites—their potential for bioremediation." *Environmental Science and Technology* 23(10): 1197-1201.

Mueller, J. G., P. J. Chapman, and P. H. Pritchard. 1989b. "Action of a fluoranthene-utilizing bacterial community on polycyclic aromatic hydrocarbon components of creosote." *Applied and Environmental Microbiology* 55(12): 3085-3090.

U.S. EPA and U.S. Air Force. 1993. *Remediation Technologies Screening Matrix and Reference Guide.* EPA 542-B-93-005, U.S. Environmental Protection Agency.

Aromatic Hydrocarbon Degradation in Hydrogen Peroxide- and Nitrate-Amended Microcosms

Barry J. Christian, Lucy B. Pugh, and Bruce H. Clarke

ABSTRACT

Fifty microcosms were constructed using aquifer materials from a former coal gasification site and divided into four groups: poisoned control, nutrient-free control, hydrogen peroxide-amended, and nitrate-amended microcosms. Each microcosm contained site soil and groundwater in a 1.2-L glass media bottle. When depleted, hydrogen peroxide and sodium nitrate were injected into the microcosms. Microcosms were periodically sacrificed for analysis of polycyclic aromatic hydrocarbons (PAHs); monocyclic aromatic hydrocarbons (benzene, toluene, ethylbenzene, and xylenes [BTEX]); total petroleum hydrocarbons (TPH); and heterotrophic plate counts (HPCs). BTEX and two- and three-ringed PAHs were degraded in microcosms receiving electron-acceptor additions compared to poisoned controls. Four-, five-, and six-ringed PAHs were not significantly degraded during this study. Except in poisoned controls, significant amounts of dissolved oxygen (DO) or nitrate were utilized, and microbial populations increased by 3 to 5 orders of magnitude compared to site soils used to assemble the microcosms (i.e., baseline samples).

INTRODUCTION

During evaluation of alternatives to remediate aquifer materials at a former coal gasification site, conventional technologies typically applied for removal of PAHs and BTEX were determined to be prohibitively expensive. A preliminary assessment of in situ bioremediation (ISB) revealed that, if feasible, it would be comparatively cost effective. Aerobic biodegradation of PAHs was demonstrated previously in outdoor test plots (Wang et al. 1990). Also, several review articles (Lee et al. 1988, Mueller et al. 1989, and Sims et al. 1989) summarized the literature on PAH and BTEX biodegradation and, in some instances, provided data showing biodegradation of these compounds under a variety of field and laboratory conditions. A microcosm study was performed to determine the feasibility of ISB at this 5-acre (2-ha) site where aquifer conditions are generally amenable to ISB; underneath several feet of surface fill is a layer of fine- to medium-grained glacial sand extending to clay at depths ranging from 50 to 70 ft (15 to 21 m) below grade. Groundwater is encountered at a depth of approximately 20 ft (6 m).

EXPERIMENTAL PROCEDURES AND MATERIALS

In October 1991, soil and groundwater samples were collected from the site in ethanol-rinsed, 5-gal (19-L) polyethylene containers. Fifty microcosms were assembled consisting of four groups: hydrogen peroxide-amended (20), nitrate-amended (18), poisoned control (6), and nutrient-free control (6) microcosms. Hydrogen peroxide-amended (i.e., oxygen-amended) microcosms and nitrate-amended microcosms were assembled with nitrogen (NH_4Cl) and phosphorus (Na_2HPO_4) in concentrations yielding a COD:N:P ratio of 100:10:2, where chemical oxygen demand (COD) was determined by summing groundwater COD (U.S. EPA Method 410.4) and soil COD (as approximated by volatile solids analysis). Hydrogen peroxide and sodium nitrate were replenished as oxygen and nitrate were depleted.

Poisoned control microcosms contained nutrients as described above, 1,500 mg/L of mercuric chloride (Baker) as a biocide, and received dosages of hydrogen peroxide when oxygen was depleted. Nutrient-free control microcosms received periodic additions of hydrogen peroxide when oxygen was depleted, but contained no supplemental nitrogen or phosphorus.

Microcosm DO concentrations were monitored with a YSI Model 58 meter and standard biochemical oxygen demand probe. When the DO was depleted below 2.0 mg/L in the oxygen-amended, nutrient-free control, and poisoned control microcosms, 140 µL of hydrogen peroxide (FMC Corp., 50% technical grade) was dosed to yield 80 mg/L of hydrogen peroxide and approximately 40 mg/L of DO (one mole of hydrogen peroxide decomposes to yield approximately one-half mole of molecular oxygen). When nitrate was depleted in the nitrate-amended microcosms (as determined by collecting 1-mL samples for analysis by U.S. EPA Method 353.2), sodium nitrate (Baker) was added as a 1-mL dosage yielding 70 mg/L of nitrate.

Each microcosm contained 300 g of site soil and 1 L of site groundwater in a 1.2-L Wheaton glass medium bottle, including a polycarbonate screw-on cap with a polyethylene liner. Site soils were first homogenized and sieved using a 4.7-mm sieve. Sieving was conducted to increase the accuracy and precision of the analytical data. Microcosms were maintained at room temperature (approximately 21°C) and mixed twice per week by inversion, resulting in soil displacement from the bottom of the media bottle to the top and back. Headspace volume in each microcosm was minimized to less than 1 mL throughout the study to minimize loss of volatile target compounds. All microcosms were handled in a uniform manner throughout the study. To track the progress of biodegradation, one, two, or three microcosms from each group were periodically sacrificed for analysis of PAHs and BTEX in both aqueous and soil phases by U.S. EPA, gas chromatography/mass spectrometry (GC/MS) methods 8270 and 8240, respectively. TPH was analyzed in the soil phase by U.S. EPA method 418.1.

HPCs in the microcosms also were tracked. As per the method (Page et al. 1982), 10 g of soil was shaken for 10 minutes in 95 mL of diluent (mineral salts in distilled water at pH 7.3). Seven to 10 subsequent serial dilutions were made and 3 mL of each dilution were plated in triplicate (1 mL each) onto sterile liquefied BBL Standard Method Agar (Belton Dickinson) plates and incubated for 7 days at 26°C, and then counted.

RESULTS

Characterization of Microcosms

Baseline soil sampling (data not provided) indicates the following: of the total volatile solids in the soils (3,700 mg/kg), only 7% was composed of PAHs and BTEX compounds (260 mg/kg and 1.0 mg/kg, respectively); and site soils with 3,700 mg/kg volatile solids and 1,240 mg/kg TPH contained 12 times more organic matter and 80 times more TPH on a weight basis than the groundwater with 302 mg/L COD and 15 mg/L TPH, respectively. Sampling of the microcosms indicated the following: 99% of the PAHs (170 mg/kg) and 50% of the BTEX compounds (0.010 mg/kg) were partitioned in the soil phase; six of the 15 PAHs measured were 2- and 3-ringed, while the remainder were 4-, 5-, and 6-ringed PAHs; and concentrations of metals in site soils used to assemble the microcosms (data not provided) were generally in the range observed for natural Michigan sands (MDNR 1988) and thus were not expected to be inhibitory to microcosm organisms.

Heterotrophic Plate Count, BTEX, and PAH Data

After 35 weeks of incubation the baseline HPC concentration increased from 9,000 colony forming units per dry weight gram (CFUs/g) to 2×10^6, 6×10^6, and 6×10^7 CFUs/g in the nitrate-amended, nutrient-free controls, and oxygen-amended microcosms, respectively, but were <1 CFU/g in the poisoned controls. These data indicate the following: (1) baseline soil samples contained low initial HPC concentrations; (2) aerobic microcosms supported large HPC increases; and (3) use of mercuric chloride provided effective poisoned controls, containing no microorganisms (this finding is supported by the fact that two dosages of 80 mg/L of hydrogen peroxide into the poisoned control microcosms resulted in elevated DO concentrations throughout the 97-week study). No HPC analyses were conducted after week 35. The baseline and week-35 HPC data fall within the range of values reported elsewhere for shallow, sandy soils (Wilson et al. 1983).

Microcosm soils were obtained from the vadose zone at the site and thus were not initially in equilibrium with the groundwater used to assemble the microcosms. Consequently, the two microcosm phases were allowed to equilibrate for 72 h prior to "time zero" sample collection. Approximately 75% of the total BTEX mass in each microcosm (0.36 mg) was contributed initially by the soil phase (0.28 mg). As shown in Table 1, aqueous-phase concentrations in all microcosms declined to levels below the analytical detection limit after 12 weeks, except in the poisoned control series microcosms where the BTEX concentrations were observed to be constant over time. Aqueous-phase BTEX analyses of microcosms were not conducted after week 12. Loss of BTEX by volatilization from the microcosms was effectively minimized as evidenced by BTEX retention in the aqueous phase of poisoned controls.

Soil-phase BTEX concentrations (data not provided) were analyzed in baseline soil samples (triplicate), time-zero microcosms (triplicate), week-35 microcosms (triplicate), and week-97 microcosms (quadruplicate). These data indicate that soil-phase BTEX concentrations were detected from time to time in the microcosms throughout the 97-week study, but were in low concentrations in all microcosm groups (including the poisoned controls) compared to baseline soil samples, indicating that BTEX partitioned to the aqueous phase.

Only soil-phase PAH data are presented, since over 99% of the entire mass of PAHs existed in the soil phase within the microcosms. Figure 1 compares the 95% confidence interval around the mean of 2- and 3-ringed PAHs in the oxygen-amended microcosms at "time zero" with the data after 35 weeks and after 97 weeks of incubation. Figure 1 shows that each of the six 2- and 3-ringed PAHs decreased in concentration compared to poisoned controls (data for poisoned controls are presented in Figure 2). A two-sample t-test (Walpole and Myers 1985) was performed to show that the mean for each PAH compound in the oxygen-amended microcosms is statistically different from the mean for the poisoned control microcosms after 35 and 97 weeks. A conservative assumption of variance inequality was made. The analysis showed that the p-values

TABLE 1. Total BTEX concentrations in microcosm water.

Microcosm Identification	Total BTEX Concentration (mg/L) at Indicated Time					
	Baseline	Time = 3 weeks	Time = 6 weeks	Time = 8 weeks	Time = 10 weeks	Time = 12 weeks
Poisoned Control	0.085	0.107	NA	0.086	NA	0.112
Nitrate-Amended	0.085	0.019	0.037	0.022	0.019	<0.004
Nutrient-Free Control	0.085	0.108	NA	<0.004	NA	<0.004
Oxygen-Amended	0.085	<0.004	<0.004	<0.004	<0.004	<0.004

NA = Not analyzed.
Note: "Time zero" data represent the average of the analytical results from 3 microcosms, whereas the data for all subsequent times represent the results from single microcosms.

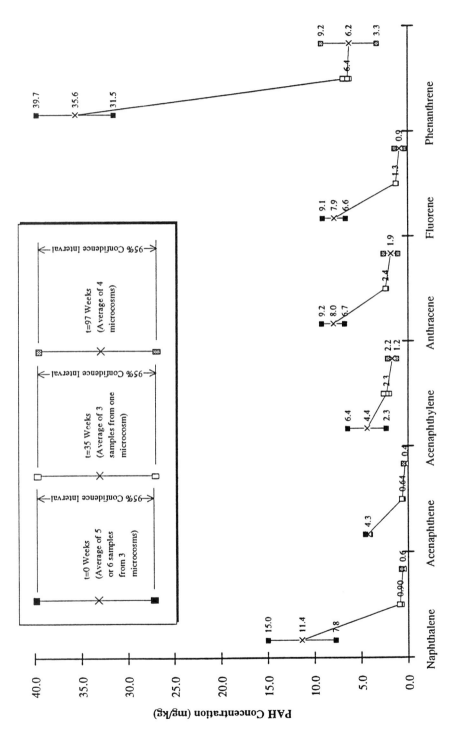

FIGURE 1. Concentrations of 2- and 3-ringed PAHs in oxygen-amended microcosm soils.

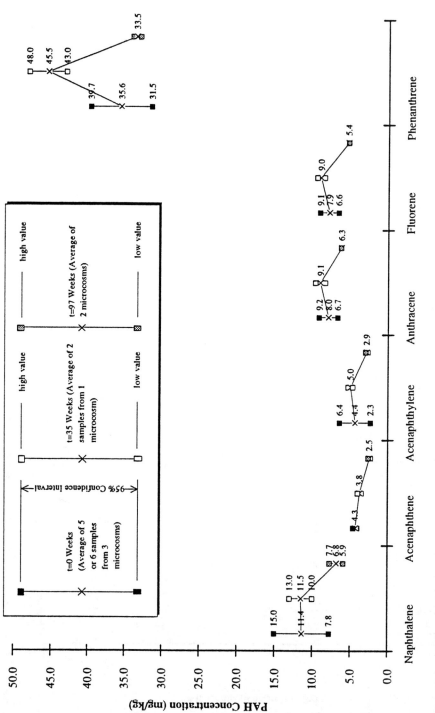

FIGURE 2. Concentrations of 2- and 3-ringed PAHs in poisoned control microcosm soils.

were less than 0.05, indicating that there is a 95% confidence level that the mean concentration of each 2- and 3-ringed PAH compound in the oxygen-amended microcosms was different than the mean concentration for each PAH compound in the poisoned control microcosms at both week 35 and week 97.

The soil-phase PAH data for the nutrient-free control and nitrate-amended microcosms are compared with the oxygen-amended and poisoned control data in Table 2. This comparison shows that removal of 2- and 3-ringed PAHs in nutrient-free control microcosms was nearly as complete as in the oxygen-amended microcosms indicating that nutrient addition did not enhance the rate of removal. Concentrations of 2- and 3-ringed PAHs in the nitrate-amended microcosms were not dramatically different than in the poisoned controls. Apparent losses in poisoned controls were most likely due to volatilization from the microcosms that occurred when caps were removed for nutrient addition.

Review of data in Figure 3 shows that 4-, 5-, and 6-ringed PAHs were not removed from oxygen-amended microcosms. In the poisoned control microcosms (data not provided) the concentrations of these PAHs increased slightly over time. Although no data exist in support of this, it is hypothesized that the PAHs became more extractable over time due to the reduction in average particle size in the microcosms, which resulted from abrasion during microcosm mixing and from degradation of the coal tar matrix to which the PAHs were partitioned. Except in poisoned control microcosms, aqueous-phase PAH concentrations were consistently below cleanup criteria (MDNR 1992) up through week 12 (despite rigorous microcosm agitation) and, therefore, were not subsequently analyzed.

TPH concentrations decreased from approximately 1,240 mg/kg at "time zero" to approximately 270 mg/kg after 97 weeks in the oxygen-amended microcosms, and decreased to lesser extents in the nitrate-amended microcosms

TABLE 2. PAH concentrations in nutrient-free, nitrate-amended, and poisoned control soils.

Microcosm I.D.	2- and 3-ringed PAHs (mg/kg) at Indicated Time						4-, 5-, and 6-ringed PAHs (mg/kg) at Indicated Time					
	Time 0	Week 8	Week 12	Week 14	Week 35	Week 94	Time 0	Week 8	Week 12	Week 14	Week 35	Week 94
Poisoned Control	72	100	NA	NA	84	57	96	187	NA	NA	136	114.8
Nitrate-Amended	72	17	72	103	NA	48.2	96	29	129	169	NA	104
Nutrient-Free Controls	72	18	NA	NA	NA	15	96	78	NA	NA	NA	86.3
Oxygen-Amended	72	66	35.5	25.6	13.9	11.7	96	145	147	130	107	93.2

NA = Not analyzed.

Note: "Time zero" data are the averages of 5 or 6 samples. Data presented for weeks 8, 12, and 14 are the result of analysis of one sample. Week 35 data are averages of duplicate and triplicate samples in the poisoned control and oxygen-amended microcosms, respectively. Week 97 data are averages of 4, 3, and 2 samples, respectively, for the nutrient-free, nitrate-amended, and poisoned control microcosms.

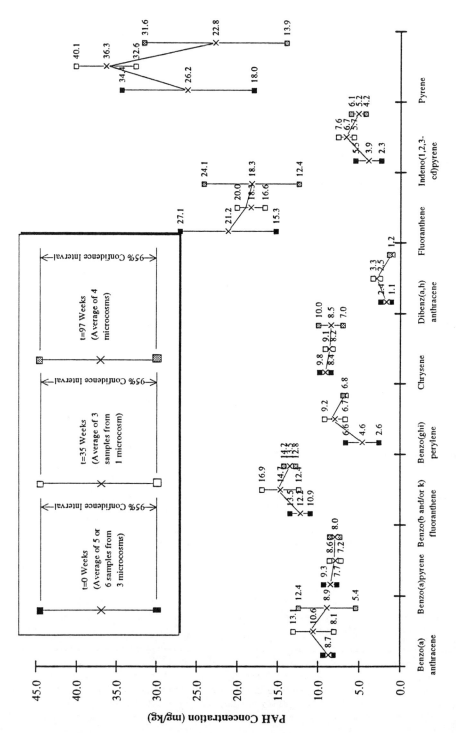

FIGURE 3.   Concentrations of 4-, 5-, and 6-ringed PAHs in oxygen-amended microcosm soils.

(750 mg/kg), nutrient-free microcosms (310 mg/kg), and poisoned controls (664 mg/kg).

DISCUSSION

These results indicate that BTEX and 2- and 3-ringed PAHs can be degraded in coal tar-impacted materials under oxygen-enhanced, laboratory microcosm conditions. The reaction mechanism is most likely biodegradation. Loss by evaporation from the microcosms of even the most volatile PAH compound, naphthalene, did not occur as evidenced by the presence of naphthalene in the soil phases of the poisoned control microcosms after 97 weeks. Half-lives of several 2- and 3-ringed PAHs were calculated and observed to be within the ranges observed by other researchers (Heitkamp et al. 1987 and McGinnis et al. 1991).

There remains a possibility that Fenton's reaction could account for some of the removal of 2- and 3-ringed PAHs since the poisoned controls received lesser amounts of hydrogen peroxide. Poisoned controls could not be given amounts of hydrogen peroxide equal to the oxygen-amended microcosms due to concerns that the evolution of oxygen gas would increase pressure in the microcosm and crack the medium bottles (as occurred in previous work). The reaction between hydrogen peroxide and ferrous iron (i.e., Fenton's reaction) can occur in the microcosms since ferrous iron exists naturally in soil and groundwater used to assemble the microcosms. Hydroxyl radicals are released from this reaction and are strong oxidizers capable of degrading environmental contaminants such as pentachlorophenol (Watts et al. 1990) and may oxidize PAH compounds. While it is difficult to separate removal caused by biodegradation from that caused by Fenton's reaction, it seems unlikely that Fenton's reaction contributed significantly to the removal of BTEX and 2- and 3-ringed PAHs considering that soluble iron is required for Fenton's reaction (Watts et al. 1990) and there were approximately 10 times as many moles of PAHs removed as total soluble iron available in the microcosms (pH conditions were not conducive for cycling insoluble iron to soluble iron [Watts et al. 1990]). In addition, significant increases in HPC populations support the evidence for biodegradation of PAHs.

The data further indicate that 4-, 5-, and 6-ringed PAHs were not dramatically biodegraded or oxidized by Fenton's reaction under the test conditions. Limited degradation of these PAHs could be related to the availability of alternative but uncharacterized carbon and energy sources contained within the coal tar. In addition, the 4-, 5-, and 6-ringed PAHs may be recalcitrant and not amenable to attack by enzyme systems associated with indigenous microorganisms or the hydroxyl radical. These compounds are also strongly adsorbed within the soil matrix or contained within the coal tar matrix (Lewis et al. 1992) and, while susceptible to solvent extraction, may be inaccessible to agents that promote biodegradation or to the hydroxyl radical.

The study results indicate that ISB may be a useful tool for remediation of former coal gasification sites. BTEX and 2- and 3-ringed PAHs, which are relatively soluble and therefore relatively mobile, were shown to be significantly degraded. Although shown to be less degradable, the 4-, 5-, and 6-ringed PAHs were shown to be strongly adsorbed to the soil phase (more than 99% of the total PAH concentrations in the microcosms existed in the soil phase), and did not leach into groundwater in concentrations exceeding the cleanup criteria. It is possible that 4-, 5-, and 6-ringed PAHs were solubilized at a low rate, resulting in immeasurably low decreases in soil concentration, while being bio-degraded at the rate of solubilization such that aqueous-phase concentrations did not increase.

REFERENCES

Heitkamp, M. A., J. P. Freeman, and C. E. Cerniglia. 1987. "Naphthalene Biodegradation in Environmental Microcosms: Estimates of Degradation Rates and Characterization of Metabolites." *Applied and Environmental Microbiology* 53(1): 129-136.

Lee, M.D., J. M. Thomas, R. C. Borden, P. B. Bedient, C. H. Ward, and J. T. Wilson. 1988. "Biorestoration of Aquifers Contaminated with Organic Compounds." *CRC Critical Reviews in Environmental Control* 18(1): 29-89.

Lewis, R. F. (U.S. EPA). 1992. "Innovative Biological Treatment Processes for Soils Contaminated with Polynuclear Aromatic Hydrocarbons." Paper presented at Fourth Annual Symposium on Emerging Technologies for Hazardous Waste Management. Atlanta, GA, September 21-23.

McGinnis, G. D., H. Borazjani, D. F. Pope, D. A. Strobel, and L. K. McFarland. 1991. *On-Site Treatment of Creosote and Pentachlorophenol Sludge and Contaminated Soil.* EPA/600/2-91/019. Project #Cr-811498.

MDNR (Michigan Department of Natural Resources). 1988. "How Clean is Clean?" Draft Guidance Document.

MDNR (Michigan Department of Natural Resources). 1992. *Michigan Environmental Response Act Operational Memorandum #8 (Revision 1).*

Mueller, J. G., P. J. Chapman, and P. H. Pritchard. 1989. "Creosote Contaminated Sites — Their Potential for Bioremediation." *Environmental Science and Technology* 23(10): 1197-1201.

Page, A. L., R. H. Miller, and D. R. Keeney (Eds.). 1982. *Methods of Soil Analysis*, Part 2 — Chemical and Microbiological Properties, 2nd ed., Chapter 37, American Society of Agronomy, Inc. and Soil Science Society of America, Inc., Madison, WI.

Sims, J. L., R. C. Sims, and J.E. Mathews. 1989. *Bioremediation of Contaminated Surface Soils.* EPA/600/9-89/073.

Walpole, R. E., and R. H. Myers. 1985. *Probability and Statistics for Engineers and Scientists*, 3rd ed. Macmillan Publishing Company, New York, NY.

Wang, X., X. Yu, and R. Bartha. 1990. "Effect of Bioremediation on Polycyclic Aromatic Hydrocarbon Residues in Soil." *Environmental Science and Technology* 24(7): 1086-1089.

Watts, R. J., M. D. Udell, P. A. Rauch, and S. W. Leung. 1990. "Treatment of Pentachlorophenol-Contaminated Soils Using Fenton's Reagent." *Hazardous Waste & Hazardous Materials* 7(4): 335-345.

Wilson, J. T., J. F. McNabb, D. L. Balkwill, and W. C. Ghiorse. 1983. "Enumeration and Characterization of Bacteria Indigenous to a Shallow Water-Table Aquifer." *Ground Water* 21(2): 134-142.

Biodegradation of Polycyclic Aromatic Hydrocarbons in a Two-Liquid-Phase System

Peter Vanneck, Marit Beeckman, Nancy De Saeyer,
Sigfried D'Haene, and Willy Verstraete

ABSTRACT

The use of a two-liquid-phase system consisting of silicone oil and water for biodegrading polycyclic aromatic hydrocarbons (PAHs) was investigated. Biomass determinations indicated that the cells were mainly growing at the silicone oil-water interface. In shaken and aerated systems with PAHs and inoculum, 97% and 80%, respectively, of the total biomass was attached to the silicone phase. PAH concentrations in the silicone phase dropped by a factor 2 to 100 when microorganisms were present. Biodegradation rates in these systems varied from 3.6 to 5 mg PAH-C/L reactor·d. In the shaken systems at 28°C, the measured CO_2 production rate was equal to 9.1 mg CO_2/L reactor·d and corresponded to a 50% conversion to CO_2. In the aerated systems at 10°C, however, only 25% of the PAH-C was converted to CO_2, resulting in a CO_2 production rate of 0.5 mg CO_2/L reactor·d.

INTRODUCTION

PAHs constitute a class of hazardous chemicals consisting of two or more fused benzene rings that originate from oil, tar, wood-preserving creosote, or incomplete combustion processes (Sims & Overcash 1983). Microorganisms capable of degrading certain low-molecular-weight PAHs have been described, and the basic mechanisms of the metabolic pathways have been elucidated (Cerniglia 1992). However, microbial degradation of sparingly soluble compounds such as PAHs is often restricted by the solubilization rate of the PAHs (Volkering et al. 1992; Stucki & Alexander 1987) or by the toxicity of the parent compound or the metabolites formed (Donnelly et al. 1992, IARC 1983).

Recently there has been considerable interest in two-liquid-phase (aqueous-hydrophobic) systems that provide a biocompatible environment for the microorganisms. In two-liquid-phase systems, the substrate diffuses from the hydrophobic phase to the aqueous phase, and the microorganisms transform or degrade the substrate at the interface and/or in the aqueous phase (Rezessy-Szabo et al. 1987).

In the present study, PAH biodegradation was investigated using a two-liquid-phase system in which silicone oil acted as the hydrophobic phase. The performance of the two-liquid-phase system was assessed by determining the CO_2 production and the degradation rate of the PAHs and by measuring the biomass fractions in the two phases at the end.

MATERIALS AND METHODS

Two-Liquid-Phase System

The two-liquid-phase system consisted of a mineral salts medium supplemented with an inert silicone oil (Janssen Chimica, Beerse, Belgium). The silicone phase was dosed with some of the following PAHs: naphthalene, fluorene, phenanthrene, anthracene, fluoranthene, pyrene, and benzo(a)pyrene. The mineral salts medium (MSM) from Ascon-Cabrera and Lebeault (1993) was used. The inoculum was taken from a contaminated soil near a tar disposal pit at a coke manufacturing plant (Ghent, Belgium). After shaking (150 rpm on a rotary shaker) this contaminated soil in physiological solution (9 g/L NaCl) at a 1/9 ratio (weight/volume) for 2 h at 28°C, a fraction of the settled-out supernatants was used as inoculum.

PAH Biodegradation in Shaken
Two-Liquid-Phase Systems

PAH biodegradation was investigated using four different two-liquid-phase systems in duplicate (Table 1): B, BI, BP, and BIP (B = biphasic, I = inoculum, P = PAHs). Each system consisted of a 100-mL glass Schott bottle with screwcap

TABLE 1. Systems used to investigate PAH biodegradation.

Two-liquid-phase system	Inoculum	PAHs
BIP	+	+
BP	–	+
BI	+	–
B	–	–

B = biphasic or two-liquid-phase.
I = inoculum.
P = PAHs added to the silicone phase.
+ = present.
– = absent.

containing 40 mL of aqueous phase and 15 mL of silicone phase. The aqueous phase was obtained by mixing 35 mL of MSM with 5 mL of inoculum or 5 mL of physiological solution (9 g/L NaCl). The nonsterile two-liquid-phase systems were placed at 28°C on a rotary shaker operating at 70 rpm, and the glass bottles were opened daily to replace the head space and to supply oxygen.

A mixture of five PAHs (naphthalene, phenanthrene, fluoranthene, pyrene, and benzo(a)pyrene) was added to the silicone phase at a concentration of 200 mg/L each. Biomass production was assessed only in the aqueous phase by measuring the optical density at 610 nm (UVIKON 930, Kontron Instruments). After 30 days of incubation, the CO_2 production during 24 h in the different two-liquid-phase systems was determined. The solution of each system was transferred to a glass container with airtight seal, provided with a glass vial filled with 10 mL 0.1 N KOH. The CO_2 captured during 24 h in the 0.1 N KOH was determined with a 0.1 N HCl solution using an automated CO_2-titrator (665 Dosimat with a 686 Titroprocessor, Metrohm, Herisau, Switzerland).

After 40 days, the silicone phase was separated from the aqueous phase using a separatory funnel. The silicone phase was transferred to a centrifuge tube and extracted with 3 × 10 mL of a solvent mixture (acetone-ethanol-chloroform, 10:10:2, vol/vol). This solvent mixture was chosen because of its ability to extract also the biomass associated with the silicone phase (Ascon-Cabrera & Lebeault 1993). The extraction was performed by shaking vigorously for 1 min and by centrifuging for 10 min at 200 g (Sorvall, RC-5B). The solvent mixture was separated from the silicone phase and centrifuged at 8820 g (Sorvall, RC-5B) for 30 min to remove all biomass. The solvent mixture was concentrated to 1 mL of acetonitrile. The efficiency of this method to extract PAHs from silicone oil was first determined in quadruplicate by extracting 10 mL of the silicone oil immediately after adding the PAHs. The aqueous phase was first centrifuged at 8820 g (Sorvall, RC-5B) for 30 min to remove all biomass and then extracted three times with an equal volume of dichloromethane. All dichloromethane fractions were combined and concentrated to 1 mL acetonitrile. After centrifugation the biomass pellets were washed three times with distilled water and the dry weight of the microorganisms was determined after oven-drying at 110°C for 24 h.

PAH Biodegradation in Aerated
Two-Liquid-Phase Systems

In this setup, three two-liquid-phase systems in duplicate were used: BI, BP, and BIP (Table 1). Each two-liquid-phase system consisted of a 500-mL Erlemeyer flask containing 200 mL of aqueous phase and 50 mL of silicone phase. When an inoculum was added, the aqueous phase was made up by mixing 50 mL of mineral salts medium (4 times concentrated) with 150 mL of inoculum. A mixture of five PAHs (fluorene, phenanthrene, anthracene, fluoranthene, and pyrene) was added to the silicone phase at a concentration of 200 mg/L each. Each nonsterile Erlemeyer flask was closed with a rubber stopper that was perforated by two glass tubes, one reaching all the way down through the silicone phase into the aqueous phase and the other one just reaching to the neck of the flask. The first glass tube was used as the air inlet, the second glass tube served as the outlet, which was connected to an air pump (KNF Neuberger, Freiburg, Germany) in the suction mode.

The nonfiltered incoming and outgoing air from each two-liquid-phase system was let through washing bottles containing 100 mL of 1 N KOH to make the incoming air CO_2 free and to trap all the CO_2 produced in the two-liquid-phase system. Every 3 days, the CO_2 concentration in the KOH solutions was determined. Biomass production and PAH concentrations in the silicone phase were determined as mentioned earlier.

PAH Analyses and Chemicals

PAH concentrations in all 1-mL acetonitrile samples were determined with an HPLC-DAD 440 (Kontron Instruments), equipped with a Vydac C18 reversed-phase column (Merck) and a MSI T-660 autosampler (Kontron Instruments). Naphthalene (98%), fluorene (98%), anthracene (97%), fluoranthene (98%), benzo-(a)pyrene (98%), and dibenzothiophene (98%) were obtained from Aldrich (Bornem, Belgium). Phenanthrene (98+%) and pyrene (98+%) were purchased from Janssen Chimica (Beerse, Belgium). All solvents used were high-performance liquid chromatography grade and were bought from UCB (Leuven, Belgium). All other solutions and chemicals used were of the highest purity commercially available.

RESULTS

PAH Biodegradation in Shaken Two-Liquid-Phase Systems

The optical density in the aqueous phase of the BIP-system during the first 14 days was not different from that of the BI-system. However, at day 14 the optical density of the BIP-system dropped from 0.5 to 0.2, whereas the optical density of the BI-system remained constant. At day 30, the CO_2 production during 24 h in the BIP- and BI-systems differed significantly (P < 0.05), and were respectively 1.2 and 1 times the CO_2 produced in the control system. After 40 days, PAH concentrations in the silicone phase and the amount of biomass in the silicone and the aqueous phase were determined. Final PAH concentrations in the silicone phase are given in Table 2. All PAH concentrations were adjusted for their respective extraction efficiencies. The concentrations of all five PAHs were significantly lower (P < 0.05) in the BIP-system compared to the BP-system. No PAHs were detected in the different aqueous phases. The total biomass produced in the BIP-system was 7.5 times that produced in the BI-system. In the latter 38% and 62% of the total biomass were found, respectively, in the aqueous and the silicone phase. However, in the BIP-system, 97% of the total biomass was attached to the silicone phase.

TABLE 2. Final PAH concentrations in the silicone phase.

Systems	PAH concentration (mg/L) in the silicone phase						
	Naphthalene	Fluorene	Phenanthrene	Anthracene	Fluoranthene	Pyrene	Benzo(a)pyrene
Shaken systems (day 40):							
BP	155 ± 7[a]	INP[c]	106 ± 6	INP[c]	127 ± 4	211 ± 5.0	236 ± 19
BIP	ND[b]	INP[c]	2.31 ± 0.01	INP[c]	0.91 ± 0.01	1.56 ± 0.04	3.10 ± 0.51
Aerated systems (day 40):							
BP	INP[c]	116 ± 7	36.9 ± 0.8	19.8 ± 2.1	67.8 ± 22.7	121 ± 11	INP[c]
BIP	INP[c]	81.9[d]	2.65 ± 0.95	18.3[d]	11.02[d]	109 ± 10	INP[c]

Initial PAH concentration = 200 mg/L each.
(a) Mean ± standard deviation of duplicate treatments.
(b) Not detected.
(c) Initially not present.
(d) Single measurement.

PAH Biodegradation in Aerated
Two-Liquid-Phase Systems

The purpose of this setup was to supply excess oxygen to the two-liquid-phase systems by aeration and to monitor continuously the CO_2 production as a measure of the mineralization rate of the PAHs. No clear difference in optical density of the aqueous phase between the BIP- and BI-systems was observed; however similar to the shaken systems, the optical density in the BIP-system dropped from 0.5 to 0.2 as degradation proceeded. Airflow rates through the different two-liquid-phase treatments varied throughout the experiment from 0.36 L/min to 0.76 L/min. The CO_2 production in the BIP-system was 1.65 times higher and significantly different (P <0.05) than in the other systems (BI and BP). After 40 days, the PAH concentrations in the silicone phase and the amount of biomass in the aqueous and the silicone phase were determined. Final PAH concentrations in the silicone phase are presented in Table 2. All PAH concentrations were adjusted for their respective extraction efficiency. Concentrations of fluorene, phenanthrene, and fluoranthene decreased in the BIP-system compared to the BP-system, but the difference was only significant (P<0.05) for phenanthrene. No PAHs were detected in the different aqueous phases. The amount of biomass produced in the BI- and BIP-systems was, respectively, 1.9 and 3.2 times that produced in the BP-system. In the BP-system, growth occurred at the end due to the nonsterile conditions: 8% of the total biomass was present in the aqueous phase and 92% was attached to the silicone phase. In the BI- and BIP-systems, respectively, 32% and 20% of the total biomass were found in the aqueous phase and 68% and 80% were associated with the silicone phase.

DISCUSSION

In all BIP-systems, the optical density in the aqueous phase showed a significant (P <0.05) drop after 14 days. This decrease in optical density may suggest that many of the cells are attaching to the silicone oil-water interface. Efroymson and Alexander (1991) reported that the number of *Arthrobacter* sp. cells in the aqueous phase decreased with time in the presence of heptamethylnonane. However, the adherence of *Arthrobacter* sp. to the hydrophobic phase was not related to the presence of a C-source for the bacteria. This observation was confirmed by the biomass distribution among the different phases. Variation coefficients for the biomass determinations in the aqueous and silicone phase were 21% and 4%, respectively. In both the shaken and aerated systems, the total biomass production was the highest in the BIP-systems, respectively, 7.3 and 1.6 times the biomass produced in the BI-systems, and significantly different (P <0.05) from that in the BI systems. In the BIP-systems, 80 to 97% of the total biomass was attached to the silicone phase, whereas this was only 62 to 68% in the BI-systems. Ascon-Cabrera and Lebeault (1993) examined the mineralization of chlorinated and nonchlorinated xenobiotic compounds in a biphasic system consisting of silicone oil and water and also observed that the microorganisms were primarily growing at the interface. However, the distribution of the micro-organisms between the two phases depends on the type of bacterial strain used: surface-attached microorganisms may utilize the substrate directly from the hydrophobic phase, while strains that grow exclusively in the aqueous phase take up dissolved substrate from the aqueous phase. Köhler et al. (1994) showed that *Pseudomonas aeruginosa* AK1 did not attach to the lipophilic surface of a heptamethylnonane-water system and utilized dissolved phenanthrene only in the aqueous phase as a C-source.

PAH analyses of the silicone phase in the shaken systems (Table 2) showed the positive effect of the inoculum: in the BIP-system, the PAH concentrations decreased to 3 mg/L or less; however in the BP-system, the PAH concentrations stayed above 100 mg/L. Naphthalene, phenanthrene, and fluoranthene concentrations in the latter system decreased somewhat probably due to physical-chemical transformations such as evaporation, light-oxidation and polymerization; however pyrene and benzo(a)pyrene concentrations remained constant. When only taking volatilization into account and assuming that the gas phase is replaced daily when opening the bottles for supplying oxygen, the remaining PAH concentrations in the liquid phase after 40 days would be 93 mg/L for naphthalene, 138 mg/L for phenanthrene, 198 mg/L for fluoranthene, and 200 mg/L for pyrene and benzo(a)pyrene. Half-lives for the five PAHs based on the observed data are all in the order of a few weeks (1 to 4), whereas Cerniglia (1992) reported half-lives of phenanthrene, pyrene, and benzo(a)pyrene in the aqueous phase of several months to years. Ascon-Cabrera and Lebeault (1993) also demonstrated that a two-liquid-phase system permitted acceleration of acclimation and an increase in degradation of chlorinated and nonchlorinated compounds, compared to a monophasic aqueous system. The amount of PAH-C biodegraded in the shaken systems was 202 mg/L reactor, resulting in an overall biodegradation rate of 5 mg PAH-C/L reactor·d (at 28°C). However, because no in-between PAH data are available, the assumed linearity of this biodegradation process cannot be assessed. When taking the daily CO_2 production at day 30 as an average, the measured mineralization rate is 9.1 mg CO_2/L reactor·d, which corresponds well to a 50% conversion of the PAH-C to CO_2 (9.3 mg CO_2/L reactor·d).

In the aerated systems, the aeration rate ranged from 86 L/L reactor·h to 182 L/L reactor·h and caused a serious stripping effect of the PAHs from the two-liquid-phase systems, as indicated by the low PAH concentrations in the BP-system (Table 2). However, the concentrations of fluorene, phenanthrene, and fluoranthene in the BIP-system were much lower than those in the BP-system, but the difference was only significant ($P < 0.05$) for phenanthrene. The concentrations for anthracene and pyrene were similar in both systems (BP and BIP). In the shaken two-liquid-phase systems the concentrations of all five PAHs, including pyrene, decreased significantly in the presence of the inoculum. Mueller et al. (1989) also showed that a fluoranthene-utilizing bacterial community could degrade other PAHs such as anthracene and pyrene. It is possible that the microorganisms responsible for the degradation of anthracene and pyrene were suppressed by those responsible for the degradation of fluorene, phenanthrene, and fluoranthene. PAH-C biodegradation rates in the aerated systems at 10°C were 0.6 mg PAH-C/L reactor·d. Based upon the trapped CO_2, the measured mineralization rate equaled 0.5 mg CO_2/L reactor·d, indicating that in these systems only 25% of the PAH-C biodegraded is converted to CO_2.

These studies indicate that PAHs can be degraded in a two-liquid-phase bioreactor consisting of silicone oil and water. Microorganisms grow primarily at the silicone oil-water interface and are degrading the PAHs directly from the silicone phase. Such two-liquid-phase bioreactors can be a novel bioremediation technology for treating PAH-contaminated water.

REFERENCES

Ascon-Cabrera, M., and J. M. Lebeault. 1993. "Selection of xenobiotic-degrading microorganisms in a two-liquid-phase aqueous-organic system." *Applied and Environmental Microbiology* 59: 1717-1724.

Cerniglia, C. E. 1992. "Biodegradation of polycyclic aromatic hydrocarbons." *Biodegradation* 3: 351-368.

Efroymson, R. A., and M. Alexander. 1991. "Biodegradation by an Arthrobacter species of hydrocarbons partitioned into an organic solvent." *Applied and Environmental Microbiology* 57: 1441-1447.

Donnelly, K. C., C. S. Anderson, J. C. Thomas, K. W. Brown, D. J. Manek, and S. H. Safe. 1992. "Bacterial mutagenicity of soil extracts from a bioremediation facility treating wood-preserving waste." *Journal of Hazardous Materials* 30: 71-81.

International Agency for Research on Cancer (IARC). 1983. *Polynuclear aromatic compounds part 1: chemical, environmental and experimental data.* Vol. 32. Lyon, France.

Köhler, A., M. Schüttoff, D. Bryniok, and H. J. Knackmuβ. 1994. "Enhanced biodegradation of phenanthrene in a biphasic culture system." *Biodegradation* 5: 93-103.

Mueller, J. G., P. J. Chapman, and P. H. Pritchard. 1989. "Action of a fluoranthene-utilizing bacterial community on polycyclic aromatic hydrocarbon components of creosote." *Applied and Environmental Microbiology* 55: 3085-3090.

Rezessy-Szabo, J. M., G.N.M. Huijberts, and J.A.M. de Bont. 1987. "Potential of organic solvent in cultivating microorganisms on toxic water-insoluble compounds." In C. Laane, J. Tramper, and M. D. Lilly (Eds.), *Biocatalysis in organic media.* Elsevier, Amsterdam, Netherlands. pp. 295-302.

Sims, R. C., and M. R. Overcash. 1983. "Fate of polynuclear aromatic compounds (PNAs) in soil-plant systems." *Residue Reviews 88*: 1-68.

Stucki, G., and M. Alexander. 1987. "Role of dissolution rate and solubility in biodegradation of aromatic compounds." *Applied Environmental Microbiology* 53: 292-297.

Volkering, F., A. M. Bruere, A. Sterkenburg, and J. G. van Andel. 1992. "Microbial degradation of polycyclic aromatic hydrocarbons: effect of substrate availability on bacterial growth kinetics." *Applied Microbiology and Biotechnology* 36: 548-552.

Evaluation of an Integrated Treatment System for MGP Site Groundwaters

O. Karl Scheible, Gary M. Grey, and Joy A. Maiello

ABSTRACT

Initially studied at bench scale, process sequences comprising dissolved air flotation (DAF), aerobic biological oxidation, air stripping, filtration, and carbon adsorption were demonstrated at pilot scale at a manufactured gas plant (MGP) site in New Jersey. Benzene, toluene, ethylbenzene, and xylenes (BTEX), and polycyclic aromatic hydrocarbons (PAHs) were the primary organics in the groundwater, ranging from levels of 2 to 8 mg/L and 0.3 to 27 mg/L, respectively; chemical oxygen demand (COD) levels were from 60 to 4,500 mg/L. Significant levels of dense, emulsified, and nonaqueous tars and oils were present in the more highly contaminated waters and were effectively removed by DAF. Carbon-based fluidized-bed biological treatment of the DAF subnatant at COD loadings between 2 and 4 g/L-d yielded effluent-soluble COD levels between 40 and 60 mg/L, with both residual BTEX and PAH concentrations ranging from nondetect levels to 0.1 mg/L. Subsequent polishing by filtration and carbon adsorption resulted in additional COD removals and nondetect levels of volatiles and semivolatiles. Air stripping was effective in lieu of the biological process for both volatile organic compound (VOC) and PAH removal.

INTRODUCTION

Investigations at a former manufactured gas plant site in New Jersey identified PAHs and VOCs as the primary contaminants in the soils and groundwaters. The remediation of the site will consist of site capping, on-site groundwater control and treatment, and off-site groundwater monitoring. In a project sponsored by Jersey Central Power and Light Company (JCP&L), New Jersey Natural Gas Company (NJNG), and the Electric Power Research Institute (EPRI), bench- and pilot-scale investigations were conducted to develop and demonstrate an integrated treatment system for the contaminated groundwaters (HydroQual, Inc. 1994). The studies were conducted over a range of groundwater contaminant levels representative of the range that might be encountered over the operating life of a full-scale pump-and-treat system at the site, and reflective of the variable conditions that exist at the many MGP sites across the country.

EXPERIMENTAL METHODS AND MATERIALS

The first phase of the study focused on bench-scale evaluations of applicable technologies, using groundwater composites developed from samples taken from the site. These conducted April through October 1993. Based on the results of the bench-scale work, a pilot facility was designed and assembled at the site; this facility was operated on groundwaters drawn continuously from selected locations over the period from April through July 1994.

The conceptual approach to the bench-scale studies is shown on Figure 1. The first task involved jar-test evaluations of chemical conditioning via coagulation, flocculation, pH adjustment, and/or chemical oxidation to enhance the separation of nonaqueous solids and their subsequent removal by settling or DAF. Bench-scale tests were then conducted with the preconditioned groundwaters to study gravity settling and DAF processes. The secondary element included evaluations of air stripping, carbon adsorption, and biological processes for their ability to remove dissolved organics, primarily BTEX and PAHs. Carbon adsorption, evaluated by isotherm procedures, was also investigated as a tertiary process following air stripping or biological treatment. Bench-scale, continuous flow reactors configured for the conventional activated sludge and fluidized-bed processes were used to study aerobic biological treatment. In the case of the fluidized bed, both sand and activated carbon were used as the biomass support medium. Preliminary estimates of residuals production were also developed for the primary (DAF) and secondary (biological) treatment processes.

A schematic of the pilot treatment systems is shown on Figure 2. Two process trains were evaluated. The first included DAF, a biological granular activated carbon-based fluidized bed, filtration, and granular activated carbon (GAC) adsorption. The DAF process, supplied by Komline-Sanderson, consisted of a flash-mix tank, flocculation tank, compressor, and flotation tank. The 1.4-m^2 DAF unit was operated at 65 to 80 m^3/d (with 100% recycle) semicontinuously to develop batches of subnatant for the subsequent processes. Polymer was added to the flash mix tank; float sludge was continuously withdrawn and held in drums. Two 27-m^3 tanks were used to store the subnatant. The DAF unit and tanks were closed and vented to a 9-m stack.

The skid-mounted, carbon-based fluidized-bed bioreactor (CFB) was provided by Envirogen, Inc. It was equipped with a 30-cm-diameter by 4.6-m-high reactor, microprocessor control, feed and recycle pumps, chemical feed equipment, oxygenator, and recycle tank. This system was operated continuously at flows of 4 to 12 L/min. The air stripper was a Shallow Tray™ system provided by Environmental Restoration Systems Inc. It was comprised of three trays, with a capacity of 4 to 160 L/min and a blower capacity of 8.5 m^3/min. The filter and GAC columns following the biological system were 23.5-cm-diameter acrylic columns,

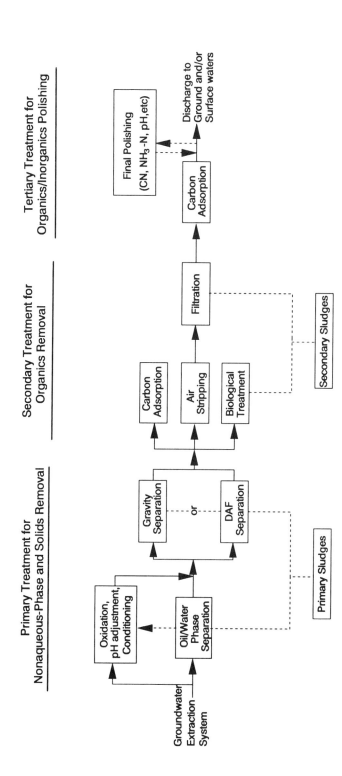

FIGURE 1. Conceptual approach to bench-scale studies.

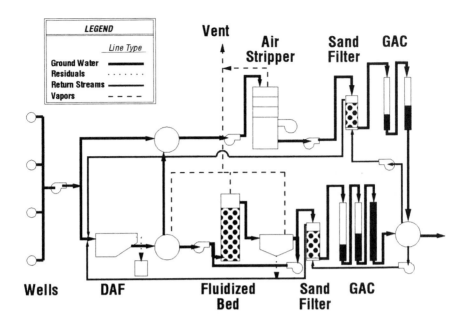

FIGURE 2. Schematic of MGP site groundwater pilot treatment plants.

1.2 m deep; these were operated at loading rates of 120 to 350 m^3/m^2-d. Those following the air stripper were 10-cm-diameter units, capable of handling flows up to 2.7 m^3/d. The filtration medium in both cases was greensand (zeolite), chosen for its ability to remove iron and ammonia. Calgon Filtrasorb 300 was used for the GAC columns.

The pilot experimental program consisted of three "campaigns" conducted over a period of approximately 3 months. After an initial acclimation and startup period for the biological system, the DAF — CFB — GAC process train was operated first with the higher strength groundwater, and then with moderate- and low-strength waters. The operation of the DAF was discontinued with the low-strength waters. The air stripper-GAC treatment sequence was tested with only the moderate- and low-strength waters.

RESULTS

Table 1 presents a summary of the groundwater characteristics. The high-strength waters were highly colored and had a distinctive coal tar odor. They contained an emulsion of dense and light, nonaqueous-phase tars and oils (DNAPLs and LNAPLs). These constituted a significant fraction of the PAHs in the sample; simple filtration would bring the sample to levels comparable to the moderate-strength groundwaters, which were primarily in a dissolved state. The moderate-strength groundwater also had a distinctive odor and a

TABLE 1. Summary of MGP site groundwater characteristics.

	Contamination Levels		
	High	Moderate	Low
Conventionals (mg/L)			
BOD	19		7.5
COD — Total	4,500	120	64
COD — Soluble	190	108	58
TSS	600	16	13
Grease & Oil	900	6.1	4.6
Volatiles (µg/L)			
Benzene	2,600	1,300	540
Total BTEX	7,800	4,100	1,600
Styrene	6,600	<100	<20
Total Volatiles	14,400	4,100	1,600
Semivolatiles (µg/L)			
Naphthalene	10,000	2,300	48
Carcinogenic PAH	1,700	<100	<5
Total PAH	27,000	3,000	270

slight sheen, without significant levels of DNAPL. The low-strength groundwater did not have an odor or sheen, but did contain significant contaminant levels. The volatiles were comprised almost exclusively of BTEX compounds; the high-strength groundwater also contained significant levels of styrene. The semivolatiles in the groundwaters were principally base/neutral extractable PAHs. Of these, naphthalene was a primary constituent, measured at levels as high as 10,000 µg/L; carcinogenic PAHs were also present in the high-strength groundwater at a level of 1,700 µg/L.

Bench-Scale Evaluations

The raw groundwaters, particularly the high-strength samples, exhibited little tendency to naturally settle or separate into distinct phases. Hydrogen peroxide, potassium permanganate, and diffused air oxidation were ineffective at various pH levels in enhancing separation. Coagulant addition did not improve performance, except at elevated pH (pH 10 to 12), particularly with the addition of a cationic polymer. Settling was enhanced at an elevated pH only (pH 11). DAF tests revealed that the solids could be floated fairly easily and efficiently at ambient conditions (pH 7.1), with the addition of 10 mg/L cationic polymer. Because pH adjustments were not needed before and after the process, DAF was selected for primary treatment, and was used to generate pretreated groundwaters for subsequent studies.

All three biological systems were able to successfully treat the DAF sub-natants. The carbon-based fluidized bed produced the best effluent, with both VOCs and PAHs at nondetectable levels (1 to 10 µg/L). The carbon medium retains the larger, less degradable compounds (such as the higher-ring-structured PAHs) beyond the system's hydraulic detention time, affording a greater oppor-tunity for breakdown than in the sand-based fluidized-bed system, or in a con-ventional suspended-growth system. In both of these configurations, at similar loadings, residual PAHs were still evident (Grey et al. 1995). Residual soluble COD levels also were lower, ranging from 30 to 40 mg/L.

Bench-scale air stripping tests confirmed that efficient removal of the VOCs could be achieved in the MGP site groundwater matrix. BTEXs were nondetect-able, while PAHs were removed by 30 to 40%. Regarding GAC adsorption, car-bon requirements were greater for samples containing residual VOCs, as expected, confirming that carbon usage can be minimized by first removing the VOCs. The studies indicated that nondetectable organic levels can be achieved at moderate carbon loadings (130 mg TOC/g GAC).

Pilot-Scale On-Site Treatment

Table 2 presents a summary of treatment performance for the process sequence comprising dissolved air flotation, fluidized bed, filtration, and granular activated carbon adsorption (DAF-CFB-GAC). The DAF system was operated with addition of 10 mg/L cationic polymer at hydraulic loading rates between 65 and 80 m^3/m^2-d. Six to 8 gal (23 to 30 L) of float sludge (30 to 40% solids) was generated with the high-strength groundwaters; centrifuge tests suggest that this volume can be further reduced, with a centrate of less than 0.5% solids.

The high- and moderate-strength DAF subnatants were similar with respect to COD and grease and oil levels, suggesting that the soluble contaminant levels were similar. However, the VOC and PAH concentrations in the high-strength subnatant were higher by a factor of approximately eight. The low-strength groundwater (without DAF pretreatment) had VOC and PAH levels similar to the moderate-strength DAF subnatant, but lower COD. The fluidized bed was operated at loadings between 2 and 4 g COD/L-d. The hydraulic flux, set by the required level of fluidization (20 to 30% bed expansion), was kept at a relatively constant level of 500 m^3/m^2-d. Effectively complete removals of VOCs and PAHs were accomplished, with effluent concentrations ranging from non-detect levels to approximately 100 µg/L, depending on the quality of the DAF subnatant. Soluble effluent COD levels ranged between 40 and 60 mg/L.

Filtration was effective in capturing the fluidized-bed effluent solids, although backwashing was frequent, particularly during periods when significant sloughing was occurring within the fluidized-bed reactor. Provision of a roughing filter or clarification (possibly with polymer addition) may be appropriate. Substantial COD reductions were accomplished through carbon adsorption, yielding levels from 5 to 20 mg/L at an empty bed contact time (EBCT) of 4 to 17 min. In all cases, VOCs and PAHs were nondetectable in the GAC effluents.

TABLE 2. Summary of DAF — CFB — GAC treatment performance.

Analytes	Campaign One: High-Strength Groundwaters				Campaign Two: Moderate-Strength Groundwaters				Campaign Three: Low-Strength Groundwaters		
	Raw	DAF	CFB Bio	GAC	Raw	DAF	CFB Bio	GAC	Raw	CFB Bio	GAC
Conventionals (mg/L)											
BOD	19	22	6.8	4	—	10	10.8	2.4	7.5	7.8	3.9
COD — Total	4,500	140	80	17	120	100	60	5	64	53	16
COD — Soluble	190	110	60	14	108	80	43	4	58	37	17
TSS	600	11	15	2	16	19	25	1.1	13	32	2.8
Grease & Oil	900	8.7	<5	<5	6.1	6.8	<5	<5	4.6	<5	<5
Volatiles (µg/L)											
Benzene	2,600	1,900	6.4	<2	1,300	960	<2	<2	540	<2	<2
Total BTEX	7,800	5,200	66	<2	4,100	2,500	<2	<2	1,600	<2	<2
Styrene	6,600	12,000	2	<2	<100	<40	<2	<2	<20	<2	<2
Total Volatiles	14,400	17,200	68	<2	4,100	2,500	<2	<2	1,600	<2	<2
Semivolatiles (µg/L)											
Naphthalene	10,000	610	<5	<5	2,300	3.5	<5	<5	48	<5	<5
Carcinogenic PAH	1,700	275	44	<5	<100	<5	<5	<5	<5	<5	<5
Total PAH	27,000	1,800	92	<5	3,000	210	<5	<5	270	<5	<5

Table 3 presents a summary of the pilot plant performance for the dissolved air flotation-air stripping-carbon adsorption sequences. During Campaign 2, the DAF was able to reduce PAH levels to approximately 200 µg/L and VOCs to approximately 2,500 µg/L. The Shallow Tray™ air stripper reduced both VOCs and PAHs to levels less than detection, at a maximum hydraulic loading of 180 m³/m²-d and a minimum gas to liquid ratio of 30. In Campaign 3, the air stripper was again able to remove VOCs and PAHs to levels less than detection, similar to the system performance during Campaign 2. In both cases, carbon adsorption provided a further reduction of COD, yielding levels less than 10 mg/L at an EBCT of approximately 12 min.

SUMMARY

The DAF process is effective in removing nonaqueous-phase tars, oils, and PAHs. It affords good protection of downstream processes and produces a float sludge that can be further dewatered by centrifugation. High leachable benzene levels in the final sludge may cause the residual to be classified as hazardous.

TABLE 3. Summary of DAF — air stripping — GAC performance.

Analytes	Campaign Two: Moderate-Strength Groundwaters				Campaign Three: Low-Strength Groundwaters		
	Raw	DAF	AS	GAC	Raw	AS	GAC
Conventionals (mg/L)							
BOD	—	10	—	1.4	7.5	—	0.9
COD	120	100	81	4	64	35	8
TSS	16	19	8.8	1.5	13	18	3.5
Grease & Oil	6.1	6.8	—	5.3	4.6	—	<5
Volatiles (µg/L)							
Benzene	1,300	960	<2	<2	540	<2	<2
Total BTEX	4,100	2,500	<2	<2	1,600	<2	<2
Styrene	<100	<40	<2	<2	<20	<2	<2
Total Volatiles	4,100	2,500	<2	<2	1,600	<2	<2
Semivolatiles (µg/L)							
Naphthalene	2,300	3.5	<5	<5	48	<5	<5
Carcinogenic PAH	<100	<5	<5	<5	<5	<5	<5
Total PAH	3,000	210	<5	<5	270	<5	<5

A high degree of treatment can be provided by process trains comprising DAF, either biological or air stripping of the subnatant, followed in the end by carbon adsorption. The performance of the two process trains under moderate- and low-strength groundwater conditions was similar; their selection may be dictated by capital, operating, and maintenance life-cycle costs. Air stripping, although a very efficient removal process for VOCs (and PAHs to a more limited extent), may require off-gas treatment. On the other hand, biological treatment, which may require greater operator attention, is a destructive process that eliminates the need for handling organics transferred to another medium. Carbon adsorption may be required only for polishing effluents from treatment of higher strength groundwaters. Air stripping and biological processes were both found capable of complete treatment for low to moderately contaminated waters.

ACKNOWLEDGMENTS

The authors are all with HydroQual, Inc., Mahwah, New Jersey, which conducted this study under subcontract to Foster Wheeler Environmental Corp., Lyndhurst, New Jersey. The work was overseen by the EPRI Steering Committee, comprised of Dr. I. Murarka (EPRI), Mr. C. R. Sweeney (JCP&L), Mr. E. Sawicki (NJNG) and Mr. A. Jain (ERM, Inc.). Mr. Jain served as the Project Manager for the EPRI Steering Committee.

REFERENCES

Grey, G. M., O. K. Scheible, J. A. Maiello, W. Guarini, and P. Sutton. 1995. "Biological Fluidized-Bed Treatment of Groundwater from a Manufactured Gas Plant Site." In R. E. Hinchee, G. D. Sayles, and R. S. Skeen (Eds.), *Biological Unit Processes for Hazardous Waste Treatment*, Battelle Press, Columbus, OH.

HydroQual, Inc. 1994. "Bench- and Pilot-Scale Treatability Studies for Contaminated Ground Waters from the Toms River Former Manufactured Gas Plant Site." Draft report prepared under subcontract to Foster Wheeler Environmental Corporation, Lyndhurst, NJ for the Electric Power Research Institute, Jersey Central Power and Light, and New Jersey Natural Gas.

Biological and Abiotic Dechlorination of Highly Chlorinated Dioxins in Biphasic Microcosms

Andrei Barkovskii and Peter Adriaens

ABSTRACT

A novel experimental approach to help increase the rates and extent of reductive dechlorination of polychlorinated dibenzo-p-dioxins (PCDD) is presented. Biphasic microcosm emulsions containing eluted micro-organisms derived from historically contaminated Passaic River (New Jersey) sediments, and 4% (v/v) of decane, were spiked with (5.25 ±0.25) mg/L of octaCDD. The microcosms were amended separately with three polyphenolic compounds—catechol, resorcinol, and 3,4-di-hydroxybenzoate—to help improve electron transfer during reductive dechlorination. Abiotic controls containing phenolic compounds only, and pasteurized cells were monitored along with the active microcosms. Lesser-chlorinated congeners were observed in all treatments, including killed cells, indicating the potential not only for biological and abiotic, but also biogenic dechlorination mechanisms. After 3 months of incubation, tetraCDD isomers were produced in biological incubations only, and up to 30% of the spiked octaCDD was removed. Polyphenolic compounds first appear to transiently complex with the dioxins prior to further dechlorination, and did not increase the dechlorination rates over unamended cells. Whereas the 2,3,7,8-/1,4,6,9-substitution ratio of hepta-chlorinated congeners increased in all treatments, 2,3,7,8-substituted hexaCDDs congeners were identified mainly in active cell incubations. Further isomer-specific analysis may thus enable distinction between abiotic and biotic dechlorination processes in anaerobic sediments.

INTRODUCTION

The widespread distribution of polychlorinated dibenzo-p-dioxins and dibenzofurans in the hydrosphere, atmosphere, and biosphere has been recognized. Yet little is known about their fate, in particular with respect to their susceptibility to microbial transformation in anaerobic soils and sediments. Previously, we reported on the microbially mediated reductive dechlorination of highly chlorinated PCDD in anaerobic soils and sediments (Adriaens & Grbic-Galic 1994; Adriaens et al. 1995a, 1995b). *Peri*-dechlorination was found to be preferred, resulting in the transient accumulation of 2,3,7,8-substituted congeners. The rates of dechlorination were on the order of 10^{-4} d^{-1}, and up to three chlorines were removed per molecule, depending on the spiked congener.

Although biological and abiotic dechlorination of PCDD/F are still largely unexplored, these systems merit attention as they could be stimulated in situ in PCDD/F-contaminated environments, once the electron transfer mediator(s) has (have) been identified and can be stimulated, and when bioavailability can be enhanced. Recently, the issue of bioavailability of sparingly soluble compounds was investigated by dissolving the organic in a nonaqueous-phase liquid (NAPL) such as *n*-hexadecane and 2,2,4,4,6,8,8-heptamethylnonane. It was found that the aerobic mineralization rate of phenanthrene exceeded the rates of spontaneous partitioning in water, and that thus improved mass transfer rates into the aqueous phase increased the bioavailability (Ascon-Cabrera & Lebeault 1991; Efroymson & Alexander 1994a; 1994b; Köhler et al. 1994).

Polyphenolic compounds are ubiquitous in the environment and influence soil biological processes through several interactions, in particular due to oxidation-reduction reactions coupled to metals (Deiana et al. 1992; Lehmann et al. 1987) or organic compounds (Flaig et al. 1975; Lehmann & Cheng 1988). Similarly, the presence of quinonic compounds as electron transfer mediators was shown to increase the reducing activity to nitroaromatic compounds in homogeneous aquatic systems (Dunnivant et al. 1992; Schwartzenbach et al. 1990). Recently, it was demonstrated that polyphenolic compounds can be used both as organic intermediates for electron transfer to the chemical terminal electron acceptor, and to support bacterial respiration as a terminal electron acceptor under oxygen-limiting conditions (Barkovskii et al. 1994b).

Because highly oxidized and sparingly soluble compounds such as PCDDs could be appropriate acceptors for electrons flowing via polyphenolic compounds from bacterial cells, the objectives of this study were to evaluate the effects of polyphenolic compounds and a NAPL on reductive dechlorination of PCDDs under anaerobic conditions.

MATERIALS AND METHODS

Medium and Inoculum

The anaerobic medium, containing a filter-sterilized mixture of aliphatic and organic acids to augment bacterial activity, and the characteristics of the Passaic River inoculum have been described earlier (Adriaens & Grbic-Galic 1992, 1994; Bopp et al. 1991). Microbial cell suspensions were prepared from acclimated sediment which was incubated with anaerobic medium (1:3 w/w), and supplied with a mixture of primary substrates (100 mg/L). After 2 months, the sediment was allowed to settle and the supernatant, containing the cells, was transferred and supplied with the primary substrate mixture (100 mg/L).

Three subsamples (5 mL each) were introduced into 20-mL vials and hermetically sealed with Teflon™-coated rubber stoppers. The headspace in the vials was analyzed for the presence of methane on a biweekly basis. When the amount of methane reached 200 to 300 nM/flask, the acclimated sediment-derived cells were used as inoculum.

Three polyphenolic compounds were chosen based their ability to form a quinoid structure, and accept electrons from a convenient electron donor. Catechol (CAT) and resorcinol (RES) represent the *ortho-* and *meta-*conformations, respectively, of ubiquitous polyphenolic molecules in the environment. The third compound, 3,4-dihydroxybenzoic acid (DHBA), is a common intermediate from aerobic degradation of aromatic compounds, and was recently demonstrated to serve as a respiratory terminal acceptor for bacteria. Decane (4%, v/v) was chosen as the solubilizing agent for octaCDD in the microcosm, based on its minimal effect on methanogenic activity.

Experimental Setup

A conceptual model, representing the carbon and electron flow in the experimental set-up, is illustrated in Figure 1. Microcosms were established in 20-mL serum bottles sealed with Teflon™-coated butyl rubber stoppers containing 5 mL

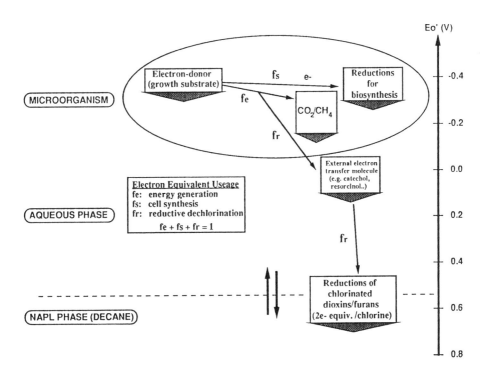

FIGURE 1. Conceptual model of carbon and electron flow in the experimental setup.

of prereduced medium and sediment-derived cells (30 ± 4 µg/L of protein). The microcosms were spiked with octaCDD dissolved in decane as second (organic) phase to a final concentration of $5,250 \pm 250$ µg/L. This standard was found to be contaminated with trace amounts of 1,2,3,4,6,7,9- and 1,2,3,4,6,7,8-heptaCDD (278 ± 20 and 173 ± 17.2 µg/L, respectively), and two unidentified hexaCDD (9.1 ± 0.1 and 8.2 ± 0.05 µg/L). The final concentrations of the contaminant dioxins in the microcosm were 11.2 ± 0.95 µg/L and 6.99 ± 0.69 µg/L for the heptaCDD, and 0.365 ± 0.66 µg/L and 0.328 ± 0.26 µg/L for the hexaCDD, respectively.

Triplicate microcosms were established for the following treatments: (1) unamended and amended active cells with 0.1 mM (final concentration) of resorcinol (RES), catechol (CAT) and with 3,4-dihydroxybenzoic acid (DHBA); (2) biological controls pasteurized at 70°C for 30 min on 4 consecutive days; and (3) chemical controls spiked with octaCDD and one of the polyphenolic compounds. The experimental matrix is shown in Table 1. The microcosms were incubated at 30°C in the dark with agitation (270 rpm), and were supplied monthly with 50 µL from a filter-sterilized substrate stock solution (100x) to maintain microbial activity and a steady supply of electrons generated from anaerobic oxidation during the time of incubation.

Analytical Procedures

After 0, 1, and 3 months, triplicates of each treatment were sacrificed for PCDD analysis. To correct for PCDD recovery efficiencies 1,2,3,4-tetraCDD (5 mg/L) was used as an internal standard. Samples were extracted manually three times with toluene. Extracts were dried under nitrogen at 45°C, redissolved in 100 µL of toluene, and used for gas chromatography/mass spectroscopy (GC/MS) analysis. Prior to analysis, the GC vials were kept at −5°C. Recovery efficiencies, based on substrate recovery corrected with 1,2,3,4-tetraCDD were consistently 80 to 90%.

The analytical method was adapted from Adriaens and Grbic-Galic (1994), with modified SIM (M+, (M+1)+, (M+2)+, (M+3)+, (M+4)+, dwelling of 100 msec per

TABLE 1. Experimental matrix for PCDD dechlorination in biphasic microcosms.

Amendments (n = 9/amendment)[a]		
Active Microcosms	**Biological Controls**	**Chemical Controls**
1. Cells w/o polyphenols	5. Pasteurized cells	6. Media + resorcinol
2. Cells + resorcinol		7. Media + catechol
3. Cells + catechol		8. Media + 3,4-DHBA
4. Cells + 3,4-DHBA		

(a) At each time point $8 \times 3 = 24$ microscosms were sacrificed.

ion). Quantitation was based on a six-level calibration established for a custom-made mixture of dioxin congeners, representing the five PCDD homolog groups monitored. The congeners were: 1,2,3,4-T_4CDD, 1,2,4,7,8-P_5CDD, 1,2,3,4,7,8-H_6CDD, 1,2,3,4,6,7,9-H_7CDD, and 1,2,3,4,6,7,8,9-OCDD. Whereas congener identification was based on coelution with a known standard, the method allows for automatic updating when unknown congeners within a specific homolog group would appear during sequential dechlorination.

RESULTS AND DISCUSSION

OctaCDD was dechlorinated in all amendments during the 3-month incubation study. Yet the patterns of dechlorination differ markedly between the different treatments, based on both the appearance of lesser-chlorinated congeners and the apparent mechanism of dechlorination. To differentiate between the different processes in each of the treatments, the appearance of lesser-chlorinated metabolites should be evaluated.

The congener-specific patterns for heptaCDD appearance in all treatments are compared in Figure 2, A-D. The initial concentrations of each congener represent to the contamination in the octaCDD standard, and are higher for the 2,3,7,9-congener than for the 2,3,7,8-substituted congener (see Materials and Methods). The general trends for each congener are essentially the same for the 3-month time period sampled. Two clearly distinctive patterns can be observed depending on the treatment. Whereas all the incubations containing polyphenolic compounds show an immediate decrease of the contaminant hepta-CDD prior to a net increase of both congeners, the congener concentration in unamended live and killed cells increases dramatically at the 3-month sampling point. This difference is postulated to be due to a rapid chemical interaction between the polyphenolic compounds and the dioxin congeners, as the pattern is observed in chemical controls as well (Figure 2, C-D). Under reduced conditions, radicals formed as metastable configurations of the polyphenolic compounds due to one-electron reductions in either microbial or chemical systems may be the driving force behind this complexation or condensation mechanism (March 1988; Dunnivant et al. 1992). The redox conditions in the anaerobic medium are favorable for these reactions, as they occur immediately (time 0).

Heptachlorinated congeners appear in all treatments, including the controls with dead cells, suggesting a biogenic mechanism involving a biological co-factor which is released upon pasteurization (Dybas et al. 1995; Assaf-Anid 1992). Vitamin B_{12}-mediated dechlorination of highly oxidized molecules, such as polychlorinated biphenyls, dibenzodioxins, and dibenzofurans, has been described earlier (Assaf-Anid et al. 1992; Barkovskii et al. 1994a; Adriaens et al. 1995b). The net production of heptaCDD congeners is similar between chemical controls, dead cells, and unamended treatments. Even though polyphenols are capable of mediating dechlorination of octaCDD to heptaCDD in chemical control microcosms, they do not enhance dechlorination in amended active cell incubations; amendments with resorcinol even decrease heptaCDD production. The ratio of

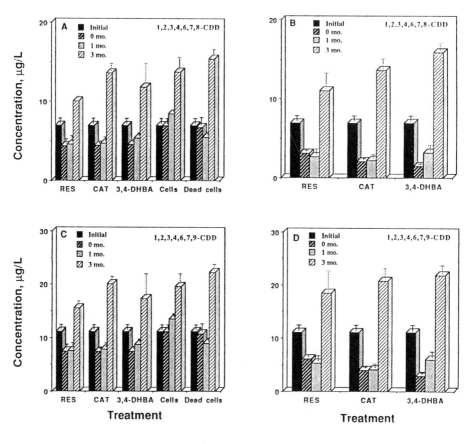

FIGURE 2. **Isomer-specific pattern of heptaCDD appearance in biological (A and C) and chemical (B and D) treatments.**

both heptaCDD congeners produced clearly shows that the appearance of the 2,3,7,8-substituted congener increases over time (Figure 3, A). The increase in ratio is particularly prominent in the chemical controls (Figure 3, B).

The appearance of hexachlorinated dioxins mirrors that observed for the heptachlorinated congeners such that an initial decrease (of the contaminant hexa-CDD, see Materials and Methods) followed by a net increase is observed only in polyphenol-amended cells and chemical controls. Direct evidence is presented for a rapid chemical reaction between the phenolic compounds and the hexaCDD, resulting in complete disappearance of the dioxin up to 1 month of incubation. After 3 months, the net production of hexaCDD was significant only in incubations with pasteurized cells, unamended cells, and 3,4-DHBA treatments (Figure 4). All other treatments resulted in the production of hexaCDD at or below the initial contaminant concentrations. Interestingly, whereas the bulk of all hexaCDD formed (>90%) are not 2,3,7,8-substituted, the appearance of the latter was observed only with unamended cells and treatments amended with

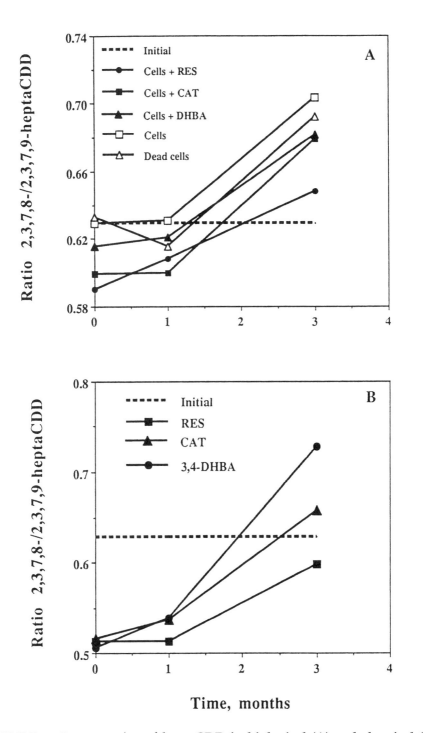

FIGURE 3. Isomer ratios of heptaCDD in biological (A) and chemical (B) treatments.

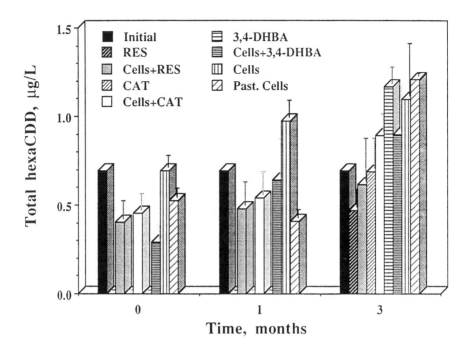

FIGURE 4. Appearance of the hexaCDD in biological and chemical treatments.

polyphenols. Whether this observation is indicative of preferential dechlorination of the 1,4,6,9-substituents via an abiotic mechanism, as was observed for hepta-CDD, will depend on further isomer-specific identification of the isomers formed.

Whereas only trace concentrations of pentachlorinated congeners were found, the appearance of tetrachlorinated dioxins was significant after 3 months of incubation. Concentrations on the order of 2 to 8 μg/L were recovered in amended and killed microcosms, as well as in live cells amended with resorcinol and catechol (Figure 5). No tetraCDD was found in any of the chemical controls. This observation concurs with the dechlorination of penta- through heptaCDD/F in anaerobic sediments, where tetraCDD/F was often the predominant congener group present (Adriaens and Grbic-Galic 1994; Adriaens et al. 1995b). No isomer-specific analyses were performed to quantify the 2,3,7,8-substituted isomer.

No attempt was made to establish a mass balance; the total concentration of lesser chlorinated dioxins recovered constitutes approximately 5 to 10% of the total OCDD removed from the system. Yet, it was demonstrated that the biphasic system and the amendment with polyphenolic compounds resulted in the formation of lesser chlorinated congeners at rates which are orders of magnitude higher than those observed in sediments (Adriaens & Grbic-Galic 1994). It was also implied that reductive dechlorination of PCDD mediated by organic electron shuttles in reduced anaerobic sediments may contribute significantly to particularly the initial transformation and immobilization of highly

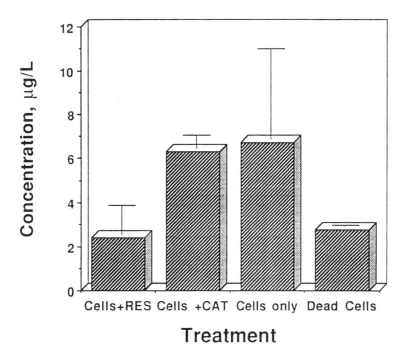

FIGURE 5. Formation of tetraCDD after 3 months of incubation.

chlorinated PCDD. Microbially mediated dechlorination reactions appear to become more dominant once lesser chlorinated PCDD are present. Congener fingerprinting may thus be useful as a tool to differentiate between abiotic and biologically mediated dechlorination of dioxins in particular, or of highly oxidized contaminants in general.

REFERENCES

Adriaens, P., and D. Grbic-Galic. 1992. *Biotransformation of Highly Chlorinated Dibenzo-p-Dioxins and Dibenzofurans in Methanogenic River Sediments.* EPA/600/R-92/126.

Adriaens, P., and D. Grbic-Galic. 1994. "Reductive Dechlorination of PCDD/F by Anaerobic Cultures and Sediments." *Chemosphere 29*: 2253-2259.

Adriaens, P., Q. Fu, and D. Grbic-Galic. 1995a. "Bioavailability and Transformation of Highly Chlorinated Dibenzo-*p*-Dioxins and Dibenzofurans in Anaerobic Soils and Sediments." *Environ. Sci. Technol.* In Press.

Adriaens, P., P. R.-L. Chang, and A. L. Barkovskii. 1995b. "Dechlorination of Chlorinated PCDD/F by Organic and Inorganic Electron Transfer Molecules in Reduced Environments." *Chemosphere.* In review.

Ascon-Cabrera, M., and J.-M. Lebeault. 1993. "Selection of Xenobiotic-Degrading Microorganisms in a Biphasic Aqueous-Organic System." *Appl. Environ. Microbiol. 59*: 1717-1724.

Assaf-Anid, N. 1992. "Reductive Dechlorination of Chlorinated Pollutants by Metals and Organometallics." Ph.D. Dissertation, The University of Michigan.

Assaf-Anid, N., L. Nies, and T. M. Vogel. 1992. "Reductive Dechlorination of a Polychlorinated Biphenyl Congener and Hexachlorobenzene by Vitamin B12." *Appl. Environ. Microbiol.* 58: 1057-1060.

Barkovskii, A., Q. Fu, and P. Adriaens. 1994a. "Biological and abiotic dechlorination of highly chlorinated PCDD/F." *Conf. Proceed. 14th Internat. Symp. of Chlorinated Dioxins, PCBs and Related Compounds*, Kyoto, Japan.

Barkovskii, A. L., M.-L. Bouillant, and J. Balandreau. 1994b. "Polyphenolic Compounds Respired by Bacteria." In T. A. Anderson, and J. R. Coats (Eds.), *Bioremediation Through Rhizosphere Technology*. ACS Symposium Series, Washington, DC. pp. 28-42.

Bopp, R. F., M. L. Gross, H. Tong, H. J. Simpson, S. J. Monson, B. L. Deck, and F. C. Moser. 1991. "A Major Incident of Dioxin Contamination: Sediments of New Jersey Estuaries." *Environ. Sci. Technol.* 25: 951-956.

Deiana, S., C. Gessa, B. Manunza, M. Marchetti, and M. Usai. 1992. "Mechanism and Stoichiometry of the Redox Reaction Between Iron (III) and Caffeic Acid." *Plant & Soil* 145: 287-294.

Dunnivant, F. M., R. P. Schwarzenbach, and D. L. Macalady. 1992. "Reduction of Substituted Nitrobenzenes in Aqueous Solutions Containing Natural Organic Matter." *Environ. Sci. Technol.* 26: 2133-2141.

Dybas, M. J., G. M. Tatara, and C. S. Criddle. 1995. "Localization and Characterization of the Carbon Tetrachloride Transformation Activity of *Pseudomonas* sp. strain KC." *Appl. Environ. Microbiol.* 61: 758-762.

Efroymson, R. A., and M. Alexander. 1994a. "Role of Partitioning in Biodegradation of Phenanthrene Dissolved in Nonaqueous-Phase Liquids." *Environ. Sci. Technol.* 28: 1172-1179.

Efroymson, R. A., and M. Alexander. 1994b. "Biodegradation in Soil of Hydrophobic Pollutants in Nonaqueous-Phase Liquids." *Environ. Toxicol. Chem.* 13 (3): 405-411.

Flaig W., H. Beutelspacher, and E. Rietz. 1975. "Chemical Composition and Physical Properties of Humic Substances." In: J. E. Gieseking (Ed.), *Soil Components, Organic Components*. Springer Verlag, Berlin. pp. 1-211.

Köhler, A., M. Schüttoff, D. Bryniok, and H.-J. Knackmuß. 1994. "Enhanced Biodegradation of Phenanthrene in a Biphasic Culture System." *Biodegradation* 5: 93-103.

Lehmann R. G., H. H. Cheng, and J. B. Harsh. 1987. "Oxidation of Phenolic Acids by Soil Iron and Manganese Oxides." *Soil Sci. Soc. Am. J.* 51: 352-356.

Lehmann, R. G., and H. H. Cheng. 1988. "Reactivity of Phenolic Acids in Soil and Formation of Oxidation Products." *Soil Sci. Soc. Am. J.* 52: 1304-1309.

March, J. 1985. *Advanced Organic Chemistry*, 3rd ed. John Wiley, New York, NY.

Schwartzenbach, R. P., R. Stierly, K. Lanz, and J. Zeyer. 1990. "Quinone and Iron Porphyrin Mediated Reduction of Nitroaromatic Compounds in Homogeneous Aqueous Solution." *Environ. Sci. Technol.* 24: 1566-1574.

Biodegradation of PCP in Soil

Ulrich Karlson, Rona Miethling, Kirsten Schu,
Susanne Schiøtz Hansen, and Jussi Uotila

ABSTRACT

Survival and activity of the pentachlorophenol (PCP)-mineralizer
Mycobacterium chlorophenolicum after inoculation into a natural soil were
studied at bench and field scales. The PCP-concentration was 30 mg kg^{-1}
soil. Inoculation levels were 10^6 to 10^9 cells g^{-1} soil. The effect of immobilization of the inocula on polyurethane foam (PUF) was investigated
at bench scale. PCP-mineralization was measured using ^{14}C-PCP at
bench scale and by soil chemical analyses in the field. The population
dynamics of the degrader were followed by immunofluorescence microscopy using specific immunoprobes. It was shown that *M. chlorophenolicum* was still present at 1 order of magnitude below inoculation level
with both immobilized and nonimmobilized cells after up to 1 year.
However, inoculation at 10^8 and 10^6 cells g^{-1} soil did not raise the PCP-
mineralization rate over the background level when cells had not been
immobilized. Inoculation with PUF-immobilized cells significantly
stimulated the rate of biodegradation at bench and field scales. Formation of chloroanisols or chloroveratroles through biomethylation of
chlorophenols was observed upon inoculation with nonimmobilized
cells at field scale.

INTRODUCTION

Pentachlorophenol (PCP) has been used worldwide as a wood preservative and fungicide. It has caused serious pollution of soil and water due to
its persistence in the environment (Crosby 1981). Numerous studies on PCP-
mineralization in liquid culture, wastewater, and soil have been conducted at
laboratory scale (Edgehill 1994; Crawford and Mohn 1985; Mikesell and Boyd
1988; Middeldorp et al. 1990a; Resnick and Chapman 1994; Radehaus and Schmidt
1992), but reports of successful bioremediation at field scale are scarce (Salkinoja-
Salonen et al. 1989). Significant effects of bioaugmentation have not yet been
demonstrated in the field. Immobilization is known to improve the survival of
inoculated bacteria (van Elsas and Heijnen 1990; Briglia et al. 1994a). It can also
affect the mineralizing activity of the bacteria by controlling bioavailability of

the pollutant (Hu 1994; Apajalahti and Salkinoja-Salonen 1984). This study consisted of two parts: (1) a laboratory investigation of the mineralization activity of *M. chlorophenolicum* PCP-1 in a natural soil at a low PCP level dependent upon immobilization on PUF, and (2) field application of immobilized *M. chlorophenolicum* CP-2 to the same soil. The field results are compared with previous work using the same bacterium, but without immobilization.

MATERIALS AND METHODS

Soil

The soil used for laboratory (bench-scale) and field experiments was from a PCP-contaminated site at Skalborg, Denmark, which had been excavated for containment purposes. Soil characteristics were as follows: origin (horizon), subsoil; texture, sand; water content (field moist), 87 g kg^{-1}; water-holding capacity, 250 g kg^{-1}; total organic carbon, 3.4 g kg^{-1}; total N, 0.25 g kg^{-1}; pH (1:1), 6.5; average PCP concentration, 30 mg kg^{-1}. For bench-scale experiments, a portion of soil with 0.2 mg PCP kg^{-1} was spiked with 30 mg kg^{-1} of ^{14}C-labeled PCP. For a field demonstration, 3500 metric tons of soil were treated on site in 2-m-high piles (case A).

Soil Inoculation

The PCP-degrading bacterium used for the laboratory study was *M. chlorophenolicum* PCP-1 (DSM 43826; Apajalahti and Salkinoja-Salonen 1986; Häggblom et al. 1994; Briglia et al. 1994b). Cells were grown at room temperature in the dark in liquid sorbitol medium (Apajalahti et al. 1986) to a density of ca. 10^9 mL^{-1}. After induction of the PCP-dehalogenase with 8 mg L^{-1} PCP (Uotila et al. 1991), PCP-mineralizing activity and purity of the culture were confirmed and cells were harvested by centrifugation and stored at 4°C until use. For fieldwork, *M. chlorophenolicum* CP-2 (Häggblom et al. 1988; Häggblom et al. 1994) was cultivated commercially (ALKO, Rajamäki, Finland) on DSM-65 medium. Strain CP-2 is taxonomically very close to strain PCP-1, except it is easier to grow on glucose, a cheap carbon compound. After an induction and activity/purity check, cells were stored in the fermentation broth at 4°C for up to 4 weeks until used. During storage, induction was repeated at 10-day intervals. Laboratory soil was inoculated at 10^6 to 10^9 cells g^{-1} soil by adding a cell suspension in mineral medium and mixing with a glass rod. Immobilized cells had been stirred overnight in mineral medium amended with "Bayvitec C 2200" iron oxide-coated PUF (Bayer, Leverkusen, Germany) that had been cut into ca. 1-mm pieces, applying 1 mg PUF g^{-1} soil in the laboratory and 100 g per metric ton of soil in the field (case A). Field soil was inoculated with 10^6 to 10^7 cells g^{-1} soil (10^7 to 10^8 cells g^{-1} PUF) by thoroughly mixing the appropriate quantity of immobilized bacterial cells with the contaminated soil using a payloader and a Combi-Screen ML-2000-S mixing machine (Vedbysønder Maskinfabrik International A/S, Slagelse, Denmark). In the mixing process, soil aggregates disintegrated to ≤0.5-cm size.

Mineralization of PCP

In the laboratory, 10-g soil samples were incubated in tightly closed 50-mL flasks at 15°C and 40% water-holding capacity. Mineralized ^{14}C-PCP was trapped as ^{14}C-CO$_2$ in 1 M KOH and quantified by liquid scintillation counting. In parallel experiments without CO$_2$ traps, entire flasks were sacrificed for cell counts at intervals of up to 205 days (Figure 1). In the field soil, PCP was determined by taking profile composite samples representing the 0 to 1, and the 1- to 2-m-depth of the piles. Subsamples (50 g) were spiked with 2,4,6-tribromophenol and

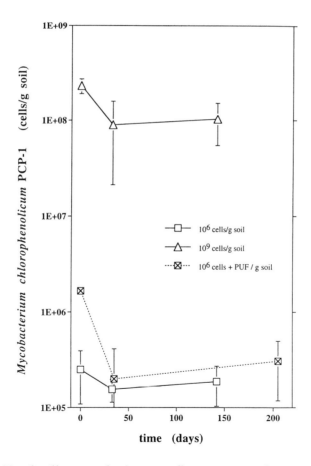

FIGURE 1. Total cell counts by immunofluorescence microscopic counts of *M. chlorophenolicum* PCP-1 inoculated into PCP-contaminated soil. Bacterial cells were added as bacterial suspension or immobilized on PUF, as indicated.

2,4,6-tribromoanisol as internal standards, and extracted in a soxhlet apparatus with acetone. The acetone extract was acetylated overnight, after which chloro-phenols, chloroanisols, and chloroveratroles were extracted using pentane. Pentane extracts were concentrated to 100 µL under N_2 and analyzed by gas liquid chromatography-mass spectrometry (Hewlett-Packard models 5970 and 5890) using selective ion monitoring. Quantification was performed using the brominated internal standards and authentic model compounds.

Bacterial Cell Counts

M. chlorophenolicum cells from laboratory and field experimental soil were extracted after Dobbins and Pfaender (1988) with modifications (78% ± 13% recovery). Rabbit antiserum for strains PCP-1 and CP-2 (Dako A/S, Copenhagen) was used without further purification. Bacterial extracts were stained with the antiserum and fluorescein-conjugated swine-anti-rabbit antibodies (Dako A/S, Copenhagen) according to standard procedure (Hobbie et al. 1977). *M. chloro-phenolicum* cells were counted by epifluorescence microscopy. The limit of detection was 10^5 cells g^{-1} soil. Immunofluorescence microscopy also was used to determine purity and cell density of bacterial inocula.

RESULTS AND DISCUSSION

PCP Mineralization in the Laboratory

The $^{14}CO_2$-measurements are presented in Figure 2. Inoculation with non-immobilized strain PCP-1 at 10^6 g^{-1} soil did not raise the PCP mineralization rate above the rate of the noninoculated control (nonsterile soil), and inoculation at 10^8 g^{-1} soil increased the activity about 2.5-fold only during the first 20 days. After the initial 35-day period, inoculated mineralization rates began to level off, much in contrast to the noninoculated soil. The noninoculated soil mineralized 55.7% of the added PCP in 85 days, whereas the inoculated nonsterile soil mineralized only 39.1% at 10^8 and 37.1% at 10^6 cells g^{-1} soil. That the noninoculated soil produced more CO_2 than the nonsterile inoculated soil cannot be explained at this point. However, one could speculate that in the inoculated treatment ^{14}C-labeled carbon was incorporated into cells of strain PCP-1, whereas the undisturbed indigenous microbiota transformed PCP more efficiently to CO_2. This would suggest that the inoculation inhibited the activity of the indigenous microbiota. Another possible explanation might be a decrease in strain PCP-1 cell numbers with time. However, the results of immunofluorescent counts (Figure 2) suggest that strain PCP-1 survived under experimental conditions, with only a minor decrease in cell numbers immediately after the start of the experiment, regardless of immobilization. The level of cellular activity of *M. chlorophenolicum* is not measured by epifluorescence microscopy. Hence nonactive cells of strain PCP-1 may have acted as a "sink" for available PCP, which was, therefore, not further degraded by the indigenous degraders. We conclude that inoculation with

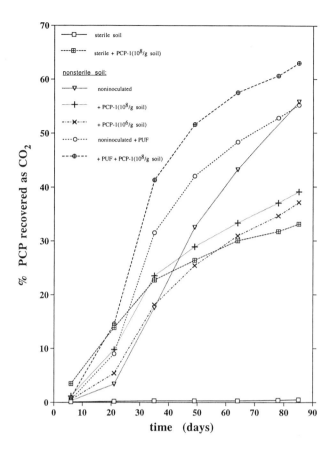

FIGURE 2. ^{14}C-PCP-mineralization rate in soil dependent upon inoculation with *M. chlorophenolicum* PCP-1, amendment with polyurethane foam (PUF), and inoculation with *M. chlorophenolicum* PCP-1 immobilized on PUF. Points represent averages of 4 replicates.

nonimmobilized strain PCP-1 does not result in improved biodegradation in an already active soil, regardless of inoculation density.

Soil amendment with PUF resulted in a strong increase in mineralization rate during the first 35 days, which slightly leveled off thereafter, such that the nonamended soil had produced the same amount of CO_2 by the end of the experiment (about 55%). Inoculation with strain PCP-1 immobilized on PUF caused the highest mineralization rate and resulted in an overall transformation to CO_2 of 63% in 85 days. Improved survival of strain PCP-1 due to the immobilization cannot have been a factor since insufficient survival was not an issue. Growth of strain PCP-1 was also not observed (Figure 2). Increased availability of the contaminant or improved maintenance of cellular activity are possible explanations of this occurrence. Availability of the PCP to the indigenous degraders may also have been affected, which would explain the effect of PUF addition in the

PUF-only treatment, where the initial mineralization rates were almost as high as with strain PCP-1 on PUF. Availability of soil PCP to the bacterial cells might be modulated by the hydrophobic surface properties of the PUF, together with its iron-oxide coating, causing adsorption of PCP (Kung and McBride 1991). This was shown in liquid systems, where reduction of PCP toxicity to a PCP-degrading *Flavobacterium* was observed (Hu et al. 1994). A similar effect was reported for bacteria immobilized on bark chips (Apajalahti and Salkinoja-Salonen 1984). Adsorption of PCP on the PUF surfaces could also facilitate diffusion of PCP from the bulk soil to the site of the bacterial cells. This might lead to enhanced mineralizing activity of the cells at decreasing soil PCP concentrations, through continued induction of the PCP-dehalogenase system (Uotila et al. 1991). The results presented here agree with results of an earlier study using a different soil (Briglia et al. 1994a).

PCP Degradation in the Field

Bioremediation was the preferred cleanup option for a PCP-contaminated site (case A), the goal being a rapid and complete degradation of the contaminant in large piles without the need for further aeration by, e.g., turning over of the soil. *M. chlorophenolicum* has been shown to be microaerophilic during PCP degradation (Apajalahti et al. 1987). Furthermore, it has been reported to perform dehalogenation even under exclusion of O_2 (Uotila et al. 1992). For these considerations and based on the results of the laboratory study, application of strain CP-2 immobilized on PUF was chosen as the most promising option for the bioremediation of case A.

PCP degradation proceeded rapidly to about 5 mg/kg during the first 2 weeks after inoculation (Figure 3). A residual concentration level of about 0.5 mg/kg was reached after 30 weeks, after which no additional degradation was measurable. The noninoculated control showed slow degradation during the initial summer and fall months, but arrived at the same residual PCP level through increased activity during the spring of the following year. It appears that the thorough mixing, which was also applied to the noninoculated control, stimulated the indigenous degraders, most likely through the addition of oxygen. Data showing degradation in undisturbed soil are not available from this field experiment.

Formation of chloroanisoles through biomethylation by the indigenous microbiota had been observed in an earlier study where inoculation with strain PCP-1 had prevented this undesired reaction (Middeldorp et al. 1990b). In case A, chloroanisoles were only detected at the µg kg^{-1} level (data not shown), regardless of inoculation. Cells of strain CP-2 were detected in the inoculated field soil throughout this study (Figure 3). After an initial drop in bacterial numbers, cell counts stabilized between 10^5 and 10^6 g^{-1} soil, which parallels the findings from the laboratory study.

Samples were obtained and analyzed from earlier field applications of non-immobilized strain CP-2 to chlorophenol-contaminated soil composts at the same inoculation density as in our field study (cases B, C, D, E; details courtesy of

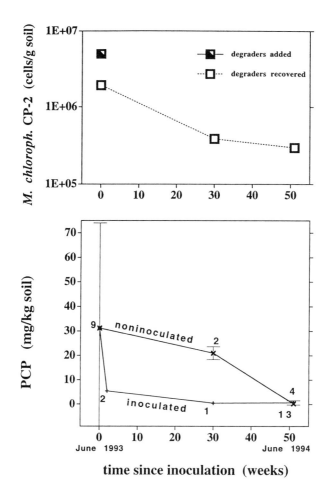

FIGURE 3. Biodegradation of PCP in soil piles (3,500 metric tons; case A) dependent upon inoculation with *M. chlorophenolicum* CP-2 immobilized on PUF, and total cell counts of the degrader by immunofluorescence microscopic counts. Numbers next to data points denote number of samples analyzed. Bars denote standard deviations. Total chloroanisoles were below 0.05 mg kg^{-1} in all cases; chloroveratroles were not determined.

Ecolution Oy, Helsinki, and Finnish Ministry of the Environment). Our analyses (Table 1) indicate that, although degradation proceeded slowly, significant amounts of chloroanisols, and in some cases chloroveratroles, were formed. Their formation through biomethylation of PCP may be caused by the higher concentrations of chlorophenols in those cases, and partially due to the heavier soil textures, compared with case A. However, it is striking that strain CP-2, which was applied without a carrier, was detected by immunofluorescence microscopy in only one of the cited cases (Table 1), and this only when allowing for

TABLE 1. Chlorinated aromatic compounds in soil piles inoculated with nonimmobilized *Mycobacterium chlorophenolicum* CP-2, and recovery of the degraders by immunofluorescence microscopic counts.

Date		Case B Loam	Case C Clay	Case D Sawdust	Case E Clay
9/91	inoculation, cfu/g[a]	10^6-10^7	10^6-10^7		
9/91	Cl-phenols, mg/kg[a]	1762	375		
9/92	inoculation, cfu/g[b]			10^6-10^7	
9/92	Cl-phenols, mg/kg[b]			15413	
10/92	Cl-phenols, mg/kg[a]	661	240		
8/93	Cl-phenols, mg/kg[b]			15940	
8/93	Cl-phenols, mg/kg[a]	604	180		
8/93	turned, mixed, + H_2O[b]			+	
11/93	inoculation, cfu/g[b]				10^6-10^7
11/93	Cl-phenols, mg/kg[b]				1024
5/94	turned, mixed[b]				+
5/94	Cl-phenols, mg/kg			5148	1008
5/94	Cl-anisols, mg/kg			2.9	14.6
5/94	Cl-veratroles, mg/kg			8.0	1.5
5/94	degraders, cells/g			$<<10^5$	
8/94	turned, mixed	+	+		
8/94	Cl-phenols, mg/kg	342	116		
8/94	Cl-anisols, mg/kg	1.7	19.2		
8/94	Cl-veratroles, mg/kg	<0.1	<0.1		
8/94	degraders, cells/g	$<<10^5$	$<<10^5$		
11/94	Cl-phenols, mg/kg				406
11/94	Cl-anisols, mg/kg				14.6
11/94	Cl-veratroles, mg/kg				1.5
11/94	degraders, cells/g				$3*10^4$

(a) Data courtesy of P. Hongisto, Ecolution Oy, Helsinki.
(b) Data courtesy of Finnish Ministry of the Environment.

counts below the statistically significant detection limit of 10^5 g^{-1} soil. Insufficient survival of the inoculant after a 2-year period, possibly caused by the lack of a carrier, may, therefore, be the indirect reason for the formation of methylated chlorophenols.

We conclude that inoculation of contaminated soil with *M. chlorophenolicum* can enhance the degradation rate of chlorophenols in the field, provided that the bacteria are immobilized on a suitable matrix. Biomethylation of chlorophenols does not seem to occur under such conditions, but may occur upon inoculation with nonimmobilized cells. Although immobilization seems to improve long-term survival of the inoculant in the field, the initial stimulation of mineralization activity by addition of the carrier is not due to increased survival of the inoculum, but to stimulation of cellular activity, most likely by control of contaminant bioavailability.

ACKNOWLEDGMENTS

We thank Mirja Salkinoja-Salonen for valuable advice. This work was supported by the European Commission's R&D program ENVIRONMENT (DG-XII/D-1), contract no. EV5V-CT93-0250 and by the Nordic Environmental Biotechnology Program 1990-1993 of the Nordisk Industrifond.

REFERENCES

Apajalahti, J.H.A., and M. S. Salkinoja-Salonen. 1984. "Adsorption of pentachlorophenol (PCP) by bark chips and its role in microbial degradation." *Microbial Ecology*. 10:359-367.

Apajalahti, J.H.A., and M. S. Salkinoja-Salonen. 1986. "Degradation of polychlorinated phenols by *Rhodococcus chlorophenolicus*." *Appl. Microbiol. Biotechnol.* 25:62-67.

Apajalahti, J.H.A., P. Kärpänoja, and M. S. Salkinoja-Salonen. 1986. "*Rhodococcus chlorophenolicus* sp. nov., a chlorophenol-mineralizing actinomycete." *Int. Journ. Syst. Bacteriol.* 36(2):246-251.

Apajalahti, J.H.A., and M. S. Salkinoja-Salonen. 1987. "Dechlorination and para-hydroxylation of polychlorinated phenols by *Rhodococcus chlorophenolicus*." *J. Bact.* 169:675-681.

Briglia, M., P.J.M. Middeldorp, and M. S. Salkinoja-Salonen. 1994a. "Mineralization performance of *Rhodococcus chlorophenolicus* strain PCP-1 in contaminated soil simulating on site conditions." *Soil Biol. Biochem.* 26(3):377-385.

Briglia, M., H.I.L. Eggen, J. D. van Elsas, and W. M. de Vos. 1994b. "Phylogenetic evidence for transfer of pentachlorophenol-mineralizing *Rhodococcus chlorophenolicus* PCP-1 to the genus *Mycobacterium*." *Int. J. Syst. Bacteriol.* 44(3):494-498.

Crawford, R. L., and W. W. Mohn. 1985. "Microbiological removal of pentachlorophenol from soil using a *Flavobacterium*." *Enzyme Microb. Technol.* 7(Dec):617-620.

Crosby, D. G. 1981. "Environmental chemistry of pentachlorophenol, IUPAC reports on pesticides (14)." *Pure and Applied Chemistry*. 53(5):1051-1080.

Dobbins, D. C., and F. K. Pfaender. 1988. "Methodology for assessing respiration and cellular incorporation of radiolabeled substrates by soil microbial communities." *Microb. Ecol.* 15:257-273.

Edgehill, R. U. 1994. "Pentachlorophenol removal from slightly acidic mineral salts, commercial sand, and clay soil by recovered *Arthrobacter* strain ATCC 33790." *Appl. Microbiol. Biotechnol.* 41:142-148.

Häggblom, M. M., L. J. Nohynek, and M. S. Salkinoja-Salonen. 1988. "Degradation and O-methylation of chlorinated phenolic compounds by *Rhodococcus* and *Mycobacterium* strains." *Appl. Environ. Microbiol.* 54(12):3043-3052.

Häggblom, M. M., L. J. Nohynek, N. J. Palleroni, K. Kronqvist, E.-L. Nurmiaho-Lassila, M. S. Salkinoja-Salonen, S. Klatte, and R. M. Kroppenstedt. 1994. "Transfer of polychlorophenol-degrading *Rhodococcus chlorophenolicus* (Apajalahti et al. 1986) to the genus *Mycobacterium* as *Mycobacterium chlorophenolicum* comb. nov." *Int. J. Syst. Bacteriol.* 44(3):485-493.

Hobbie, J. E., R. J. Daley, and S. Jasper. 1977. "Use of nuclepore filters for counting bacteria by fluorescence microscopy." *Appl. Environ. Microbiol.* 33(5):1225-1228.

Hu, Z.-C., R.-A. Korus, W. E. Levinson, and R. L. Crawford. 1994. "Adsorption and biodegradation of pentachlorophenol by polyurethane immobilized *Flavobacterium.*" *Environ. Sci. Technol.* 28(3):491-496.

Kung, K-H.S., and M. B. McBride. 1991. "Bonding of chlorophenols on iron and aluminum oxides." *Environ. Sci. Technol.* 25:702-709.

Middeldorp, P.J.M., M. Briglia, and M. S. Salkinoja-Salkonen. 1990a. "Biodegradation of pentachlorophenol in natural soil by inoculated *Rhodococcus chlorophenolicus.*" *Microb. Ecol.* 20:123-139.

Middeldorp, P.J.M., M. Briglia, V. Kitunen, R. Valo, and M. S. Salkinoja-Salkonen. 1990b. "Biodegradation and -transformation of polychlorinated phenols in soil." Presented at Proceedings of the 5th European Congress on Biotechnology, Copenhagen, July 8-14.

Mikesell, M. D., and S. A. Boyd. 1988. "Enhancement of pentachlorophenol degradation in soil through induced anaerobiosis and bioaugmentation with anaerobic sewage sludge." *Environ. Sci. Technol.* 22(12):1411-1414.

Radehaus, P. M., and S. K. Schmidt. 1992. "Characterization of a novel *Pseudomonas* sp. that mineralizes high concentrations of pentachlorophenol." *Appl. Environ. Microbiol.* 58(9): 2879-2885.

Resnick, S. M., and P. J. Chapman. 1994. "Physiological properties and substrate specificity of a pentachlorophenol-degrading *Pseudomonas* species." *Biodegradation* 5(1):47-54.

Salkinoja-Salonen, M. S., P. Middeldorp, M. Briglia, R. Valo, M. Häggblom, and A. McBain. 1989. "Cleanup of old industrial sites." *Adv. Appl. Biotechnol.* 4:347-367.

van Elsas, J. D., and C. E. Heijnen. 1990. "Methods for the introduction of bacteria into soil: a review." *Biol. Fertil. Soils* 10:127-133.

Uotila, J. S., M. S. Salkinoja-Salonen, and J.H.A. Apajalahti. 1991. "Dechlorination of pentachlorophenol by membrane bound enzymes of *R. chlorophenolicus* PCP-1." *Biodegradation* 2(1):25-31.

Uotila, J. S., V. H. Kitunen, J.H.A. Apajalahti, and M. S. Salkinoja-Salonen. 1992. "Environment-dependent mechanism of dehalogenation by *R. chlorophenolicus* PCP-1." *Appl. Microbiol. Biotechnol.* 38:408-412.

Composting Systems for the Bioremediation of Chlorophenol-Contaminated Land

Kirk T. Semple and Terry R. Fermor

ABSTRACT

Chlorophenols are among the most environmentally hazardous and intractable xenobiotic compounds. Particular attention is being paid to pentachlorophenol (PCP). The microflora found in PCP-augmented compost and PCP-contaminated soils from Finland and the United States were screened for their putative PCP-degradative abilities using both liquid and solid-state enrichment techniques. These putative degraders were isolated on R8 agar containing PCP. The capabilities of the putative PCP-degrading organisms, several of which are actinomycetes, were characterized in liquid culture using high-performance liquid chromatography and radiorespirometry. These isolated organisms have been found to colonize composts, and radiorespirometric studies are under way using [universally labeled (U)-^{14}C]PCP to determine the ability of these composts to mineralize the ^{14}C-label to $^{14}CO_2$ in small-scale composting units (500 g). The next stage will investigate mixes of chlorophenol-contaminated soils with enriched composts (up to 50 kg), thus evaluating the potential to remediate chlorophenol-contaminated soils.

INTRODUCTION

The long-term aim of this European Community-funded project, which has partners in Belgium, Germany, Finland, and the United Kingdom, is exploring composting technology for environmental benefit. It will examine the application of composting for treating of wastes and sites contaminated with chlorinated organic compounds. Chlorophenols have been used extensively in agriculture, industry, and public health because of their wide-spectrum biocidal properties (Apajalahti and Salkinoja-Salonen 1986). As a result of their widespread usage, chlorophenol contamination of the environment is extensive (Valo et al. 1984; Kitunen et al. 1985). One of the most widely used compounds of this class is PCP, which is considered a priority toxic pollutant by the U.S. Environmental Protection Agency (Sittig 1981). In 1985 the worldwide industrial production of PCP was over 100,000 tonnes, with approximately 80% being used for wood preservation (Wild et al. 1992). Intensive worldwide application and spills from industries using or manufacturing PCP have resulted in extensive pollution of some sites. These sites have been shown to contain gram quantities of chlorophenols per kilogram of soil (Valo et al. 1984; Kitunen et al. 1985). In the environment, organic pollutants are often biodegraded by microorganisms, and, normally, this is an extremely effective process. However, in nature, PCP is recalcitrant to microbial attack because of the toxicity of the compound, insufficient populations of PCP-degrading microorganisms, or certain environmental chemicals (e.g., low available carbon and nitrogen levels) or physical conditions (e.g., temperature, pH, salinity) inhibiting the indigenous degradation processes (Valo et al. 1985). We aim to overcome these problems by the production and addition of nutrient-rich mushroom composts, containing large populations of PCP-degrading actinomycetes, to PCP-contaminated soils on site.

EXPERIMENTAL PROCEDURES AND MATERIALS

Liquid Enrichment

Pentachlorophenol-contaminated soils (5 g) from Finland containing the wood preservative Ky-5, and from the United States containing Penta and Phase 2 mushroom compost (substrate for mushroom cultivation) produced at HRI (5 g) were mixed with 50 mL mineral salts medium (Saber and Crawford 1985) containing 10 mg/L PCP and incubated at 25°C and 100 rev/min for 14 d. Samples were taken from each of the enrichments, inoculated onto actinomycete isolation medium R8 (Amner et al. 1989) containing 10 mg/L PCP, and incubated at either 25°C or 40°C to improve the isolation of both putative PCP-degrading mesophilic and thermophilic microorganisms.

Solid-State Enrichment

Laboratory-made composts, prepared as described by Fermor et al. (1991), and Phase 2 mushroom composts were mixed with a range of PCP concentrations (10 to 50 mg/kg) to reduce the bacteriocidal effects of the chlorophenol on the compost microflora. These composts were placed in 2-L multiadaptor flasks for controlled-environment composting (Figure 1) for 21 d (Smith and Wood 1990). Samples of the composts (10 g) were then mixed with 90 mL NaCl (0.9% w/v) and 100 μL Tween 80 surfactant (20%) and shaken for 1 h. After sedimentation, a dilution series (10^{-1} to 10^{-4}) was made of the supernatant. Samples (100 μL) from each dilution series were plated onto R8 medium containing PCP (10 mg/L) and incubated at either 25°C or 40°C.

Radiorespirometry

A modified screw cap 250 mL Erlenmeyer flask with a center well (Figure 2) was used to conduct experiments involving [U-^{14}C]PCP. The CO_2-trapping

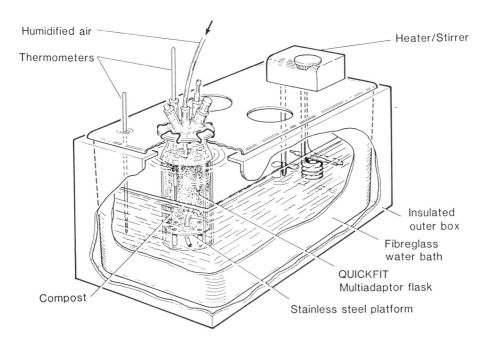

FIGURE 1. Schematic diagram of small-scale (500-g) composting unit (Smith and Wood 1990, p. 303, reprinted with permission of Chapman and Hall Ltd., Andover, United Kingdom).

efficiency was tested using $Na_2^{14}CO_3$, which was acidified with 5 M H_2SO_4 liberating $^{14}CO_2$, which was trapped in the center well containing 5 M KOH and a fluted filter paper. The system was found to be >98% efficient. Biological testing of the apparatus was carried out using [U-^{14}C]glucose with some 70% mineralization, and the rest of the ^{14}C-label was accounted for in the growth medium and cell fractions. All samples were analyzed by liquid scintillation.

Nine laboratory composting systems have been constructed. They can each produce 500 g fresh weight mushroom compost starting from chopped raw ingredients in 10 days. Aeration, moisture content, and temperature of composts can be carefully controlled, thus allowing replication of compost batches. Laboratory-made, Phase 2, and waste Phase 2 (cooked-out) mushroom composts, some spiked with PCP (50 mg/kg), were incubated for 3 weeks, at which point 10 g samples were incubated with [U-^{14}C]PCP (50 mg/kg) at 37°C, which was found to be the best temperature for PCP removal in composts (Watts et al. 1995). The KOH-trapped $^{14}CO_2$, plus the filter paper, was sampled at regular intervals, mixed with 10 mL of Optiphase High Safe 3 scintillation fluid, and analyzed using scintillation counting. The composts were prevented from drying out by placing a filter paper disk on top of the compost sample, which was moistened with 1 mL of sterile double-distilled H_2O once a week. All incubations were in duplicate and each sample was analyzed twice by liquid scintillation.

RESULTS

Isolates obtained from the enrichment techniques were initially selected for their ability to grow on R8 medium containing 10 mg/L PCP, which was slowly increased to 50 mg through successive subcultures. This allowed the isolation of approximately 50 putative PCP-degrading microorganisms, including *Pseudomonas* spp., *Flavobacterium* spp., *Bacillus* spp., and a number of actinomycetes including *Thermomonospora* spp., *Streptomyces* spp., and *Saccharomonospora* spp. In previous studies, not discussed in this paper, several of the actinomycete isolates were found to remove PCP (10 mg/L) from R8 growth medium, but none of the putative PCP-degrading isolates tested, so far, have shown any significant ability to mineralize $[U\text{-}^{14}C]$PCP in liquid medium. However, actinomycete isolates are still being analyzed for their degrading capabilities.

Laboratory-made, Phase 2, and waste-pasteurized mushroom composts, previously spiked with PCP, were incubated with $[U\text{-}^{14}C]$PCP (50 mg/kg). In all three sets of incubations, the composts, previously spiked with PCP, mineralized the ^{14}C-label to a greater extent and at a faster rate than the composts not previously exposed to PCP (Figures 3 to 5). Phase 2 compost mineralized the $[U\text{-}^{14}C]$PCP at a faster rate and to a greater extent than that either laboratory-made or waste-pasteurized composts.

DISCUSSION

Composting is an aerobic process that allows the rapid proliferation of a variety of microbial groups such as aerobic actinomycetes, bacilli, and fungi.

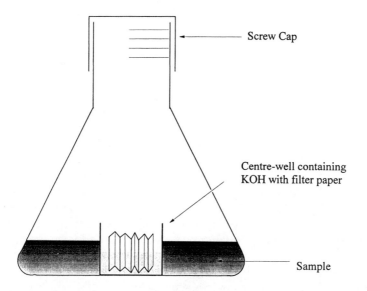

FIGURE 2. Schematic diagram of the radiorespirometric flask.

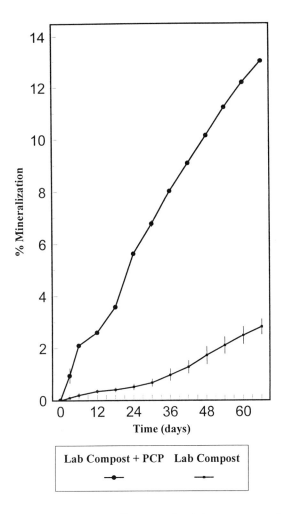

FIGURE 3. Mineralization of [U-^{14}C]PCP by laboratory-made compost.

Mushroom compost is being investigated in this study because it is readily available; at the end of cropping waste compost generally is landfilled. The role of
mushroom compost is to act as a carrier for the xenobiotic-degrading microorganisms and as a nutrient source for the inoculant and indigenous microorganisms in contaminated soils. In this project, actinomycetes have been targeted
as potential PCP-degraders because there are large populations of both mesophilic and thermophilic actinomycetes present in Phase 2 and waste mushroom
composts that have the ability to colonize the composts by mycelial growth, and
have been shown to catabolize a wide range of aromatic compounds using either
endo- or extracellular enzymes (Apajalahti and Salkinoja-Salonen 1986, 1987;
McCarthy and Williams 1992).

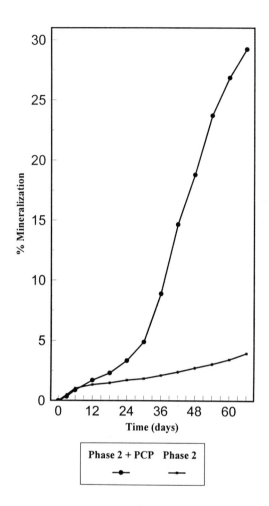

FIGURE 4. Mineralization of [U-¹⁴C]PCP by Phase 2 mushroom compost.

From this study, it is clear that PCP can be mineralized in composts by the populating microflora after exposure to the chlorophenol. However, none of the isolates tested so far have shown any mineralizing capabilities in vitro, despite their ability to remove PCP from liquid medium. This suggests that the organisms responsible for PCP mineralization have not been isolated from the composts, or there is a consortium of microorganisms involved in the complete degradation of PCP, or the physical relationship between the compost solid substrate and the populating microflora may play an important role in PCP mineralization.

The aim of this project is to bioremediate chlorophenol-contaminated soils, which implies the removal of the contaminant(s) from that environment. The most effective way of achieving this is to convert the pollutant(s) to harmless

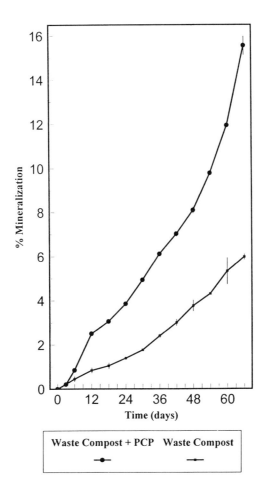

FIGURE 5. Mineralization of [*U*-^{14}C]PCP by waste pasteurized mushroom compost.

compounds such as CO_2 and H_2O. Ongoing studies will establish to what degree PCP is mineralized by these induced composts, and if 100% mineralization is not achieved, identify which compost fractions contain the residual ^{14}C-label. The next stage is to introduce either PCP-spiked or PCP-contaminated soils to the enriched composts (500 g) and to monitor the degradation of the chlorophenol using HPLC and GC. The bioremediated soils will be tested for their phyto-toxicity to measure the effectiveness of the cleanup treatment. The process will then be scaled up to 40 kg soil-compost mixes where the fate of the chlorophenol will, once again, be monitored. Large-scale investigations also are being carried out by the Finnish partners in this project, with a view to applying this tech-nology to chlorophenol-contaminated timber treatment sites.

ACKNOWLEDGMENTS

This work was supported by EC (contract no. EV5V-CT92-0240) and coordinated by Dr. A. J. McCarthy, Liverpool University, UK.

REFERENCES

Amner, W., C. Edwards, and A. J. McCarthy. 1989. "Improved Medium for Recovery and Enumeration of Farmer's Lung Organism, *Saccharomonospora viridis.*" *Applied and Environmental Microbiology 55*: 2669-2674.

Apajalahti, J. A., and M. S. Salkinoja-Salonen. 1986. "Degradation of Polychlorinated Phenols by *Rhodococcus chlorophenolicus.*" *Applied Microbiology and Biotechnology 25*: 62-67.

Apajalahti, J. A., and M.S. Salkinoja-Salonen. 1987. "Complete Dechlorination of Tetrachlorohydroquinone by Cell Extracts of Pentachlorophenol-Induced *Rhodococcus chlorophenolicus.*" *Journal of Bacteriology 169*: 5125-5130.

Fermor, T. R., D. A. Wood, S. P. Lincoln, and J. S. Fenlon. 1991. "Bacteriolysis by *Agaricus bisporus.*" *Journal of General Microbiology 137:* 15-22.

Kitunen, V., R. Valo, and M. S. Salkinoja-Salonen. 1985. "Analysis of Chlorinated Phenols, Phenoxyphenols and Dibenzofurans around Wood-Preserving Facilities." *International Journal of Environmental Analytical Chemistry 20*: 13-28.

McCarthy, A. J., and S. T. Williams. 1992. "Actinomycetes as Agents of Biodegradation in the Environment — A Review." *Gene 115*: 189-192.

Saber, D. L., and R. L. Crawford. 1985. "Isolation of and Characterization of *Flavobacterium* Strains that Degrade Pentachlorophenol." *Applied and Environmental Microbiology 50*: 1512-1518.

Sittig, M. 1981. *Handbook of Toxic and Hazardous Chemicals.* Noyes Publications, Ridge Park, NJ.

Smith, J. F., and D. A. Wood. 1990. "Cultivation of edible fungi on plant residues." In M. P. Coughlan and M. T. Amaral Collaço (Eds.), *Advances in Biological Treatments of Lignocellulosic Materials*, pp. 297-310. Elsevier Applied Science, London.

Valo, R., V. Kitunen, M. S. Salkinoja-Salonen, and S. Räisänen. 1984. "Chlorinated Phenols as Contaminants of Soil and Water in the Vicinity of Two Finnish Saw-Mills." *Chemosphere 13*: 835-844.

Valo, R., J. A. Apajalahti, and M. S. Salkinoja-Salonen. 1985. "Studies on the Physiology of Microbial Degradation of Pentachlorophenol." *Applied Microbiology and Biotechnology 21*: 313-319.

Watts, N. U., T. R. Fermor, and A. J. McCarthy. 1995. "Effect of Chlorophenol Addition on the Microbial Population of Compost." *Abstracts of the 130th Meeting of the Society of General Microbiology*, Leicester, UK, p. 23.

Wild, S. R., S. J. Harrad, and K. C. Jones. 1992. "Pentachlorophenol in the UK Environment: (I) A Budget and Source Inventory." *Chemosphere 24*: 833-845.

Wood Preserving Waste-Contaminated Soil: Treatment and Toxicity Response

Scott G. Huling, Daniel F. Pope, John E. Matthews,
Judith L. Sims, Ronald C. Sims, and Darwin L. Sorenson

ABSTRACT

Soils contaminated with pentachlorophenol (PCP) and creosote were treated without bioaugmentation at field scale in two 1-acre land treatment units (LTUs) at Libby, Montana. The concentration of indicator compounds of treatment performance, i.e., PCP, pyrene, total carcinogenic polycyclic aromatic hydrocarbons (TCPAHs) were monitored in the soil by taking both composited and discrete soil samples. A statistically significant decrease in PCP, pyrene, and TCPAH concentration occurred at field scale, and first-order degradation rate kinetics, from discrete and composited soil samples, satisfactorily represented the chemical loss for these compounds. Detoxification, as measured by the Microtox™ assay, occurred in the soil; and toxicity reduction corresponded with PCP, pyrene, and TCPAH disappearance. No increase in toxicity in the lower treated soil layers (lifts) of the LTUs was observed with time, while the upper, more recently applied lifts were highly contaminated and toxic. This indicated that vertical migration of soluble contaminants had little effect on microbial activity in lower lifts of treated soil.

INTRODUCTION

Evaluation of soil treatment in a prepared-bed system was conducted at the Champion International Superfund Site in Libby, Montana (Libby Site). The system consisted of two 1-acre (0.4-ha) LTUs. The field performance evaluation was designed and conducted by personnel at Utah State University (USU) with support from the U.S. Environmental Protection Agency (EPA) Robert S. Kerr Environmental Research Laboratory. Champion International maintains and operates the LTUs at the Libby Site. The indicator compounds evaluated in this study were PCP, pyrene, and TCPAHs. TCPAHs consist of benzo(a)anthracene, chrysene, benzo(b)fluoranthene, benzo(k)fluoranthene, benzo(a)pyrene, dibenzo-(a,h)anthracene, benzo(ghi)perylene, and indeno(1,2,3-cd)pyrene. Screened soil was taken from the waste pit area, placed in the LTUs in ≈ 6-in. (15.2-cm) layers (lifts), and irrigated to maintain soil moisture content (40 to 70% field capacity) to optimize treatment and to control dust. Nutrients were added as needed, and the LTUs were tilled weekly when weather conditions permitted. New lifts were not applied until contaminant cleanup levels had been achieved.

EXPERIMENTAL APPROACH
AND MATERIALS AND METHODS

The concentrations of indicator compounds in discrete and composited soil samples were measured to evaluate treatment performance and to relate to toxicity measurements. There were 32 randomly selected sampling points per LTU (Mason 1983); at least 30 discrete samples were collected during each sampling event; a subset, usually consisting of 18 to 20 samples, was analyzed. Soils were extracted with 1:1 methylene chloride:acetone (U.S. EPA Method 3550) using sonication. PCP was analyzed using gas chromatography (GC) (EPA Method 8040), and poly-cyclic aromatic hydrocarbon (PAH) compounds using GC/mass spectrometry (U.S. EPA Method 8270). Concurrently, Champion International conducted routine operational monitoring of the LTUs using composited soil samples. Composited soil samples were collected and analyzed by Champion International from 4 locations within each of the 4 quadrants of the LTUs. An equal volume of soil was taken from the top 6 in. (15.2 cm) and mixed in a stainless steel bowl, for a total of 4 samples per sampling event. PAHs and PCP were analyzed using EPA Method 8100 and 8040, respectively (Woodward-Clyde Consultants 1990).

The toxicity of water extracts of soil was evaluated using the Microtox™ assay. The Microtox™ assay is a simple, standardized, acute toxicity assay that measures the microbial response of an aqueous sample, as bioluminescent light output. The assay utilizes a suspension of marine luminescent bacteria (*Photobacterium phosphoreum*) as the bioassay organism and a "challenging" solution (Sims 1990). *P. phosphoreum*, which produce light from a complex chain of biochemical reactions, are used to generate a dose-response curve to determine the effective concentration (EC50) of test sample required to cause a 50% reduction in light output (Microbics Corp. 1992). The reduction in light output represents physiological inhibition, not just mortality, indicating the presence of toxic constituents in the test sample. These data, when correlated with the degradation of the chemical compounds, can provide a general indicator of toxicity reduction (Matthews and Bulich 1984; Matthews and Hastings 1987; Loehr 1989; Sims et al. 1986b; Symons and Sims 1988). However, this assay is not intended to provide information on toxicity from a human health or safety perspective.

RESULTS

Linear regression analyses of the indicator compound concentrations in composited samples (Figure 1) were used to calculate the coefficients of determination

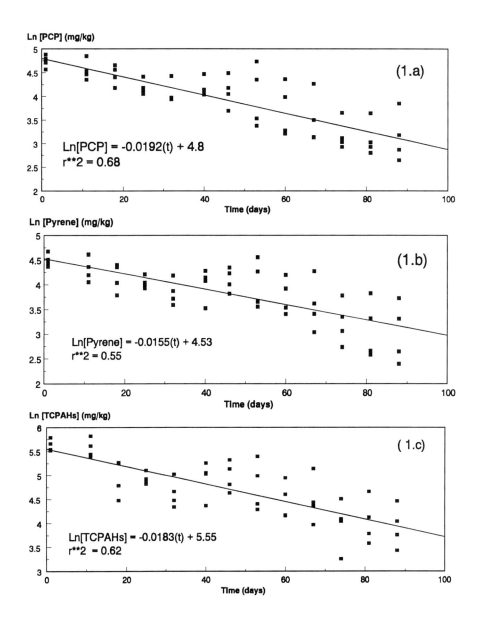

FIGURE 1. (a) PCP, (b) pyrene, and (c) TCPAHs concentrations in soil, LTU 2, Lift 1, first-order degradation kinetics, composited samples.

(r^2), degradation rate constants (k), and half-lives ($t_{1/2}$) (Table 1). Half-life values for the composited samples were calculated from the first-order degradation rate constant. Half-life values for the discrete samples, were calculated assuming first-order degradation rate kinetics using two points in time (days 1 and 54) (Table 2). Although the concentration data are variable, half-lives based on the

TABLE 1. Summary of chemical and kinetic data, LTU 2, Lift 1, composite and discrete samples.

| Contaminant | Coefficient of Determination (Quadrants 1, 2, 3, 4) | Composite Samples | | |
		Initial Mean Concentration (mg/kg) (n=4)	Degradation Rate Constant k (1/day)	Half-life (days)
PCP	0.97, 0.90, 0.70, 0.76	115.1	0.0192	36
Pyrene	0.88, 0.90, 0.60, 0.48	90.5	0.0155	45
TCPAHs	0.88, 0.90, 0.70, 0.58	279.5	0.0183	38
		Discrete Samples		
PCP	0.97, 0.90, 0.70, 0.76	101.4	0.0163	43
Pyrene	0.88, 0.90, 0.60, 0.48	84.9	0.0125	55
TCPAHs	0.88, 0.90, 0.70, 0.58	204.0	0.0127	55

mean values for PCP, pyrene, and TCPAH, using data sets derived from different sampling approaches, differ only by 15%, 19%, and 31%, respectively. Low values for r^2 indicate that time of treatment explains 55 to 68% of the concentration change. Variability is believed to be attributed to (1) hot spots from the heterogeneous placement of highly contaminated soil in the LTU from the wastepit area (stockpile), and (2) to the presence of highly contaminated soil-waste oil conglomerates. Contaminant concentration variability is also observed in the

TABLE 2. Mean concentration[a] of PCP, pyrene, and TCPAHs, LTU 2, Lift 1, composite and discrete soil samples.

		PCP	Pyrene	TCPAHs
Day 1	Composite (7/26/91)	115.1 (100.1–130.1)	90.5 (71.0–110.0)	279.5 (223–336)
Day 1	Discrete (7/27/91)	101.4 (66.1–136.7)	84.9 (65.0–105.0)	204.0 (153.3–255)
Day 60	Composite (9/23/91)	45.9 (5.4–86.4)	43.4 (17.1–69.7)	92.5 (34.6–150.4)
Day 54	Discrete (9/18–19/91)	42.7[b] (26.4–59.0)	43.2 (0–94.0)	103 (1.2–204)

(a) Mean concentration in mg/kg (95% confidence interval for the mean), PCP day 1, n = 20; day 54, n = 19, pyrene and TCPAH day 1, n = 20; day 54 n = 6.
(b) Excluding one value in quadrant 2 (1,168 mg/kg).

discrete sample data (Table 2). For example, in quadrant 3, the mean PCP concentration was greater at day 54 (73.6 mg/kg) than at day 1 (71.3 mg/kg), whereas the overall mean in LTU 2 showed a statistically significant decrease with time.

Based on the mean PCP concentration in discrete samples collected in LTU 2, PCP decreased by 58% during 53 days; PCP decreased in the composited samples by 60% during 60 days. The estimated half-lives were generally greater than those reported by Dasappa and Loehr (1991), and fall within the range reported for a variety of soils by McGinnis et al. (1991). Laboratory results using radio-labeled PCP and soil from LTU 1 and 2 indicated that a portion of PCP was mineralized to CO_2 and water by indigenous soil microorganisms at temperatures and moisture contents representative of the field conditions in LTUs 1 and 2 (Sims et al. 1994). Further, irreversible incorporation of radiolabeled PCP into the soil solid phase (humification) was significant relative to the mineralized fraction. Abiotic controls were not used, and therefore it is unclear whether humification can be entirely attributed to biotic processes. Volatilization of PCP was minor based on losses observed in laboratory "flowthrough" soil microcosms. Overall, the loss of PCP at field-scale, as indicated by both the discrete and composite data, can be attributed to both biotic and abiotic processes, although biotic processes are believed to be the primary process.

Pyrene decreased 49% during 54 days, based on the mean concentration in discrete samples, and by 52% during 60 days in the composited samples. The estimated half-lives are consistent with other values reported for soil (Symons and Sims 1988; Bulman et al. 1985; McGinnis et al. 1991).

TCPAHs decreased 50% during 54 days based on the mean TCPAH concentration in discrete samples in LTU 2 and the mean TCPAH concentration in composite samples was estimated to decrease 67% during 60 days. The coefficients of determination in quadrants 1 and 2 (Table 1) indicate that the PCP, pyrene, and TCPAH soil concentrations conformed relatively well to first-order degradation rate kinetics. The low coefficients of determination in quadrants 3 and 4 indicate that other factors, in addition to time of treatment, affected data variability.

Microtox™ analyses were performed on discrete soil samples collected from LTU 2, lift 1, 2 days after lift 1 was applied and on day 53, referred to as day 1 and day 53, respectively. Microtox™ results from day 1 and day 53 are presented in Table 3. The toxicity, as measured by Microtox™, decreased significantly from day 1 (mean EC50 6.6) to day 53 where 7 of the 10 soil samples had no dose response, as did background samples. Values in Table 3 (day 53), where an EC50 value was measured, reflects the variability of contaminant concentrations in LTU 2 that was observed earlier. The reduction in toxicity corresponded with reductions in PCP, pyrene, and TCPAH concentrations (Tables 1 and 2). A similar correspondence between toxicity reduction, as measured by Microtox™, and PCP soil concentration (Dasappa and Loehr 1991) and PAH soil concentration (Symons and Sims 1988; Abbott and Sims 1989; Wang et al. 1990) has been observed in laboratory microcosms.

The potential for vertical migration of the contaminants being treated or their transformation products is an important concern for LTU performance. The

TABLE 3. Soil Microtox™ toxicity (EC50) values, LTU 2, Lift 1, discrete soil
samples collected 7/27/91 (day 1) and 9/18/91 (day 53).

Day[a]	EC50	Low[b]	High[b]	Coefficient of Determination
1	3.73	2.54	5.5	0.9950
1	3.03	1.89	4.86	0.9939
1	5.09	3.51	7.37	0.9937
1	7.21	4.42	11.75	0.9840
1	8.44	6.15	11.59	0.9920
1	7.1	6.62	7.63	0.9997
1	7.84	5.7	10.77	0.9926
1	8.89	8.08	9.77	0.9992
1	6.74	3.87	11.74	0.9810
1	7.88 Mean 6.60	4.82	12.89	0.9822
53	47.5	24.1	93.7	0.9724
53	8.86	6.68	11.76	0.9933
53	9.01	7.85	10.33	0.9984

(a)   No dose response in 7 of 10 samples on Day 53.
(b)   95% confidence interval for the EC50.

Microtox™ assay represents an indirect measurement of this process, because PCP
and its decomposition products, i.e., tetra-, tri-, di-, and chlorophenols (McGinnis
et al. 1991), are very toxic to *P. phosphoreum*. The Microtox™ assay is generally
more sensitive to the toxicants involved (PCP, PAHs) than are other indicators of
soil microbial activity (Sims et al. 1986a; Sims et al. 1986b; Symons and Sims 1988;
and Aprill et al. 1990).   In this study, samples from buried lifts were assayed
by Microtox™ to determine whether vertical migration of contaminants occurred
from upper lifts. PCP, pyrene, and TCPAH concentrations in lift 5 (LTU 1), when
it was applied onto lift 4 (Table 4), were greater than values in LTU 2, lift 1,
on day 1 (Table 2), which yielded a mean EC50 of 6.6. It was assumed that the
EC50 of soil extracts from LTU 1, lift 5 would have also been high. Indicator
compound concentrations in lift 5 on two later dates (9/18/91, 9/1/92), when
lift 4 soils were sampled for the Microtox™ assay, indicate a significant loss of
these compounds in lift 5. Microtox™ EC50 values for the 14 discrete soil samples collected from the buried lift 4, 56, and 292 days after application of lift 5,
were nontoxic. Additional sets of discrete soil samples (55 total) collected over
time (5/7/91 to 9/1/92) from LTU 1, lifts 1 to 3, were nontoxic. These data indicate that loading contaminated lifts onto lifts that had previously reached the

TABLE 4. Mean concentration[a] of PCP, pyrene, and TCPAHs, LTU 1, Lift 5, discrete soil samples.

Date	PCP	Pyrene	TCPAHs
7/27/91	119.4 (93.9–145.0) n=20	135 (102–167) n=20	254 (179–328) n=20
9/18/91	40.5 (28.9–52.1) n=20	35.3 (14.5–56.0) n=20	103 (52.5–153) n=20
9/1/92	16.9 (13.3–20.5) n=20	4.3 (1.4–7.3) n=19	37.1 (23.1–51.2) n=19

(a) Mean concentration in mg/kg; 95% confidence interval for the mean.

cleanup goals had no measurable effect on the Microtox™ response in lower lifts in LTU 1. Therefore, the vertical migration of soluble contaminants from such lifts had little effect on the microbial activity in the underlying treated soil.

CONCLUSIONS

A statistically significant decrease in PCP, pyrene, and TCPAH concentrations occurred at field scale as determined by composite or discrete soil sampling and analysis. Based on mean concentrations, first-order degradation rate kinetics satisfactorily represented the chemical loss for these compounds. Good agreement was observed between composited and discrete soil sample data for the mean concentrations and half-lives of PCP, pyrene, and TCPAHs. Detoxification occurred in PCP- and creosote-contaminated soil being treated at field scale in prepared-bed land treatment units. Detoxification occurred over the same time frame as the degradation of indicator compounds, PCP, pyrene, and TCPAHs. No increase in toxicity in lower lifts of soil was observed when highly contaminated soil was applied to lifts that had previously undergone remediation. This indicated that the vertical migration of soluble contaminants had little effect on the microbial activity in the underlying treated soil.

ACKNOWLEDGMENTS

Without the cooperation and support of the Champion International staff at Libby, Montana, (specifically, Ralph Heinert, Dave Cosgriff, Gerald Cosgriff, Jim Carraway, and Jim Davidson), this project could not have been accomplished. Also, invaluable support was provided by Bert E. Bledsoe and Dr. Mary Randolph of EPA-RSKERL and Jon Ginn, a doctoral candidate in the Dept. of Environmental Engineering at Utah State University, Logan, Utah.

DISCLAIMER

Although the research described in this article has been funded by the U.S. EPA, it has not been subject to internal agency peer review and therefore does not necessarily reflect the views of the Agency and no official endorsement should be inferred.

REFERENCES

Abbott, C. and R. C. Sims. 1989. "Use of Bioassays to Monitor PAH Contamination in Soil." *Proceedings of the 10th National Superfund Conference*, Washington DC, November 27-29.
Aprill, W., R. C. Sims, J. L. Sims, and J. E. Matthews. 1990. "Assessing Detoxification and Degradation of Wood Preserving and Petroleum Wastes in Contaminated Soil." *Waste Mgmt. and Res. 8*: 45-65.
Bulman, T. L., S. Lesage, P.J.A. Fowlie, and M. D. Webber. 1985. "The Persistence of PAHs in Soil." *PACE Report No. 85-2, Petroleum Association for Conservation of the Canadian Environment.* 1202-275 Slater St., Ottawa, Ontario, Canada. November.
Dasappa, S. M. and R. C. Loehr. 1991. "Toxicity Reduction in Contaminated Soil Bioremediation Processes." *Wat. Res. 25*(9): 1121-1130.
Loehr, R. C. 1989. *Treatability Potential for EPA Listed Hazardous Waste in Soil*, U.S. EPA Technical Report EPA/6002-89/011, Robert S. Kerr Environmental Research Laboratory, Ada, OK, March.
Mason, B. J. 1983. *Preparation of Soil Sampling Protocol: Techniques and Strategies.* U.S. EPA Technical Report, EPA-600/4-83-020. Environmental Monitoring Systems Laboratory, Las Vegas, NV.
Matthews, J. E. and A. A. Bulich. 1984. "A Toxicity Reduction Test System to Assist in Predicting Land Treatability of Hazardous Organic Wastes." In J. K. Petros, Jr. et al. (Eds.), *Hazardous and Industrial Solid Waste Testing: Fourth Symposium.* Philadelphia, PAASTM/STP 886.
Matthews, J. E. and L. Hastings. 1987. "Evaluation of Toxicity Test Procedure for Screening Treatability Potential of Waste in Soil." *Toxicity Assessment. 2*: 265-281.
McGinnis, G. D., H. Borazjani, D. F. Pope, D. A. Strobel, and L. K. McFarland. 1991. *On-Site Treatment of Creosote and Pentachlorophenol Sludges and Contaminated Soil.* U.S. EPA Technical Report, EPA/600/2-91/019, Robert S. Kerr Environmental Research Laboratory, Ada, OK.
Microbics Corp., 1992. *Microtox™ Manual*, Microbics Corp., Carlsbad, CA.
Sims, R. C. 1990. "Soil Remediation Techniques at Uncontrolled Hazardous Waste Sites: A Critical Review." *Journal of the Air & Waste Mgmt. Ass. 40*(5): 703-732.
Sims, R. C., J. L. Sims, D. L. Sorensen, and L. L. Hastings. 1986a. *Waste/Soil Treatability Studies: Methodologies and Results. Volume 1.* U.S. EPA Technical Report, EPA/600/6-86/003a, Robert S. Kerr Environmental Research Laboratory, Ada, OK.
Sims, R. C., D. L. Sorensen, W. J. Doucette, and L. L. Hastings. 1986b. *Waste/Soil Treatability Studies: Methodologies and Results. Volume 2, Waste Loading Impacts on Soil Degradation, Transformation, and Immobilization.* U.S. EPA Technical Report, EPA/600/6-86/003b, Robert S. Kerr Environmental Research Laboratory, Ada, OK.

Sims, R. C., J. L. Sims, D. L. Sorensen, and J. E. McLean. 1994. *Champion International Superfund Site, Libby, Montana: Bioremediation Field Performance Evaluation of the Prepared-Bed Land Treatment System.* U.S. EPA Technical Report, R.S. Kerr Environmental Research Laboratory, Ada, OK, DRAFT.

Symons, B. D., and R. C. Sims. 1988. "Assessing Detoxification of a Complex Hazardous Waste, Using the Microtox™ Assay." *Arch. of Environ. Cont. and Tox.* 17: 497-505.

Wang X., X. Yu, and R. Bartha. 1990. "Effect of Bioremediation on Polycyclic Hydrocarbon Residues in Soil." *Environ. Sci. and Technol.* 24: 1086-1089.

Woodward-Clyde Consultants. 1990. *No Migration Petition Report: Land Treatment Units, Libby Montana.* Prepared for Champion International Corporation, Stamford, CT. May.

Effect of Soil on the Stimulation of Pentachlorophenol Biodegradation

Marie-Paule Otte, Charles W. Greer,
Yves Comeau, and Réjean Samson

ABSTRACT

The presence of soil was essential to obtain an active pentachlorophenol (PCP)-degrading consortium from a contaminated soil in a fed-batch bioreactor. The effects of whole contaminated soil, an aqueous extract of the soil, and different molecular weight fractions of the extract on the activity of the consortium were investigated. Indigenous adapted biomass was prepared by enrichment in mineral salts medium (MSM) containing 100 mg/L PCP, and activity was determined by following PCP mineralization. Using a gene probe from a PCP degradation pathway, it was shown that probe-positive bacteria increased from 27% to 89% of the total viable population after four transfers of the enrichment process. Sterile soil had the most beneficial effect on the mineralization of 300 mg/L of PCP with 75% mineralization after 10 days. At a higher concentration (400 mg/L PCP), consortium activity was maintained by the addition of soil, but was lost when either soil extract was added or no addition was made. Soil enhanced the consortium's PCP mineralization activity and its resistance to PCP, probably by serving as a contaminant and biomass adsorbent, as well as a source of nutrients. This concept could be useful for producing active biomass for soil remediation involving bioaugmentation.

INTRODUCTION

The extensive use of toxic compounds such as PCP and creosote in the wood-preserving industry has led to serious soil contamination problems. Remediation of these soils can be achieved by aerobic biopile or soil/slurry bioreactor treatment, after inoculation with an active biomass. Mixed indigenous strains were shown to have the advantage of being more resistant to extreme environmental changes and predation than pure cultures (Grosser et al. 1991, Fewson 1988).

In an experiment to produce an active consortium of PCP degraders using a fed-batch bioreactor, the presence of the soil was shown to be essential (Otte et al. 1994). This led to the present investigation of the effect of soil on the activity and resistance of the consortium, because soil may act as support for the biomass, adsorbant for the contaminant, and source of nutrients for the microorganisms.

The objective of this research was to evaluate the effect of a whole contaminated soil, an aqueous extract of the soil, and different molecular weight fractions of the extract on the mineralization of PCP at high concentrations (up to 400 mg/L) by a microbial consortium obtained from a wood-preserving site. The active biomass was obtained after four liquid transfers (10% v/v) of the initial contaminated soil slurry in MSM (Greer et al. 1990) containing 100 mg/L PCP. Bacterial selection was monitored during the liquid transfers using a *pcpB* gene probe. To evaluate the consortium activity and resistance, mineralization tests in microcosms with [U-^{14}C]PCP were performed.

MATERIAL AND METHODS

Inoculum Enrichment

The sandy loam soil, contaminated for several years with PCP, polycyclic aromatic hydrocarbons (PAHs), and heavy metals was obtained from the pole storage area of a wood-preserving plant. The PCP concentration was 34 mg/kg of dry soil. An initial 5% (w/v) contaminated soil suspension (T0) was prepared in MSM. Liquid enrichment of the biomass was made by four subsequent 10% v/v transfers (T1, T2, T3, and T4) of T0 in fresh MSM containing 100 mg/L PCP. Monitoring of PCP mineralization during enrichment was done in microcosms containing [U-^{14}C]PCP. Bacterial selection during enrichment was followed with a *pcpB* gene probe. The size of the probe is 560 base pairs, and probing was done following the technique of Greer et al. (1993). The *pcpB* gene encodes the enzyme PCP-4-monooxygenase from *Flavobacterium*, the first enzyme in the PCP degradation pathway (Orser et al. 1993).

Mineralization Experiments

Supports such as contaminated and pristine soil, mineral illite clay, the silt fraction of pristine soil, Anachemia sand, and pure humic acids were all sterilized by gamma irradiation as previously described (Otte et al. 1994). An aqueous soil extract was obtained by filtration (0.22 μm) of a 50% (w/v) suspension of the contaminated soil in distilled water. This extract was then separated by cross-flow filtration (regenerated cellulose membrane) in four decreasing molecular weight fractions: FA > 100,000 D, 100,000 D > FB > 30,000 D, 30,000 D > FC >3,000 D, and FD < 3,000 D. Mineralization studies of [U-^{14}C]PCP were conducted in 125-mL microcosms as described previously (Otte et al. 1994). Each microcosm contained 10 mL of the T4 biomass suspension and one of the following amendments: a sterile support (500 mg), humic acids (2%), soil extract (1 mL), or remained unamended ("no addition").

RESULTS AND DISCUSSION

Liquid transfers were made to obtain an active biomass free of soil particles. Monitoring of the $[U^{-14}C]PCP$ mineralization rate showed a decrease in the lag phase from 100 h in the T0 transfer to no lag period for the T4 transfer. The PCP mineralization rate remained constant at 2 mg/L/h between transfers. These results indicate that, as expected, the biomass remained active and the lag phase shortened after each liquid transfer, since biomass was taken in its exponential phase of growth (Dawes 1989).

Monitoring of PCP degraders with the *pcpB* gene probe clearly demonstrated an increase in PCP degraders compared with the total viable bacterial population. The percentage of the total colony-forming units (CFUs) capable of degrading PCP was 27% in the initial soil suspension, 68% in T1, 81% in T2, 82% in T3, and 89% in T4 (Fig. 1). After the second liquid transfer (T2), a

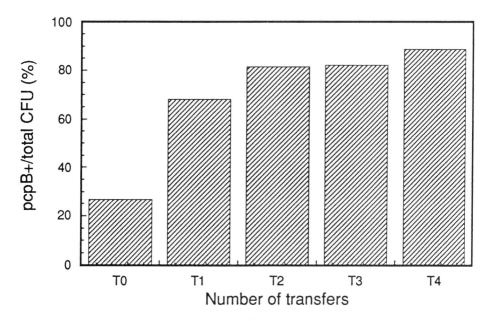

FIGURE 1. Ratio (%) of positive *pcpB* colonies (pcpB$^+$) to total CFUs during liquid enrichment in MSM containing 100 mg/L PCP. The enrichment culture transfers are labeled as follows: T0, the 5% (w/v) soil suspension in MSM containing 100 mg/L PCP; T1-T4 refer to the first, second, third, and fourth transfer (10% v/v) in MSM containing 100 mg/L PCP, respectively.

plateau was reached at about 85% indicating a selection of the PCP-degrader biomass. The PCP-degrading population appeared to be predominantly one strain (colony type), which resulted from subculturing biomass on PCP as sole carbon source. This method of monitoring the biomass proved to be very sensitive in detecting PCP-degraders, and allowed the evolution of selection in the total biomass to be followed (Orser et al. 1993).

The effect of different sterile solid supports (contaminated soil, silt fraction, clay, sand) on PCP mineralization by the biomass obtained after three liquid transfers (T3) was tested for 10 days in microcosms containing 100 mg/L PCP (Fig. 2). The presence of soil (pristine or contaminated) enhanced consortium activity to the greatest extent, the silt fraction from a pristine soil had a weaker influence, and sand had no influence at all. In the presence of sterile contaminated soil, silt fraction, and clay, a PCP mineralization level of 60% was obtained in 1.7, 2.9, and 5.8 days, respectively; in the presence of sand, no mineralization occurred. When humic acids (2%) were added to the sand, however, the PCP mineralization rate obtained was identical to that observed with clay. These findings demonstrate that some components were essential to maintain the consortium-degrading activity.

The effects of sterile soil, soil extract, and different fractions (F1 to F4) of the soil extract on PCP mineralization also were tested (data not shown). Sterile

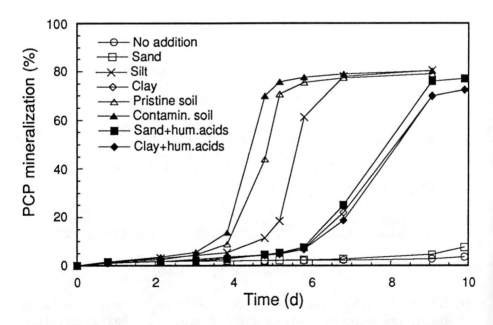

FIGURE 2. **Effect of the addition of different solid supports and humic acids on the mineralization of 100 mg/L PCP by the indigenous bacterial consortium. The active biomass used was that obtained after the third liquid enrichment transfer T3.**

soil, soil extract, and the low-molecular weight-fraction F4 (3,000 D) of the soil extract had a positive effect on 100 mg/L PCP mineralization (75% mineralization in 2.4 days). A lesser effect was also observed with the other soil extract fractions F1, F2, and F3 (75% mineralization in 6.2 days). These results confirm the important role of soil particles (adsorption of PCP and biomass) and soil extract (supplementary nutrients) on consortium-degrading activity (Holm et al. 1992). Microscopic mass-transfer processes, such as diffusion of some carbon sources and nutrients, may be important to consider in quantitative descriptions of the biodegradation of organic chemicals in soil (Scow & Alexander 1992).

The effect of soil and soil extract was also tested for higher PCP concentrations, up to 400 mg/L (Fig. 3). At 300 mg/L PCP, 75% mineralization was

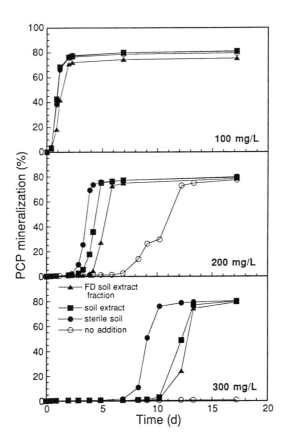

FIGURE 3. Effect of the addition of soil, soil extract, and a low-molecular-weight fraction of the soil extract (FD < 3,000 D) on PCP mineralization by an indigenous bacterial consortium in the presence of increasing concentrations. The active biomass used was that obtained after the fourth liquid enrichment transfer T4.

observed in the presence of soil after 9.6 days and in the presence of soil extract
after 13.3 days. In the absence of an amendment, no mineralization occurred
after 17 days. With the other PCP concentrations tested, the following results
were obtained: at 100 mg/L PCP, no difference in mineralization rates were
observed between sterile soil, soil extract, fraction F4 of the soil extract, and "no
addition" (75% after 2 days). At 200 mg/L PCP, 75% mineralization was reached
within 4.1, 4.8, 6.8, and 13.5 days in the presence of sterile soil, soil extract, fraction F4, and "no addition," respectively. Finally, at 400 mg/L PCP, preliminary
data reveal 25% mineralization after 17.2 days in the presence of sterile soil,
although activity was lost in the presence of soil extract, in fraction F4, and with
"no addition."

These results demonstrate the importance of soil in maintaining and
enhancing PCP degrader activity by decreasing PCP toxicity to the microorganisms and providing them protection and essential nutrients to ensure
biomass growth. Maintaining contaminated soil could be very important for
the production of active and resistant biomass in bioaugmentation processes
for soil remediation.

REFERENCES

Dawes, E.A. 1989. "Growth and survival of bacteria." In J.S. Poindexter, and E.R Leadbetter
    (Eds.), *Bacteria in Nature,* Vol. 3, pp. 67-187. Plenum Press, New York, NY.
Fewson, C.A. 1988. "Biodegradation of xenobiotic and other persistent compounds: The
    cause of recalcitrance." *Tibtech 6*: 148-153.
Greer, C.W., J. Hawari, and R. Samson. 1990. "Influence of environmental factors on 2,4-
    dichlorophenoxyacetic acid degradation by *Pseudomonas cepacia* isolated from peat."
    Arch. Microbiol. 154: 317-322.
Greer, C.W., L. Masson, Y. Comeau, R. Brousseau, and R. Samson. 1993. "Application of mol-
    ecular biology techniques for isolating and monitoring pollutant degrading bacteria." *W.
    Poll. Res. J. Can. 28*(2): 275-287.
Grosser, R.J., D. Warshawsky, and J.R. Vestal. 1991. "Indigenous and enhanced mineralization
    of pyrene, benzo[a]pyrene, and carbazole in soils." *Appl. Environ. Microbiol. 57*(12): 3462-
    3469.
Holm, P.E., P.H. Nielsen, H-J. Albrechtsen, and T.H. Christensen. 1992. "Importance of unat-
    tached bacteria and bacteria attached to sediment in determining potentials for degra-
    dation of xenobiotic organic contaminants in an aerobic aquifer." *Environ. Microbiol.
    58*(9): 3020-3026.
Orser, C.S., C.C. Lange, L. Xun, T.C. Zahrt, and B.J. Schneider. 1993. "Cloning, sequence
    analysis, and expression of the *Flavobacterium* pentachlorophenol-4-monooxygenase
    gene in *Escherichia coli.*" *J. Bacteriol. 175*(2): 411-416.
Otte, M-P., J. Gagnon, Y. Comeau, N. Matte, C. W. Greer, and R. Samson. 1994. "Activation of
    an indigenous microbial consortium for bioaugmentation of pentachlorophenol/cre-
    osote contaminated soils." *Appl. Microbiol. Biotechnol. 40*: 926-932.
Scow, K.M., and M. Alexander. 1992. "Effect of diffusion on the kinetics of biodegradation:
    Experimental results with synthetic aggregates." *Soil Sci. Soc. Am. J. 56*: 128-134.

Degradation of PCP During Acidogenesis

Peter R. Cali and
Sanjoy K. Bhattacharya

ABSTRACT

This study was conducted to assess the potential for biodegradation of organic compounds by anaerobic acidogenic organisms. The reducing environment present in acidogenic systems is conducive to the reductive dechlorination of chlorinated organic compounds, such as pentachlorophenol (PCP). Both batch and continuous reactors were used to evaluate the effects of PCP on anaerobic acidogenesis and to assess its potential for acidogenic biodegradation. It was determined that up to 7.0 mg/L PCP did not inhibit acetate formation, that PCP was degraded during acidogenesis, and that acclimation to lower doses was beneficial for acidogenic degradation of PCP.

INTRODUCTION

It has been shown that PCP can be biodegraded to end products of methane and carbon dioxide in an anaerobic reactor (Yuan 1991, Wild et al. 1993, Fox et al. 1993, Majumdar et al. 1994). To optimize the efficiency of anaerobic or combined anaerobic-aerobic treatment systems, a greater understanding of the role of the acidogenesis phase is needed. The goal of this research was to study the biodegradation of PCP during the acidogenesis phase and the effect that PCP has on the intermediates of acidogenesis, specifically volatile acids. A concomitant objective was to investigate the effect of acclimation to PCP by the acidogens. To accomplish these goals continuously fed, completely mixed reactors were operated at disparate combinations of solids retention times (SRT), 0.25 to 0.61 day, organic loading rates (OLR), 8,500 to 30,000 mg glucose/L-day, and PCP concentrations (1.4 to 7.0 mg/L PCP). The reactors were fed a mixture of glucose, PCP, and a nutrient solution. Batch reactors were used to establish the toxicity threshold and the mechanism of inhibition due to PCP toxicity in anaerobic systems.

MATERIALS AND METHODS

Batch reactors consisted of 500-mL serum bottles containing 250 mL of a glucose enrichment culture. The stock culture was enriched for a period of 1 year prior to use by feeding glucose (99.7%; Sigma Chemical, St. Louis, Missouri) as the sole carbon source and by maintaining an SRT of 50 days and an organic loading rate of 1,000 mg/L-day. Anaerobic toxicity assays (ATA) were conducted on triplicate spiked serum bottles using the method developed by Owen et al. (1979); one set of reactors was left unspiked to serve as the control. All reactors were stored at 35 ± 1C and an organic loading rate of 1,000 mg glucose/L-day was maintained throughout the experiments. Gas production was measured manometrically and compared with the stoichiometric values.

Continuous reactors consisted of 2.0 L plexiglas chemostats (1.5 L of culture). Glucose, nutrients (Haghighi-Podeh, 1991), and PCP were fed continuously via a Cassette Multichannel peristaltic pump (Manostat, New York, New York). PCP was spiked continuously through the glucose feed line. The glucose feed concentration was maintained at 5,000 mg/L, and the SRT was controlled by adjusting the pump flowrate. Because acetate utilization constitutes the rate-limiting step in the anaerobic digestion process, acetate utilizers theoretically can be washed out at an SRT of approximately 4 days. Here, acetate utilization was kinetically excluded by operating the chemostats at SRTs well below the washout SRT for acetate utilizers, between 0.3 day and 0.6 day.

Intermediate volatile acids were measured by gas chromatography (GC) with flame ionization detection (FID). Glucose and ethanol were detected using colorimetric methods (Sigma Diagnostics, St. Louis, Missouri). PCP was analyzed using a high-performance liquid chromatograph (HPLC) equipped with an ultraviolet (UV) detector. Gas chromatography with mass spectrometer (GC/MS) was used to check for possible intermediates of PCP degradation. Volatile suspended solids (VSS) was used as a measure of the biomass in the reactors and was determined according to Method 2940E (Standard Methods 1992).

RESULTS AND DISCUSSION

The results of the ATAs, presented in Figure 1, showed that inhibition of methane production occurred in all the PCP-spiked systems. Below a PCP spike concentration of 2.71 mg/L, approximately 40% methane production inhibition occurred; above 2.71 mg/L, the degree of inhibition of methane formation became dependent on spike concentration (Table 1). Methane formation is an end product of the degradation of glucose to higher volatile acids, then to hydrogen and acetate, and finally to methane and carbon dioxide. Because methane formation was inhibited in the spiked reactors, either acetate was not being formed from glucose (acidogenesis inhibition), or the acetate was formed

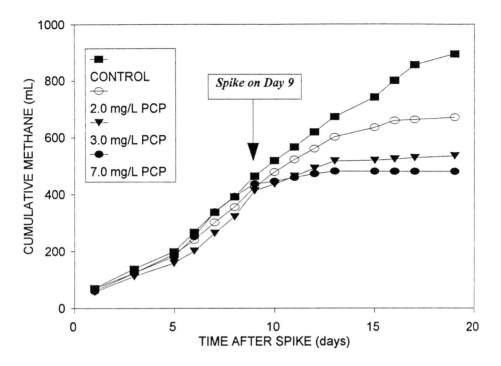

FIGURE 1. Inhibition of methane production in PCP-spiked reactors.

TABLE 1. ATA for glucose enrichment culture spiked with PCP.

Toxicant Concentration	VSS (mg/L) at Start of Experiment	Percent Inhibition of Methane Production
Control	4,720	0
0.54 mg/L PCP	4,315	24
1.50 mg/L PCP	4,085	34
1.93 mg/L PCP	5,205	44
2.33 mg/L PCP	4,450	44
2.71 mg/L PCP	4,365	40
3.06 mg/L PCP	4,310	63
3.39 mg/L PCP	4,250	78

but was not degraded (methanogenesis inhibition). Analysis of the intermediate volatile acids revealed that for spike concentrations as high as 7.0 mg/L PCP, glucose was totally degraded and acetate was still being formed.

As seen in Figures 1 and 2, acetate was degraded and methane produced in the control group, but methane production was inhibited and acetate accumulated in the PCP-spiked systems. This showed that inhibition was primarily the result of toxicity to the acetate utilizers.

In the continuous reactors, which were operated at 0.6-day maximum SRT, virtually all acetate utilization was excluded. These reactors exhibited no toxic effects on the intermediates in response to 7.0 mg/L PCP. Volatile acids formation continued in the same composition (86% acetate, 7% propionate, 6% butyrate, and 1% others) as in the control. Inhibition of acidogenesis, or the retarded production of acetate, was not observed in the continuous systems as a result of the presence of 7.0 mg/L PCP.

PCP was degraded in all spiked continuous reactors to different degrees depending on the spike concentration. Analysis of the effluent by GC/MS detected no phenolic intermediates of the PCP degradation. Table 2 shows that PCP degradation decreased from 63 to 81% for PCP spike concentrations below 2.8 mg/L to 50 to 64% for spike concentrations above 2.8 mg/L. To assess the effects of acclimation to PCP by the acidogenic organisms, some continuous systems were exposed to gradual increases in PCP concentration from 2.8 to 7.0 mg/L. Table 3 shows that those systems acclimated to lower dosages exhibited 64 to 90% degradation of the continuously spiked 7.0 mg/L PCP,

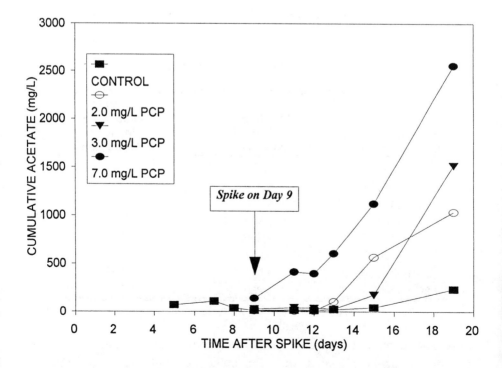

FIGURE 2. Acetate accumulation in PCP-spiked reactors.

TABLE 2. Effect of spike concentration on PCP degradation.

Chemostat No.	SRT (days)	Influent PCP (mg/L)	% Degradation
2	0.61	1.4	63
3	0.55	2.1	81
4	0.65	2.8	79
5	0.61	7.0	53
8	0.37	5.0	64
9	0.32	7.0	50

TABLE 3. Effect of acclimation on PCP degradation.

Chem No.	SRT (days)	PCP spike (mg/L)	% PCP degradation
2	0.60	1.4	63
2	0.58	2.8	68
3	0.52	2.1	81
3	0.49	5.0	82
4	0.53	2.8	79
4	0.51	7.0	90
5	0.61	7.0	53
8	0.39	5.0	64
8	0.25	7.0	64
9	0.32	7.0	50

whereas the unacclimated systems spiked with 7.0 mg/L PCP degraded only 50 to 53%.

CONCLUSIONS

The following conclusions are drawn from this research:

1. The presence of PCP inhibited acetate utilization. For spike concentrations above 2.7 mg/L PCP, the degree of inhibition increased with increased spike concentration.
2. Spike concentrations of 7.0 mg/L PCP did not inhibit acidogenesis in either batch or continuous systems.

3.   Between 50 and 90% of the spike concentration of PCP was degraded during acidogenesis, depending on spike concentration. Acclimation to lower dosages improved degradation efficiency.

ACKNOWLEDGMENTS

This research has been partially funded by DOE through the Center for Bioenvironmental Research (EM project) at Tulane University, New Orleans, LA.

REFERENCES

Cali, P. 1995. "Inhibition by Pentachlorophenol in Anaerobic Acidogenic Systems." Ph.D. Dissertation, Civil and Environmental Engineering Department, Tulane University, New Orleans, LA.

Fox, P., V. Rangachar, and N. Rhadasagar. 1993. "Reductive Dechlorination of Aromatics under Acidogenic vs Methanogenic Conditions." In *66th Water Environment Federation Proceedings*. pp. 131-143.

Haghighi-Podeh, M. R. 1991. "Fate and Toxic Effects of Cobalt, Cadmium, and Nitrophenols on Anaerobic Treatment Systems." Ph.D. Dissertation, Civil and Environmental Engineering Department, Tulane University, New Orleans, Louisiana.

Majumdar, P., S. Bandyopadhyay, and S. K. Bhattacharya. 1994. "Effects of Heavy Metals on the Biodegradation of Organic Compounds." In R. E. Hinchee, D. B. Anderson, F. B. Metting, Jr., and G. D. Sayles (Eds.), *Applied Biotechnology for Site Remediation*, pp. 354-359. CRC Press, Boca Raton, FL.

Owen, W. F., D. C. Stuckey, J. B. Healy, L. Y. Young, and P. L. McCarty. 1979. "Bioassay for Monitoring Biochemical Methane Potential and Anaerobic Toxicity." *Water Research*, 13(6).

Standard Methods for the Examination of Water and Wastewater. 1992. 18th ed., APHA, AWWA, and WPCF, Washington, DC.

Wild, S., S. Harrad, and K. Jones. 1993. "Chlorophenols in Digested U.K. Sewage Sludges." *Water Research* 27(10): 1527-1534.

Yuan, Q. 1991. "Fate and Effects of Pentachlorophenol on Combined Anaerobic-Aerobic Treatment Systems," M.S. Thesis, Civil and Environmental Engineering Department, Tulane University, New Orleans, LA.

Co-Contaminated Sites: Biodegradation of Fossil Fuels in the Presence of PCBs

Pamela J. Morris, Michael E. Shelton,
and Peter J. Chapman

ABSTRACT

Polychlorinated biphenyl (PCB)-contaminated sites are often co-contaminated with fossil fuels making biodegradation more difficult. Our current studies examine biodegradation of the fossil fuel components of two PCB-contaminated sites: (1) a former racing Drag Strip soil contaminated with Aroclor 1242 and (2) a sediment from Silver Lake contaminated with Aroclor 1260. The sandy surface soil at the Drag Strip site contains 1.9% organic carbon and 1.5% fossil fuel component. Analysis of the solvent-extractable organic fraction, by alumina column chromatography, shows the distribution of organics to be 91.2% hydrocarbons, 7.8% polars, and 1.1% asphaltenes. This oil is extremely weathered and contains few readily biodegradable components. Enrichments have yielded undefined mixed cultures of bacteria capable of extensive degradation of components of both the Drag Strip and Silver Lake site materials. One culture, enriched from a creosote-contaminated soil adjacent to a utility pole, transformed approximately 28% and 37% (by weight) of the Drag Strip and Silver Lake oils, respectively. While the presence of fossil fuels has been shown to inhibit aerobic PCB degradation, our studies show that the presence of PCBs negatively impacts fossil fuel biodegradation. Continuing studies will examine the nature of PCB inhibition of fossil fuel biodegradation.

INTRODUCTION

Fossil fuels are complex mixtures of chemicals found in crude oil and coal-derived materials and are among the most common contaminants at hazardous waste sites. Their effective bioremediation requires a physiologically diverse microbial population. In addition, the prevalence of fossil fuels in surface soils and sediments results in the presence of a complex hydrophobic phase that can potentially sequester other hydrophobic contaminants. The fossil fuel component of contaminated soils and sediments can be more effective at sequestering PCBs than natural organic matter (Sayler and Colwell 1976, Boyd and Sun 1990). Harkness and Bergeron (1990) observed inhibition of the aerobic degradation of PCBs in the presence of mineral oil, suggesting that PCBs were partitioning into the mineral oil, resulting in a decrease in availability to the PCB-degrader.

Effects of increasing PCB concentrations on crude oil degradation have been examined using cultures established by enrichment with a weathered crude oil not containing PCBs. We have also been studying microbial cultures enriched on extracted PCB-fossil mixtures. These enrichment cultures degrade not only the readily degradable components of fossil fuels found at these sites, but preliminary evidence has demonstrated that many of the more recalcitrant components, such as the triterpanes, also are degraded. Our long-term goal is to understand the biological process by which complex mixtures of fossil fuels can be degraded, and to apply that knowledge to lessen potential human health effects.

EXPERIMENTAL METHODS AND MATERIALS

Fossil Fuel and PCB-Contaminated Soils

The Drag Strip soil (Glen Falls, New York) contains 1.5% by weight of a methylene chloride-extractable fraction containing fossil fuels and PCBs. The soil is contaminated with weathered Aroclor 1242 (averaging 500 ppm) to the top 15 cm (McDermott et al. 1989) (Figure 1A). Silver Lake (Pittsfield, Massachusetts) is a 10-ha urban pond that received municipal and industrial discharges (Brown et al. 1987). Total oil recovered from the sediment was 10% by weight, with a total concentration of PCBs (Aroclors 1254 and 1260, Figure 1B) at roughly 2,300 μg/g sediment (John Quensen, personal communication).

Enrichment of Fossil Fuel-Degrading Cultures

Mixed microbial populations were enriched on methylene chloride-extractable fractions from both the Drag Strip soil and Silver Lake sediment. The extractable materials, containing both fossil fuels and PCBs, were recovered from the soil and the sediment using 24th Soxhlet extraction with methylene chloride as the solvent. The extract was concentrated to dryness to remove solvent and used as a carbon and energy source in the enrichments. Soil contaminated with leachate from a creosote-contaminated utility pole (Fairhope, Alabama) supplied the biological enrichment source. Enrichments were initiated by the addition of 1 g of soil to 25 mL of mineral medium amended with 50 mg hydrocarbon-PCB mixture from either source. Incubations were carried out at 30°C and 200 rpm for 30 days. Primary enrichments were transferred (4% v/v) for two successive 30-day periods prior to use in any experimentation. Scaled-up cultures used 200 mg quantities of the extracted hydrocarbon-PCB component added to 100 mL of mineral medium in 500-mL Erlenmeyer flasks.

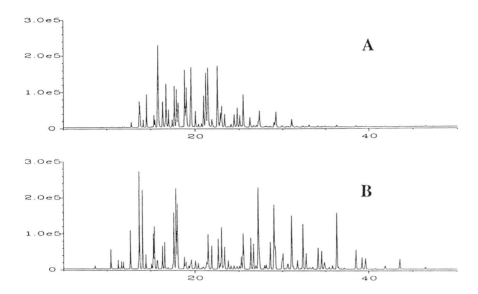

FIGURE 1. Capillary gas chromatographic (ECD) traces of the PCB congener profiles from (A) Drag Strip soil and (B) Silver Lake sediment.

The incubations were inoculated with 4% v/v of a 30-day culture pregrown under the conditions described above.

PCB-Contaminated Fossil Fuel Degradation

Studies were conducted to test whether the presence of PCBs (Aroclor 1242) in an artificially weathered crude oil (Alaskan North Slope 521 or ANS 521) influenced the biodegradability of the crude oil.   ANS 521 was spiked with the following five concentrations of Aroclor 1242: 0, 500, 2,500, 5,000, and 10,000 μg Aroclor 1242 per g of ANS 521. A mixed culture, enriched on ANS 521, was inoculated (4% v/v) into 500 mL Erlenmeyer flasks containing 200 mg of the PCB-ANS 521 mixture and 100 mL of mineral medium. Cultures were incubated for 30°C and 200 rpm for 30 days.

Fossil Fuel Extractions

The entire contents of each culture were extracted three times with 1/10 volumes of methylene chloride in a separatory funnel to recover the remaining fossil fuel-PCB component. Extracts were combined, concentrated to dryness, and weighed to determine the amount of material degraded. The recovered residues were redissolved in 10 mL *n*-hexane, shaken vigorously by hand approx. 1 min and allowed to stand overnight. Samples were centrifuged to pellet the hexane-insoluble material, and the hexane-soluble material was examined by gas chromatography (GC).

Alumina Chromatography

The methylene chloride extractable material from Drag Strip soil and Silver Lake sediments was subjected to fractionation on alumina. Alumina (4 g, Fisher Brand, 80 to 300 mesh) was activated at 300°C overnight, cooled, and stored in a vacuum desiccator at room temperature. The column was packed in chloroform/1% methanol. Residues (100 mg) were loaded onto each column in approximately 500 μL chloroform/1% methanol and then were eluted with an additional 15 μL to recover the hydrocarbon fraction and PCBs. Continued elution with an additional 10 mL of chloroform/1% methanol followed by elution with 5.0 mL methanol then 10 mL chloroform combined to elute fractions of polar and asphaltene compounds. Individual fractions were concentrated to dryness for gravimetric analysis. Typical total recovery for a 100-mg hydrocarbon-PCB component fractionation was approximately 93 to 94%. A modified alumina cleanup was employed for larger-scale incubations, where extracts were loaded onto a column in 500 μL methylene chloride, and eluted with 9.5 mL of methylene chloride. The recovered 10-mL volume was submitted for analysis.

Fossil Fuel and PCB Analysis

Samples were analyzed by GC on a Hewlett-Packard (HP) Model 5890 Series II equipped with a flame ionization detector (FID). The column used in the analysis was type HP-5 with a length of 25 M, inside diameter of 0.32 mm, and phase thickness of 0.52 μm. The carrier gas was helium, and the combustion gas was hydrogen. The injector and detector temperatures were 290 and 315°C, respectively. The temperature program begins at 50°C for 1 min and proceeds at 5°C/min to 310°C with a hold time of 20 min. Control and degraded oil samples were analyzed by mass spectrometry (MS) using an HPGC equipped with a Model 5971 Series MS Detector. Helium was the carrier gas, and the temperature settings and analysis program were identical to the GC/FID conditions. The column was an HP ULTRA-2. PCB analysis was performed by capillary GC with an HP-5 capillary column and an electron capture detector (ECD) according to Quensen et al. (1988).

RESULTS

Microorganisms enriched from a soil adjacent to a creosote-contaminated utility pole were used to study the effect of PCBs on crude oil (ANS 521) biodegradation. Increasing concentrations of PCBs (Aroclor 1242) added to ANS 521 resulted in a decrease in the extent of degradation of the crude oil. When no Aroclor 1242 was added to ANS 521, approximately 18.5% of the weight of crude oil was removed in a 30-day incubation. Over the same period of time, cultures incubated with ANS 521 containing 500, 2,500, 5,000, and 10,000 μg Aroclor 1242 per g oil degraded approximately 14.7%, 11.8%, 5.4%, and 3.7% of ANS 521, respectively. Unlike the cultures from this site enriched on Drag Strip and Silver Lake fossil fuel residues, the culture used in this study was originally enriched on ANS 521 and had no previous exposure to PCBs. Transformation of the PCBs was not observed (data not shown).

Microorganisms enriched from the same creosote-contaminated soil on PCB-containing fossil fuel components from both the Drag Strip and Silver Lake sites yielded significant biodegradation of the fossil fuel material extracted from these PCB-contaminated sites. The Drag Strip enrichment culture degraded approximately 28% of the weight of the Drag Strip residue during a 30-day incubation, while the Silver Lake enrichment culture was able to remove 37%. The controls retained an overall 93% of the starting weight after the 30-day incubation. GC-FID traces of the sterile Drag Strip control (no inoculum, Figure 2A) compared with an active (inoculum, Figure 2B) sample show that significant depletion of the unresolved complex material has occurred. For the Drag Strip fossil fuel material, the PCBs detected by FID remain at the same abundance and serve as surrogate internal standards. Considerable depletion of the unresolved complex material was also observed with the Silver Lake fossil fuel material (Figure 2, panels C and D).

Fractionation of the biodegraded and control fossil fuel materials confirms that biodegradation of the hydrocarbon fractions was significant, with the asphaltenic material remaining relatively constant (Table 1). The polar material increased slightly upon degradation of the Drag Strip fossil fuel component. Analysis of the solvent-extractable fraction of the Drag Strip material, by alumina column chromatography, showed an average distribution of organics

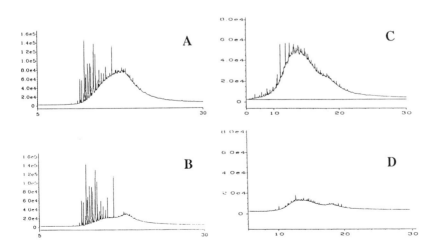

FIGURE 2. Capillary gas chromatographic traces (FID) of the (A) control (uninoculated), (B) biodegraded Drag Strip fossil fuel materials, (C) control (uninoculated), and the (D) biodegraded Silver Lake fossil fuel materials.

TABLE 1. Recovery of hydrocarbon, polar, and asphaltenic fractions of the Drag Strip and Silver Lake fossil fuels in biodegraded (inoculated) compared to control (uninoculated) samples by alumina chromatography. Numbers in parentheses are percentages of the total material represented by each of the three fractions.

	Total[a] mg	Hydrocarbons mg (%)	Polars mg (%)	Asphaltenes mg (%)
Drag Strip Control				
Replicate 1	202.2	183.4 (91.7)	14.6 (7.3)	2.2 (1.1)
Replicate 2	200.0	181.7 (90.6)	16.4 (8.2)	1.9 (1.0)
Drag Strip Inoculated				
Replicate 1	103.3	82.6 (80.0)	18.6 (18.0)	2.1 (2.0)
Replicate 2	99.3	79.6 (80.2)	17.6 (17.7)	2.1 (2.1)
Silver Lake Control				
Replicate 1	200.0	180.9 (90.5)	17.7 (8.9)	1.4 (0.7)
Replicate 2	200.7	181.1 (90.6)	17.8 (8.9)	1.8 (0.9)
Silver Lake Inoculated				
Replicate 1	63.0	45.7 (72.5)	15.5 (24.6)	1.8 (2.9)
Replicate 2	86.5	67.9 (78.5)	17.0 (19.7)	1.6 (1.9)

(a) Represents the amount of fossil fuel material that underwent fractionation by alumina column chromatography. Amounts differ due to differences in material remaining in biodegraded compared to control samples.

to be 91.2% hydrocarbons, 7.8% polars, and 1.1% asphaltenes; biodegraded material consisted of 80.1% hydrocarbons, 17.9% polars, and 2.1% asphaltenes. For the Silver Lake material, the uninoculated control showed an average of 90.6% hydrocarbons, 8.9% polars, and 0.8% asphaltenes, and biodegraded material showed 75.5% hydrocarbons, 22.15% polars, and 2.4% asphaltenes. Selective ion monitoring of different components of the fossil fuel materials showed a decrease in the abundance of a number of specific hydrocarbons, including a decrease in the abundance of the triterpanes, including the hopanes, after degradation (m/z = 191) (Figure 3). This decrease was more evident for the Silver Lake fossil fuel material where levels of tricyclic and pentacyclic triterpanes were barely detectable (Figure 3D).

DISCUSSION

We have demonstrated the microbial degradation of two different weathered fossil fuel materials extracted from PCB-contaminated soil and sediment.

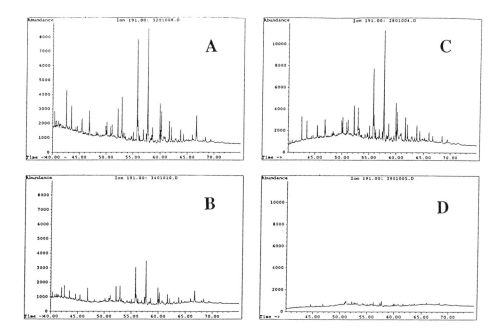

FIGURE 3. Mass fragmentography (m/z = 191) of the hydrocarbon fractions recovered from control and biodegraded materials. Panel designations are as for Figure 2. The peak at 57.5 min is 17α(H),21β(H)-hopane.

After 1 month of incubation with a fossil fuel-degrading enrichment culture, the fossil fuel component was significantly biodegraded, including portions of the unresolved complex material observed by GC. The PCB component of these materials was not degraded. Studies are continuing to examine whether biodegradation of the fossil fuel component of these PCB-contaminated fossil fuels can increase the availability of the PCBs, to facilitate their increased biodegradation.

Studies have shown that the presence of increasing concentrations of PCBs in a crude oil matrix may inhibit the biodegradation of the crude oil. Because so many waste sites are co-contaminated, we need to understand the influence that one contaminant may have on another contaminant during biodegradation. Despite demonstrated negative effects of PCBs on crude oil biodegradation, when enrichment cultures were established on PCB-containing fossil fuels from two different contaminated sites, significant degradation of fossil fuel components was shown to occur.

Preliminary evidence suggests that components of the unresolved complex material, and specific recalcitrant chemicals such as the hopanes, are biodegraded by the enrichment cultures. It has been suggested that 17α(H),21β(H)-hopane can be used as a conserved internal marker for studying oil biodegradation in the environment (Prince et al. 1994), but these preliminary results suggest that even this compound and its homologs may be transformed under laboratory conditions with certain enrichment cultures. However, more studies are required to support these preliminary results.

ACKNOWLEDGMENTS

The authors would like to extend their appreciation to Dr. John Quensen III (Michigan State University) for the Silver Lake methylene chloride-extracted fraction and to the General Electric Co. for supplying the Drag Strip soil. We would also like to acknowledge Simon Akkerman (University of Florida), Tai Hyunh (University of Florida), and Beat Blattman (Avanti Corporation) for analytical assistance. P. J. Morris has been supported at the University of Florida by the General Electric Co. and in part by U.S. Department of Energy grant DE-FG01-912EW506 to the Medical University of South Carolina. M. E. Shelton has been supported by U.S. Environmental Protection Agency Cooperative Agreement #EPA/CR820771-01-0 between the University of Minnesota and the U.S. EPA Environmental Research Laboratory (Gulf Breeze, Florida) where P. J. Chapman is a research microbiologist. The U.S. Environmental Protection Agency does not endorse the use of products mentioned in this manuscript.

REFERENCES

Boyd, S.A. and S. Sun. 1990. "Residual petroleum and polychlorobiphenyl oils as sorptive phases for organic contaminants in soils." *Environ. Sci. Technol.* 24: 142-144.
Brown, Jr., J.F., R.E. Wagner, H. Feng, D.L. Bedard, M.J. Brennan, J.C. Carnahan, and R.J. May. 1987. "Environmental Dechlorination of PCBs." *Environ, Toxicol. and Chem.* 6:579-593.
Harkness, M.R. and J.A. Bergeron. 1990. "Availability of PCBs in soils and sediments to surfactant extraction and aerobic biodegradation." *Research and Development Program for the Destruction of PCBs,* Ninth Progress Report, General Electric Company Corporate Research and Development, p. 109.
McDermott, J.B., R. Unterman, M. J. Brennan, R. E. Brooks, D. P. Mobley, C. C. Schwartz, and D. K. Dietrich. 1989. "Two strategies for PCB soil remediation: Biodegradation and surfactant extraction." *Experimental Progress* 8:46-51.
Prince, R.C., D.L. Elmendorf, J.R. Lute, C.S. Hsu, C.E. Halth, J.D. Senius, G.J. Dechert, G.S. Douglas, and E.L. Butler. 1994. "17a(H),21b(H)-hopane as a conserved internal marker for estimating the biodegradation of crude oil." *Environ. Sci. Technol.* 28: 142-145.
Quensen, J.F., III, J.M. Tiedje, and S.A. Boyd. 1988. "Reductive dechlorination of polychlorinated biphenyls by anaerobic microorganisms from sediments." *Science* 242: 752-754.
Sayler, G.S., and R.R. Colwell. 1976. "Partitioning of mercury and polychlorinated biphenyl by oil, water, and suspended sediment." *Environ. Sci. Technol.* 10: 1142-1145.

Biostimulation of PCB-Degrading Bacteria by Compounds Released from Plant Roots

John S. Fletcher, Paula K. Donnelly, and Ramesh S. Hegde

ABSTRACT

Flavonoid and coumarin compounds produced by plants supported the growth of polychlorinated biphenyl (PCB)-degrading bacteria, and the bacteria retained their PCB-degrading properties. Root leachates and washings from mulberry trees also supported the growth of a PCB-degrading bacterium. These results indicate that chemicals released by some plant roots may serve as cometabolites for PCB-degrading bacteria. Identification of the right plant species and development of appropriate cultivation practices promises to lead to an ecologically sound means to achieve sustained in situ degradation of PCBs at contaminated terrestrial sites.

INTRODUCTION

Screening and isolation studies (Bedard et al. 1986, 1987) have shown that PCB-degrading bacteria do exist in the wild, but the persistence of PCBs in the environment indicates, that under conditions prevailing at PCB sites, the indigenous bacteria are ineffective in degrading PCBs. In situ inoculation of contaminated soil with laboratory grown PCB-degrading bacteria has only been marginally successful (McDermott et al. 1989). Thus, the primary challenge for successful bioremediation of PCB-contaminated soil is to devise ways to encourage the growth and sustained PCB-degradation of select groups of microbes that are either indigenous to PCB sites or are introduced to the sites.

The successful use of biphenyl (Figure 1) as a cometabolite to enhance laboratory degradation of PCBs (Focht and Brunner 1985) suggests that in situ addition of appropriate organic substrates may enhance PCB degradation at contaminated sites. Uniform dispersal of biphenyl on a large scale at terrestrial sites is not feasible because of its low water solubility. Therefore, there is a need to identify other compounds that support the growth of PCB-degrading bacteria and to devise ways to disseminate them in a uniform and continuous

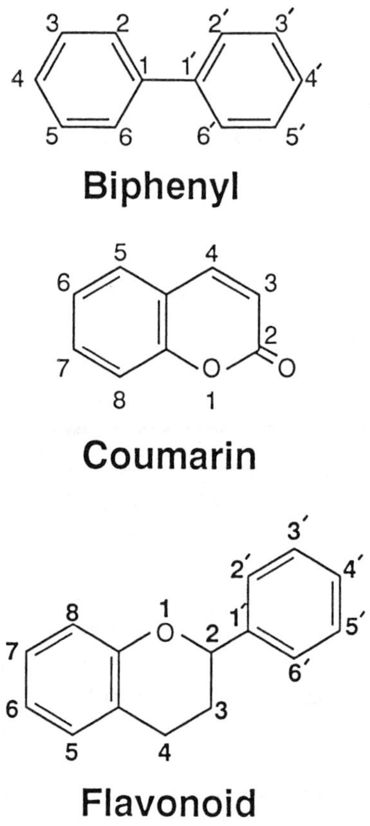

Biphenyl

Coumarin

Flavonoid

FIGURE 1. General structures of biphenyl, coumarin, and flavonoid.

manner at contaminated sites. We hypothesize that the roots of some plant species release phenolic compounds such as flavonoids and coumarins into the soil where the compounds selectively foster the growth and function of PCB-degrading bacteria, and thereby cause sustained, long-term degradation of PCBs. Results presented in this paper from microbial growth studies conducted with pure compounds and mixtures of compounds released by plant roots support this hypothesis.

METHODS

The suitability of plant flavonoids and/or coumarins to support the growth of PCB-degrading bacteria was examined by comparing the growth of three different PCB-degrading bacterial strains on biphenyl versus 14 different compounds (apigenin, catechin, chrysin, coumarin, dihydrofistin, maclurin, morin, myricetin, naringenin, naringin, phloridzin, quercetin, scopoletin, and vitexin). The compounds tested served as the sole carbon source for pure cultures grown in liquid media. The three strains of bacteria tested (*Alcaligenes eutrophus* H850, *Corynebacterium* sp. MB1, and *Pseudomonas putida* LB400) were provided by the General Electric Company, which had demonstrated the ability of each strain to degrade PCBs (Bedard et al. 1986). The PCB-degrading properties of bacteria grown on flavonoids were examined after 3 transfers in each of the compounds studied. The ability of each organism to metabolize PCBs was measured with the assay described by Bedard et al. (1986).

A screening study was conducted to identify plant species that release large amounts of phenols from their roots. Twelve different plant species were grown as duplicates in sand culture in the greenhouse for 3 to 5 months, at which time the potted plants were eluted with 3 L of H_2O, and the phenol content of the eluates was determined according to the method of Amorin et al. (1977). Two criteria were used to select the species for examination: (1) all test plants had been reported in the literature to accumulate flavonoids in some tissue (usually leaves), and (2) the plants were suitable for growth under conditions prevailing at contaminated sites. Concerning the latter criterion, species were selected that are perennial, easy to grow, have large root systems, are widely distributed in the United States, and have a high tolerance for stress conditions such as poor soil and drought.

Plant root leachates and root washings were centrifuged and the supernatant was filter sterilized using a 0.2-μm Gelman filter. One mL of the sterile sample was added as the sole carbon source to test tubes containing 4.0 mL of PAS-salts medium (Bedard et al. 1986). Each sample tube was inoculated with 50 μL of 1.0 optical density *A. eutrophus* H850. Triplet cultures were incubated at 23°C on a rotary shaker. Cultures receiving multiple additions received 100-μL additions at 6-day intervals. The optical densities of 1-mL aliquots were read at 615 nm.

RESULTS

Microbial Growth on Plant Flavonoids

Three PCB-degrading bacterial strains obtained from General Electric (MB1, H850, and LB400) have been tested for their ability to grow on 14 different plant flavonoids or coumarins (Figure 1) as the sole carbon source (Donnelly et al. 1994). All three bacterial strains grew on at least 4 of the 14 compounds, and the H850 strain grew better on six of the compounds than it did on biphenyl. The PCB-degrading properties of each of these strains were examined after three successive passes in media with one of the flavonoids serving as the sole carbon source. The LB400 strain that General Electric used in its terrestrial field test (McDermott et al. 1989) retained all of its PCB-degrading properties following growth on all three flavonoids tested (Donnelly et al. 1994). The retention of PCB-degrading properties by MB1 and H850 differed depending on the flavonoid provided. These results prove that some of the natural products produced by plants will support the growth of PCB-degrading bacteria and that the bacteria retain their ability to metabolize PCBs. The substrate specificity observed emphasizes the necessity to identify specific plant species that produce and release those compounds instrumental in fostering growth of PCB-degrading bacteria.

Release of Phenols from Plant Roots

All of the plant species examined (Table 1) released phenols from their roots. The amount of phenols present in leachates recovered from potted plants following one growing season ranged from a low of 1.0 μmoles of morin equivalents for mahonia to a high of 9.8 μmoles for barberry (Table 1). Although time in culture may have contributed to some of the differences observed, there was no clear correlation between the growth period and phenolic yield. A more important factor appears to be the physiological differences that exist between the species examined.

When the growth-promoting properties of the root leachates were screened with H850 as the test organism, only osage orange and mulberry fostered growth. Further examination of mulberry showed that pot leachates were more stimulatory towards H850 growth than root washings (Table 2). Sequential additions of either the leachate or the washing improved growth, but again the leachate preparations were more stimulatory.

DISCUSSION

We have demonstrated that PCB-degrading bacteria will grow on plant flavonoids, and the root exudates of some species. These results support the

TABLE 1. Phenolic content of leachates recovered from potted plants.

Plant Species	Growth Period, days	Total Phenols, μmoles morin
Barberry	105	9.8
Sumac	116	9.3
Perennial sunflower	92	8.8
Osage orange	139	8.3
Rose	131	8.3
Asparagus	98	8.3
Tall fescue	93	8.3
Mulberry	96	7.2
Cottonwood	175	4.7
Indian grass	160	2.6
Big bluestem	154	2.6
Mahonia	150	1.0

TABLE 2. Relative growth of *A. eutrophus* H850 on mulberry pot leachates and root washings.

	Weeks of Growth	
Culture Addition	0	4
Pot leachate	(Optical density values)	
Single Addition		
Control (pot with no plant)	0.012	0.116
Plant 1	0.011	0.155
Plant 2	0.007	0.151
Multiple additions		
Plant 1	0.010	0.519
Plant 2	0.007	0.388
Root wash		
Single addition		
Plant 1	0.009	0.026
Plant 2	0.005	0.125
Multiple addition		
Plant 1	0.001	0.076
Plant 2	0.002	0.186

hypothesis that the rhizosphere of certain plant species will favor the growth and survival of PCB-degrading bacteria. The release of phenolics into the soil by the fine network of plant roots may be thought of as a naturally occurring injection system capable of delivering desired substrates into the soil that foster the growth and action of PCB-degrading bacteria. However, it is important to recognize that the roots of all plant species do not produce and release equal amounts and kinds of phenolic compounds; therefore, the rhizosphere zone of all plants must not be considered a haven for PCB-degrading bacteria. It may be that only a few plant species may have the desired characteristics. Awareness of such species would be extremely valuable, because growing such plants at contaminated sites has the potential of selectively fostering the growth of PCB-degrading bacteria over competing organisms. The outcome could be a sustained population of PCB-degrading bacteria that would degrade PCBs over an extended time period. Thus, plant-microbe systems have the potential of providing inexpensive, ecologically stable bioremediation systems.

REFERENCES

Amorin, H. V., D. K. Dougall, and W. R. Sharp. 1977. "The effect of carbohydrate and nitrogen concentration on phenol synthesis in Paul's Scarlet rose cells grown in tissue culture." *Physiol. Plant.* 39:91-95.

Bedard, D. L., R. Unterman, L. H. Bopp, M. J. Brennan, M. L. Habert, and C. Johnson. 1986. "Rapid assay for screening and characterizing microorganisms for the ability to degrade polychlorinated biphenyls." *Appl. Environ. Microbiol.* 51:761-768.

Bedard, D. L., R. E. Wagner, M. J. Brennan, M. L. Habert, and J. F. Brown, Jr. 1987. "Extensive degradation of aroclors and environmentally transformed polychlorinated biphenyls by *Alcaligenes eutrophus* H850." *Appl. Environ. Microbiol.* 53:1094-1102.

Donnelly, P. K., R. S. Hegde, and J. S. Fletcher. 1994. "Growth of PCB-degrading bacteria on compounds from photosynthetic plants." *Chemosphere* 28:981-988.

Focht, D. D., and W. Brunner. 1985. "Kinetics of biphenyl and polychlorinated biphenyl metabolism in soil." *Appl. Environ. Microbiol.* 50:1058-1063.

McDermott, J. B., R. Unterman, M. J. Brennan, R. E. Books, D. P. Mobley, C. C. Schwartz, and D. K. Dietrich. 1989. "Two strategies for PCB soil remediation: biodegradation and surfactant extraction." *Environ. Prog.* 8:48-51.

Anaerobic Bioremediation of DDT-Contaminated Soil With Nonionic Surfactant

Guanrong You, Gregory D. Sayles,
Margaret J. Kupferle, and Paul L. Bishop

ABSTRACT

This study focused on the anaerobic bioremediation of DDT (1,1,1-trichloro-2,2-*bis*(p-chlorophenyl)ethane)-contaminated soil in slurry reactors with anaerobic culture. Effects of nonionic surfactant (Triton X-114) and oxidation-reduction potential (ORP) reducing agent additions on DDT transformation were evaluated. DDT transformation resulted in DDD (1,1-dichloro-2,2-*bis*(p-chlorophenyl)ethane) as the major product, with DDT residue remaining at 10% of the initial DDT concentration. The initial DDT transformation rate increased in reactors with Triton X-114 added. Triton X-114 addition decreased DDD production slightly; reducing agent additions decreased DDD production and increased production of less-chlorinated intermediates. When reducing agents and larger amounts of Triton X-114 were added, a greater decrease in DDD production was observed. The soil tested had a high sorption capacity for Triton X-114; thus, Triton X-114 application may result in a high residue in the environment. Anaerobic biotransformation with an appropriate surfactant and reducing agents is a promising strategy for the first phase of DDT bioremediation. Transformation rates are increased, and less DDD is produced. Complete mineralization of DDT may require subsequent aerobic treatment.

INTRODUCTION

DDT (1,1,1-trichloro-2,2-*bis*(p-chlorophenyl)ethane) is a hydrophobic chlorinated pesticide and a priority pollutant designated by the U.S. Environmental Protection Agency (U.S. EPA). Use of DDT was banned in the USA in 1972, but DDT persists in nature. Existing DDT-contaminated soil and sediment sites are a threat to the ecosystem and human health.

Extensive DDT transformation occurs mainly under anaerobic conditions (Fig. 1). With little exception, the first reductive product, DDD (1,1-dichloro-2,2-

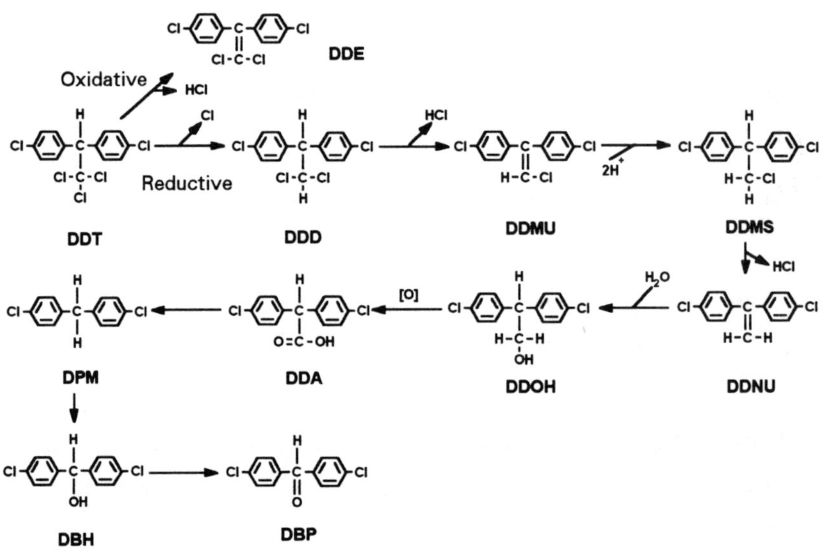

FIGURE 1. Metabolic pathway for the biotransformation of DDT. (Adapted from Rochkind and Blackburn 1986). DDT = 1,1,1-trichloro-2,2-*bis*(*p*-chlorophenyl)ethane; DDD = 1,1-dichloro-2,2-*bis*(*p*-chlorophenyl) ethane; DDMU = 1-chloro-2,2-*bis* (*p*-chlorophenyl)ethylene; DDMS = 1-chloro-2,2-*bis*(*p*-chlorophenyl)ethane; DDNU = unsym-*bis*(*p*-chloro-phenyl)ethylene; DDOH = 2,2-*bis*(*p*-chlorophenyl)ethanol; DDA = di-chlorodiphenylacetate; DPM = dichlorodiphenylmethane; DBH = di-chlorobenzhydrol; DBP = dichlorobenzophenone; DDE = 2,2-*bis*(*p*-chlorophenyl)1,1-dichloroethylene.

bis(*p*-chlorophenyl)ethane), has been observed as the major product. DDD is also a pesticide and priority pollutant; successful bioremediation of DDT requires simultaneous DDD removal.

Anaerobic dechlorination of DDT is not coupled with general biological activities; it does not require prior acclimation of microorganisms to DDT (You et al. 1994a). Aqueous-phase and DDT-spiked soil-slurry-phase studies have shown extensive DDT transformation can be encouraged and DDD production can be reduced significantly with the addition of nonionic surfactant under highly reduced conditions (You et al. 1994a, b).

The study described here applied the same methodology to the treatment of site DDT-contaminated soils. Because site-contaminated soils have been aged for decades, contaminants tend to be less available for remediation. Because surfactant may increase DDT availability by releasing DDT from soil, the methodology was expected to be useful in remediation of aged soils. Sorption of the nonionic surfactant Triton X-114 (TX-114) to the soil and its applicability as an amendment in remediation was also considered.

MATERIALS AND EXPERIMENTAL PROCEDURES

DDT-contaminated soil collected from San Francisco Bay (courtesy of Levine Fricke Consulting Engineers, Emeryville, California, USA) was sieved through a 2-mm screen and stored at 4°C. The water content for the soil is 8.9% and the organic content is 3.4%. Soil-water content was measured by drying at 105°C; total organic matter content was estimated from the ignition loss at 550°C.

An anaerobic microbial culture was taken from a culture source reactor (*E* culture in You et al. 1994a) maintained at 35°C and fed ethanol as the primary carbon source without prior exposure to DDT.

Test reactors were set up by transferring 1 g of soil to each 40-mL reactor followed by addition of chemicals such as surfactant, reducing agents (250 mg/L of cysteine hydrochloride and sodium sulfide) or mercuric chloride ($HgCl_2$). A 20-mL aliquot of anaerobic culture was transferred to each reactor; the reactor was then flushed with nitrogen and capped tightly with a Teflon™-lined septum cap. The reactors were mixed on a circular shaker (130 rpm) at 35°C. Killed control reactors were set up by adding 300 mg/L $HgCl_2$ without nitrogen flushing to ensure higher ORPs and no DDT biotransformation. An adequate number of reactors was established such that one whole test reactor was sacrificed for each sampling.

Soil extraction, DDT and intermediates analysis, and ORP measurements followed procedures described previously (You et al. 1994a). Concentrations of DDT and its intermediates are expressed in terms of "DDT equivalent concentration" (mg/kg soil), defined as the molar concentration of each compound multiplied by the molecular weight of DDT. From the DDT transformation pathway, concentrations of DDT, DDD, DDMU, DDE, DDOH, and DBP were measured. The sum of the six measured mass concentrations (defined as "total DDT products") was used for mass balance calculations. Because not all DDT transformation by-products could be measured, strict closure of the mass balance was not possible. ORPs are presented versus a standard hydrogen electrode (SHE) reference.

RESULTS AND DISCUSSION

The DDT concentration did not change significantly with time in the killed control reactors; the average concentration was 1,940 mg/kg dry soil (Fig. 2, curve 1). DDT transformation was observed in all the other reactors (Fig. 2 and 3). The initial DDT transformation rate was higher in reactors with surfactant added. After 30 days, the residual DDT concentration was 120 to 225 mg/kg, or 6% to 12% of the initial DDT concentration. The residual DDT may represent the portion of DDT strongly bound to the soil and not available for transformation.

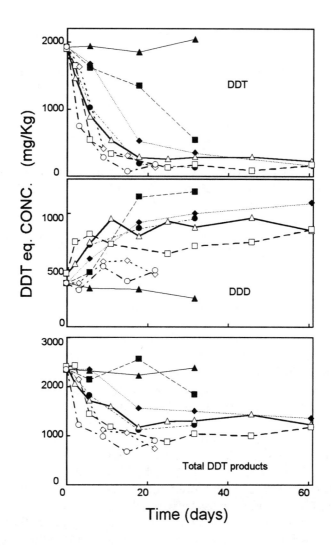

FIGURE 2. **Transformation of site DDT-contaminated soil (part 1).**

──▲── 1: control, exposed to air in setup	─ ─■─ ─ 2: no reducing agents
····◆···· 3: reducing agents	─ ·◆· ─ 4: 500 mg/L TX-114
──△── 5: 250 mg/L TX-114, reducing agents	─ ─□─ ─ 6: 500 mg/L TX-114, reducing agents
···◇··· 7: 1,164 mg/L TX-114, reducing agent	─ ·○· ─ 8: 2,811 mg/L TX-114, reducing agent

DDD concentration, which had an initial level of 382 mg/kg soil, or 20% of the initial DDT concentration, increased with time. The amount of DDD production was equivalent to 4 to 59% of the converted DDT. DDD generation was highest in tests without amendments (curve 2), where production was 59% of the converted DDT. With reducing agents (curve 3) and surfactant (curve 4) addition, the conversion ratios were 40% and 28%, respectively. With reducing

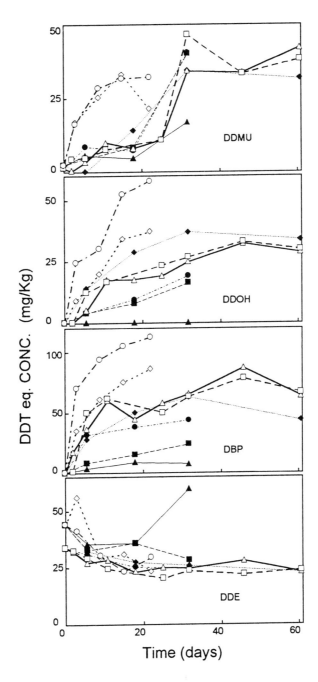

FIGURE 3. Transformation of site DDT-contaminated soil (part 2).

```
———▲——— 1: control, exposed to air in setup      — –■– — 2: no reducing agents
····◆···· 3: reducing agents                       – · –●– · –  4: 500 mg/L TX-114
———△——— 5: 250 mg/L TX-114, reducing agents     — –□– — 6: 500 mg/L TX-114, reducing agents
··· ◇ ··· 7: 1,164 mg/L TX-114, reducing agent     — · ○ · — 8: 2,811 mg/L TX-114, reducing agent
```

agents and higher TX-114 concentrations (curves 5 through 8), less DDD production was observed. With 1,164 mg/L TX-114 and reducing agents (curve 7),
DDD production was equivalent to only 4% of the DDT loss. DDD concentrations below the initial concentration were not observed, indicating that DDD
previously accumulated in the soil was not readily degradable. This observation is similar to that seen in aqueous systems. DDD originally present in DDT-
acclimated biomass was not readily degradable (You et al. 1994a).

DDMU, DDOH, and DBP accumulated with time and concentrations
increased most in reactors with added reducing agents and higher TX-114 concentrations. DDE, not a reductive product of DDT, showed partial transformation, with more than 60% of the initial concentration remaining at the end of
the experiments. The incomplete transformation of DDE was likely due to the
low availability of DDE, possibly caused by its high binding capacity to soil
matrices (You et al. 1994b).

DDT and DDD accounted for most of the total DDT products. Total DDT
products decreased with time. The decrease was greater with reducing agents
and higher TX-114 concentration, indicating increased concentrations of
unmeasured intermediates.

The pH in the reactors was in a range of 7.44 to 7.97. ORPs in killed control
reactors were in the range of –55 to +122 mV; ORPs in reactors without reducing agents were in the range of –140 to –120 mV; with reducing agents, ORPs
were between –190 to –70 mV. Variations of ORP values may be due, in part, to
the incomplete nitrogen flushing during the reactor setup or slight air contamination of the reactor contents when the reactor caps were loosened by mixing.

TX-114-soil and TX-114-anaerobic culture sorption isotherms can be
expressed as (You et al. 1994c)

soil: when $C < 48.4$ mg/L, $S = 0.274 C^{0.769}$
soil: when $C \geq 48.4$ mg/L, $S = 2.48 * 10^{-5} C^{3.17}$
anaerobic culture: $S = 0.505 (C/D)^{0.953}$

where S is the sorbed concentration of TX-114 (mg/g soil or volatile suspended
solids [VSS]), C is the aqueous concentration of TX-114 (mg/L), and D is the biomass concentration (g VSS/L). Based on these isotherms, the TX-114 distribution in the DDT transformation reactors as a function of total TX-114 concentration was calculated (Fig. 4). A minimum 78% of the added TX-114 was
sorbed by soils, whereas a maximum of 16% of the TX-114 was present in the
bulk solution. In order for TX-114 aqueous concentration to reach the critical
micelle concentration (80.3 mg/L), addition of 1,350 mg/L of TX-114 was necessary. At these concentrations, more than 93% of TX-114 was sorbed by the biomass and soil.

The high fraction of TX-114 that is sorbed may limit the use of this surfactant in full-scale remediation. TX-114 is an alkylphenol ethoxylate surfactant.
This class of surfactants is known to be recalcitrant to biodegradation (Kravetz
1981, Wagener and Schink 1987). If applied to site remediation, the high sorption capacity of TX-114 will result in a large residue of TX-114 in the environment, which is not desirable. If we assume the surfactant present in the bulk

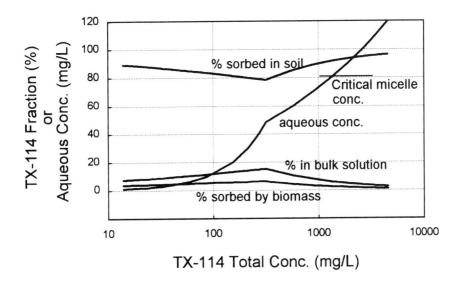

FIGURE 4. TX-114 phase distribution and aqueous concentration versus TX-114 total concentration in soil slurry systems

solution is more effective for DDT transformation enhancement, the large total concentration required to generate an effective bulk solution concentration may prove to be economically undesirable.

CONCLUSIONS

The current work shows that anaerobic bioremediation of DDT-contaminated soil supplemented with surfactant and reducing agents is a promising technology. Surfactant addition was effective in increasing the DDT transformation rate and in decreasing DDD production. The decrease in DDD production is meaningful because DDD, an environmentally regulated pesticide, is recalcitrant to aerobic transformation. Less-chlorinated intermediates (e.g., DBP, DDOH) accumulated under anaerobic conditions, but these compounds may be aerobically biodegradable. Complete mineralization may require sequential anaerobic-aerobic conditions. Further study is required to identify an environmentally and economically acceptable alternative surfactant that promotes DDT transformation.

ACKNOWLEDGMENT

This work was supported in part by U.S. EPA Cooperative Agreements 816700 and 821029, and by a cooperative research and development agreement

(CRDA 0009-92) with Levine-Fricke Consulting Engineers, Emeryville, California, USA. The authors thank Maoxiu Wang and Tiehong Huang for their laboratory assistance.

REFERENCES

Kravetz, L. 1981. "Biodegradation of nonionic ethoxylates." *J. Am. Oil Chem. Soc.* *58*:58A-65A.

Rochkind, M. L., and J. W. Blackburn. 1986. *Microbial Decomposition of Chlorinated Aromatic Compounds.* EPA/600/2-86/090. pp. 138-145.

Wagener, S. and B. Schink. 1987. "Anaerobic degradation of nonionic and anionic surfactants in enrichment cultures and fixed-bed reactors." *Wat. Res. 21*:615-622.

You, G., G. D. Sayles, M. J. Kupferle, I. S. Kim, and P. L. Bishop. 1994a. "Anaerobic DDT Biodegradation: Enhancement by Application of Surfactants and Low Oxidation Reduction Potential." Submitted to *Chemosphere.*

You, G., G. D. Sayles, M. J. Kupferle, P. L. Bishop, and D. S. Lipton. 1994b. *Anaerobic DDT Degradation of Contaminated Soil by a Mixed Consortia and Enhancement by Surfactant Addition in Slurry Reactors.* pp. 635-646. Water Environment Federation, 67th Annual Conference & Exposition, Chicago. October, 1994.

You,G., M. Wang, G. D. Sayles, M. J. Kupferle, and P. L Bishop. 1994c. "Isotherms for the Sorption of a Nonionic Surfactant onto Soils and Biomass and Implications to Environmental Remediation." In preparation.

Biological and Nonbiological Losses of the Chlorinated Hydrocarbon Endosulfan from Soil

Turlough F. Guerin

ABSTRACT

Nonbiological processes were predominantly responsible for the loss of the chlorinated hydrocarbon endosulfan I in a bioremediation feasibility study using indigenous microorganisms without organic amendments. The study highlighted the importance of minimizing losses due to volatilization and chemical hydrolyses when investigating the bioremediation potential of certain organic pollutants in soil. This is the first report of the chlorinated ring of endosulfan I being mineralized to $^{14}CO_2$.

INTRODUCTION

Several reports have described the dissipation of endosulfan applied to soil (Guerin 1993 and references cited therein). These studies, however, have not clearly demonstrated the relative contributions of biological and nonbiological processes in the loss of endosulfan I (the major isomer of the chlorinated insecticide endosulfan) in soil including those conducted by Van Dyk and Van Der Linde (1976), Martens (1977), and Doelman (1990). Although previous liquid culture studies have indicated that soil microorganisms are capable of transforming endosulfan to various degradation products in aerobic liquid culture (Miles and Moy 1979), these studies do not reflect the extent of biodegradation (as opposed to undefined dissipation) from soil. The current study attempted to determine the role of indigenous microorganisms in two different soils in the dissipation of endosulfan I and to compare this with the combined effect of nonbiological processes, specifically chemical degradation and volatilization. A further hypothesis tested was that endosulfan I, radiolabeled in the chlorinated ring, could release $^{14}CO_2$.

MATERIALS AND METHODS

Trial Design

Chemicals. The chlorinated hydrocarbon [6,7,8,9,10-^{14}C]-endosulfan I (Hoechst, specific activity 2962 MBq/g) was dissolved in 100% methanol and added to small subsamples of soil. These subsamples of soil (1 to 2 g) were ground to form a fine powder to which the methanol solution was added. Immediately after the methanol evaporated, the dried subsample was distributed evenly through the bulk of the soil (30 g) in each incubation vessel (a 250-mL glass jar).

Characteristics of the Soil and Incubation Conditions. Two soils were used in this study. One soil was collected from a cotton farming area (CFS) and the other a contaminated soil from a sullage pit (used for the disposal of agricultural chemicals), located on the same cotton farm (CS). The soil characteristics are presented in Table 1. This farm was located at Narrabri, in northwestern New South Wales, Australia. Both samples were collected at the beginning of a cotton growing season and stored at 4°C until used in the experiments. The moisture characteristics of the CFS and the CS were determined so that the water content corresponding to the field capacity (FC) water potential (–0.3 bar or –33 J.kg^{-1}) could be obtained accurately. Determining the water content at FC of both soils allowed the CFS and the CS to be compared at the same water availability. The available water for microbial growth (not the gravimetric moisture content) was considered important in the design of this trial. The pressure plate and vacuum desiccator methods described by Guerin (1993) were employed to determine the moisture characteristics of each soil.

TABLE 1. Characteristics of the soils used in the study.

Soil	pH	OC(%)[a]	OM(%)[a]	Fraction %			
				C[b]	S[c]	FS[d]	CS[e]
CFS	7.5	1.09	1.91	62.4	20.4	14.3	2.9
CS	7.8	2.20	2.11	51.1	15.9	17.0	16.0

(a) OC = organic carbon; OM = organic matter. OC was measured using the Walkley and Black method described in McLeod (1975).
(b) C = clay fraction.
(c) S = silt fraction.
(d) FS = fine sand fraction.
(e) C = coarse sand fraction. The fractionation was conducted using the Bouyacos method described in Day (1953).

The 250-mL glass jars were sealed with Teflon™ (polytetrafluoroethylene, or PTFE) tape. The lid inserts in the glass jars were lined with sterilized aluminum foil, and these were replaced at each of the sampling times. Sodium azide, NaN_3, a respiration inhibitor and previously reported to be used as a soil sterilant (El Beit et al. 1981) was added to the control soils at 0.1% (w/w). Prior to the addition of NaN_3, the control soils were autoclaved twice (on two consecutive days) at 121°C for 30 min. In the sterile controls the radiolabeled endosulfan was added to the soil after autoclaving and unlabeled endosulfan I concentrations were determined after autoclaving. The flasks were incubated at 30±1.5°C in the dark and subsamples were taken at 0, 1, 2, 3, 5, 7 and 9 weeks. All subsamples were removed aseptically from the jars at the sampling times using a laminar flow cabinet. The amounts of radioactivity (in the form of endosulfan I) and unlabeled endosulfan I added to the soil are presented in Table 2.

Analysis of Radioactive Carbon Dioxide. Wheaton vials (4-mL total capacity and 4 cm long) containing 1 mL 2M NaOH (alkaline solution), were placed on the soil surface in each of the flasks. The amount of ^{14}C captured was determined by adding the alkali solution into a liquid scintillation fluid (Hionic Fluor, Canberra Packard) designed for samples with high ionic concentrations. A further capture of $^{14}CO_2$ was conducted. The soil subsamples (1 to 2 g), removed from the sealed incubation flasks at the various sampling times for analysis, were treated with 0.1M HCl for 12 to 15 h to release $^{14}CO_2$. These soil subsamples in the 4-mL vials were incubated within larger 30-mL sealed vials so that any evolved $^{14}CO_2$ could be captured in an alkaline solution. Any radioactivity in the alkali absorbent within the larger vials was then determined.

Extraction of Radioactivity from Soil. The acidified soil samples were treated with an equal mass of distilled water as a slurry and placed into 250-mL ground glass sealed flasks. The soil-water slurry was then extracted on an orbital shaker at 160 rpm for 2 to 3 h with 40 mL hexane:acetone (3:1). After extraction, the hexane-acetone extract (HAE) (<40 mL) was decanted into a

TABLE 2. Initial concentrations of endosulfan I.

Soil	Unlabeled[a]	Labeled	Total
CFS	0.51	1.66	2.2
CS	414	3.32	417

(a) These values represent the total amounts of endosulfan I in the soil as determined by the soil extraction procedure described in Guerin (1993) and the GC-ECD analytical method described in Guerin et al. (1992)

250-mL separation funnel and partitioned with 10 mL of distilled water. The HAE component remaining after partitioning was then made up to 50 mL with hexane and radioactivity determined. The volume of the aqueous phase was then measured. This aqueous phase was separated and termed the water re-extract (WRE). The soil residue remaining from the HAE was re-extracted with 15 mL 100% methanol to provide the methanol (ME) phase. The radioactivity in each of the three phases (HAE, ME, and WRE) was then determined with a Packard liquid scintillation counter (4000 Series) (Guerin 1993).

Analyses of Total Endosulfan I. The determination of total endosulfan I in the soil was conducted using the extraction procedure described in Guerin (1993) and analyzed using the electron-capture detector/gas chromatography (ECD-GC) method as previously described in Guerin et al. (1992).

RESULTS AND DISCUSSION

Dissipation of Endosulfan I from the CFS

Due to the low water solubility (high lipophilicity) of endosulfan I (approximately 2.3 mg/L) (Guerin et al. 1992), the majority of the extracted radioactivity (76 to 93%) at the beginning of the trial was found in the HAE. The remainder of the radioactivity was unrecovered, presumably bound in the soil (Table 3). During the course of the trial, radioactivity in HAE decreased in both the sterile and nonsterile treatments. The radioactivity remaining in the nonsterile treatment was 51% (SE = 7%) of that originally applied, compared with 60% (SE = 7%) in the sterile treatment at the end of the trial (Table 3). These results indicate that there was an effect of the indigenous soil microflora on the degradation of endosulfan I. Although there were only small differences between the sterile and nonsterile treatments in the WREs, the radioactivity in the ME from the nonsterile treatment increased substantially, further indicating that microorganisms were transforming the applied ^{14}C-endosulfan I (Table 3). A previous report, which attempted to determine the specific role of microorganisms in the degradation of endosulfan in soil, indicated that biotransformation to less hydrophobic products and binding of the [5,9]-^{14}C-endosulfan (radiolabeled in the nonchlorinated carbons) can occur (Monteiro et al. 1989).

Although the increase in methanol-soluble radioactivity in this particular study was not measured, endosulfan diol and endosulfan sulfate were identified (but not quantified) as degradation products in the nonsterile treatments. Given that the solubility of endosulfan diol and sulfate are 300 and 18 mg/L, respectively (Guerin et al. 1992), compared to 2.3 mg/L for endosulfan I, it is likely these products contributed to the increase in the radioactivity in the ME and WRE phases in the current study.

The data from the sterile treatments indicated that dissipation occurred even though the soil was autoclaved and treated with NaN_3. This suggested

TABLE 3. Distribution of ^{14}C-endosulfan I in soil extracts during the trial (%).

Time (weeks)	Soil	HAE Sterile	HAE Nonsterile	ME Sterile	ME Nonsterile	WRE Sterile	WRE Nonsterile	$^{14}CO_2$ Sterile	$^{14}CO_2$ Nonsterile	Unrecovered[c] Sterile	Unrecovered[c] Nonsterile
0[a]	CFS	76	78	1.1	1.4	0.3	0.3	0	0	23	20
0	CS	93	88	0.7	0.5	0.4	0.4	0	0	5.9	11.1
9[b]	CFS	60	51	1.4	7.5	1.5	0.7	1.4	1.4	36	39
9	CS	80	73	1.8	2.5	1.3	0.8	0.4	0.4	17	23

(a) The percentage values are the mean of 2 separate analyses and the average standard errors ranged from 5 to 8% at week 0.
(b) The percentage values are the mean of 2 separate analyses and the average standard errors ranged from 7 to 14% at week 9.
(c) The ^{14}C-label in this fraction included bound endosulfan I at week 0 and bound as well as volatilized residues at week 9.

that nonbiological processes were involved in the decrease of HAE residues of endosulfan I (Table 3). Because the initial soil pH values were >7, it is likely that there were some minor losses due to alkaline hydrolysis (Goebel et al. 1982). El Beit et al. (1981) also have indicated that NaN_3 may increase soil pH. The NaN_3 may have influenced the rate of loss of endosulfan I in the current study, because small increases in soil pH (up to 0.2 pH units) were measured in the sterile but not in the nonsterile treatments.

In addition, because small amounts of water were shown to be lost from the incubation vessels (as measured at each of the sampling times), it is likely that endosulfan I losses may have occurred through volatilization, due to the volatility of this compound (Henry's constant = 1.07) (Guerin et al. 1992). The formation of endosulfan diol from alkaline hydrolysis of endosulfan I is likely to substantially reduce the volatility of endosulfan I. Although there are no data on the volatility of endosulfan diol, its relatively high water solubility compared to endosulfan I suggests that its volatility is considerably lower than that of endosulfan I. The higher water solubility of endosulfan diol would have resulted in tighter binding of this form of endosulfan to the soil particles and soil water.

Losses of endosulfan I would have also been expected to occur on opening the incubation vessels for sampling. Because of its high volatility, substantial amounts of endosulfan I would have become equilibrated with the volume of air above the soil in the incubation flasks (approximately 200 mL compared with 30 mL of soil). Each time the incubation flasks were opened for sampling, the equilibrium vapor was dissipated. It was likely that this loss was higher in the CS compared with the CFS where the concentration of applied (or "non-aged") ^{14}C-endosulfan I was relatively high.

Dissipation of Endosulfan I from the CS

The radioactivity in the HAE from the sterilized and unsterilized CS was 80 and 73% (average SE = 20% for both) respectively, of that originally applied after the 9-week incubation (Table 3). The smaller decrease in radioactivity in the HAE (compared with the CFS) over the 9-week incubation in both the sterile and nonsterile treatments suggested that chemical and physical processes were most significant in the losses of endosulfan I from this soil. Although there was only a small difference between the sterilized and nonsterilized treatments with respect to radioactivity remaining in the HAE, there was an increase in the radioactivity in the ME in the nonsterile treatments. This indicated that there was microbial conversion of endosulfan I to less hydrophobic, methanol-soluble degradation products. In the sterile treatment, there was also a slight increase in methanol-soluble radioactivity indicating that nonbiological processes contributed (at least in part), to this transformation.

Due to the difficulty in completely sterilizing soil, there was a possibility that there was a small amount of biological activity in the sterilized flasks.

In the nonsterile treatment, it was likely that volatilization and chemical hydrolysis had removed much of the endosulfan I prior to any substantial bio-transformation taking place.

The final distribution of radioactivity observed in the HAE of the CS (both sterile and nonsterile), suggested that there was a considerably lower rate of endosulfan I dissipation in this soil compared with the CFS (Table 3). This can be explained in terms of the apparently low specific activity of endosulfan I in the CS treatments. This resulted from the dilution of applied radioactive endo-sulfan I with endogenous residues present as a result of previous agricultural activities. Although twice as much radioactive endosulfan I was added to the CS treatments, the higher total concentrations of endosulfan I in the CS samples (417 μg/g) compared to 2.2 μg/g in the CFS (concentrations in both soils determined prior to the trial but after autoclaving) caused the theoretical specific activity of the endosulfan I in the CS to be approximately 100 times less than that contained in the CFS (Table 2). Although there was only a relatively small change in the measured radioactivity, the actual amount of endosulfan I dissipated from the CS was considerably higher than that from the CFS assuming the applied and endogenous endosulfan I behaved in the same way. If this was the case, then over the first 2 weeks of the trial, the amount of endosulfan I in the CS decreased by approximately 100 μg/g, whereas in the CFS the decrease was only 0.5 μg/g. However, the endogenous endosulfan I present in the CS was "aged" and consequently more strongly adsorbed to the soil. As a consequence, the applied ^{14}C-endosulfan I would have been preferentially removed from the soil by volatilization, chemical hydrolysis, and/or biodegradation, compared to the aged endogenous residues. In the CS, it was observed that the amount of ^{14}C-label remaining unrecovered was substantially less than that recorded for the CFS, indicating that there was considerably less binding of the ^{14}C-label in the CS.

The endogenous endosulfan I had been in both the CS and CFS soil for up to 10 years—the period since technical-grade endosulfan was used routinely in producing cotton in the region where the soil samples were collected (Guerin 1993). It is recognized that aging of organic compounds can occur in soils leading to increased amounts of bound residues. Hatzinger and Alexander (1995) have demonstrated that aged phenanthrene (also a hydrophobic compound), becomes increasingly resistant to biodegradation with time.

Although no studies have been reported on the dissipation of endosulfan I from soils at such high concentrations, the findings of Van Dyk and Van Der Linde (1976) corroborate the observed losses of "non-aged" endosulfan both due to volatilization in the current study. These researchers demonstrated that freshly applied endosulfan (comprised of isomers I and II in a ratio of 3:1) was rapidly volatilized when added to nonsterile soil.

Although autoclaving has the potential to influence the sorptive properties of soil, this was conducted prior to addition of radioactive endosulfan I and prior to analyses of the initial soil concentrations. Therefore this would not have contributed directly to the loss of radioactivity in the HAE phase in the sterile treatment.

Significance of Nonbiological Processes in the Design of Trials for Bioremediation Feasibility Studies

The biological contribution to endosulfan I dissipation can be difficult to predict due to the confounding effects of nonbiological processes, in particular volatilization and chemical hydrolysis. An explanation for the predominant role of volatilization in the loss of endosulfan I from soil at high concentrations can be given by comparing the expected rates of loss by this process to rates of loss by biological processes. Loss by volatilization increases linearly with concentration because loss is directly proportional to the concentration (or activity) of the compound. In a soil already containing high concentrations of endosulfan I, there are likely to be fewer sites remaining on the soil surface available for further adsorption of added endosulfan I. Thus, the radioactive endosulfan I added to the contaminated soil would have been only weakly absorbed and therefore would have volatilized more easily than the aged residues. In contrast, in the cotton farming soil, where the concentrations of endosulfan I were considerably lower, the added endosulfan I would have been more strongly bound. Under these conditions, rates of loss by biodegradation were more significant.

In soil, where the surfaces are strongly adsorbing, and particularly towards organochlorine compounds, the rate of loss will be influenced by the physical and chemical factors controlling absorption of endosulfan I onto the soil surface. These factors affecting absorption, which have been discussed in detail by Spencer and Cliath (1981), cause the effective vapor pressure of an adsorbed chemical to be reduced. The rate of loss from volatilization is therefore likely to be greater immediately after its application, before stronger adsorption to soil occurs.

Rates of loss due to biodegradation do not follow the same mode as volatilization because biological systems become saturated with increasing concentrations of applied compounds (Alexander 1977; Howard et al. 1981). Depending on the nature of the compound and the activity of the microorganisms degrading the compound, the concentration at which the rate of loss by biodegradation reaches a maximum will vary. However, this concentration generally will be less than that at which volatilization reaches a maximum rate. Thus, with increasing concentrations, losses due to volatilization may increase in a linear manner. On the other hand, loss from biodegradation typically will follow a Michaelis-Menten type curve with concentration, that is, a rectangular hyperbola, eventually reaching a maximum rate with increasing concentrations. In the current study, the microorganisms in the contaminated soil were possibly kinetically saturated with endosulfan I.

It is recommended, in studies where the potential role of microorganisms is to be determined, that a sufficiently wide range of concentrations be used. If volatilization occurs, then the contribution of microorganisms to the compound's loss would be most obvious in treatments where lower contaminant concentrations were used.

In trials where the values for soil pH are >7, or where volatilization is likely to be significant (such as high concentrations or an open system), nonbiological losses of endosulfan I will be more significant, and smaller differences between sterile and nonsterile treatments will be expected. In soil with pH values <7, biodegradation is likely to be more significant than chemical degradation in the dissipation of endosulfan I. These points should be considered at the design stage of setting up bioremediation treatability studies.

Mineralization of Endosulfan I

Over the 9-week trial period, the total amount of $^{14}CO_2$ released was equivalent to 1.4% of the original radioactivity applied. Controls demonstrated that the alkali solution did not absorb any volatile ^{14}C-endosulfan I. Although there was no significant difference between the sterile and nonsterile treatments with either of the soils, a greater proportion of $^{14}CO_2$ was released from CFS compared with the CS.

The results from the study indicate that the chlorinated cyclodiene ring structure in endosulfan I can be oxidized, at least to a small extent (by microbial and/or nonbiological processes) in these soils. A similar result was reported by Monteiro et al. (1989). These researchers incubated a mixture of endosulfan I and II (radiolabeled in the nonchlorinated carbons of the molecule *viz* [5a,9a]-^{14}C) in three different soils with pH values of <7. The respective amounts of $^{14}CO_2$ released from their sterilized and nonsterilized treatments over a period of 48 days were 0.3 and 0.4% of that originally applied as the parent compound. The results presented in the current paper are the first to report that the chlorinated ring of endosulfan I can be mineralized to $^{14}CO_2$ even though the rates of formation are low and it occurred in both sterile and nonsterile soil.

Reducing conditions may be required in soil contaminated with endosulfan I to promote initial dechlorination of the chlorinated ring and thereby expose the compound to a more extensive oxidation. Although there is no direct evidence in the literature for the dechlorination of endosulfan I under reducing conditions, previous research has demonstrated that related cyclodienes can be reductively dechlorinated (Guerin 1993 and references cited therein). Endosulfan diol also can be formed at high rates under reducing conditions (Doelman 1990, Guerin 1993).

Influence of Soil pH on the Dissipation
and Measurement of Endosulfan I

Soil pH can influence the rate of hydrolysis of endosulfan I. At the beginning of the trial, after adding ^{14}C-endosulfan I and NaN_3 and autoclaving, the pH of both soils was determined. The pH of the CFS and CS was 7.5 and 7.8, respectively. The slightly higher pH of the CS may account, at least in part, for the greater increase of radioactivity in the WRE and ME in this soil. The rate of endosulfan diol formation from chemical hydrolysis was expected to be greater in the higher pH CS.

Soil pH also influences the retention of $^{14}CO_2$. At pH values less than approximately 6.4, CO_2 is released from carbonic acid and may be absorbed by an alkali solution. Above pH 6.4, CO_2 tends to remain in the soil solution as bicarbonate (HCO_3^-) and CO_3^{-2} ions or fixed as carbonates. Bicarbonate, carbonate and carbonic acid can then be converted to CO_2 by acidification of the soil. Both soils in the experiment had pH values considerably higher than 6.4. Therefore, as a precaution to determine whether any of the originally applied radioactivity was converted to $^{14}CO_2$ and subsequently had become bound in the soil as carbonates or carbonic acid, the soil samples collected at each of the sampling times were treated with acid and the presence of any bound radioactive CO_2 was determined prior to the solvent extraction of the sample. During the acidification, the soil pH decreased to <1.5 pH units.

The efficiency of this procedure for determining the release of bound CO_2 was verified using radioactive bicarbonate in controls using the same soil (treated with the same acid) and in a dilute acid solution (positive control) (Guerin 1993). The recovery for CO_2 release was 96 to 100%. None of the treatments or controls released $^{14}CO_2$ upon acidification indicating, unexpectedly (because of the soil pH values of approximately 8 pH units), that the formation of bicarbonate in the soil was not limiting the total $^{14}CO_2$ evolved and measured.

CONCLUSIONS

Radioactive endosulfan I was dissipated from both the CFS and the CS. The dissipation was not appreciably accelerated in the nonsterile treatments, indicating that microorganisms did not contribute substantially to the observed losses of this compound under the conditions described. The specific nonbiological losses were likely to be due to chemical hydrolysis and volatilization. This explanation was proposed because (1) this compound is susceptible to alkaline hydrolysis, which was likely because the soils were alkaline, and (2) because endosulfan I is volatile.

It is apparent that nonbiological processes can be very significant in the dissipation of endosulfan I from soil. Therefore, under conditions of high soil pH, and where volatilization can occur, care should be taken regarding the interpretation of the relative role of biological processes in the degradation of endosulfan I in soil.

Special precautions should be taken when designing bioremediation feasibility trials for other volatile compounds, and those containing chemical structures susceptible to hydrolysis, to ensure that biological processes are adequately quantified. The much higher concentrations of endosulfan I in the CS probably accentuated losses by volatilization because the rate of volatilization increases in a directly proportional manner to contaminant concentration. From the comparison of rates of biological and nonbiological losses of endosulfan I from the CFS and CS, it can be concluded that under the conditions of the current study the proportion of biological contributions to the overall dissipation is greatest where endosulfan I concentrations in soil were low.

For a bioremediation process to be effective for the treatment of endosulfan I in the soils used in the current study, it is likely that organic matter amendments would be required to increase the activity of microorganisms capable of degrading endosulfan I. Organic carbon was limiting in both soils in the current study. A composting procedure, similar to that previously applied to the treatment of various organic contaminants in soil, may be an effective aerobic treatment process for endosulfan and related compounds (Rhodes et al. 1994). However, with the high chlorination level of endosulfan I, the addition of simple organic nutrient sources under reducing (anaerobic conditions) may be an appropriate way to biodegrade this compound and there is evidence that endosulfan I can be biodegraded under these conditions (Doelman 1990, Guerin 1993). Further feasibility studies and pilot-scale trials using composting and anaerobic treatments will be required before recommendations on field-scale treatment can be made.

In the current study, a small proportion of the ^{14}C-endosulfan I was converted to $^{14}CO_2$. The radioactive endosulfan I used in these experiments contained the ^{14}C-label entirely in the chlorinated ring carbons which, in related compounds, has been shown to be highly recalcitrant to degradation in the environment. Nevertheless, this is the first report of $^{14}CO_2$ release from the chlorinated ring with this insecticide.

ACKNOWLEDGMENTS

I. R. Kennedy from the Department of Agricultural Chemistry and Soil Science, University of Sydney (where the study was conducted) is acknowledged for discussions in the design and implementation stages of the project and feedback on early drafts. Funding from the Cotton Research Development Corporation and the support and interest provided by Hoechst (Melbourne and Frankfurt) is gratefully acknowledged.

REFERENCES

Alexander, M. 1977. *Introduction to Soil Microbiology.* 2nd ed. John Wiley and Sons, New York, NY.

Day, P. R. 1953. "Experimental Confirmation of Hydrometer Theory." *Soil Science* 74: 181-186.

Doelman, P. 1990. "Microbial Degradation of Hexachlorocyclohexane Isomers in Mineral Soil and of Endosulfan Isomers in Organic Soil in Connection with Soil and Water Quality: Ecotoxicological Research." *Report of the Netherlands Organisation for Applied Scientific Research (TNO)* 23: 73-91.

El Beit, I. O. D., J. V. Wheelock, and D. E. Cotton. 1981. "Pesticide-Microbial Interaction in the Soil." *International Journal of Environmental Studies* 16: 171-180.

Goebel, H., S. G. Gorbach, W. Knauf, R. H. Rimpau, and H. Huttenbach. 1982. "Properties, Effects, Residues and Analytics of the Insecticide Endosulfan." *Residue Reviews 83*: 1-122.

Guerin, T. F. 1993. "The Relative Significance of Biodegradation and Physico-chemical Dissipation of Endosulfan from Water and Soil." Ph.D, University of Sydney, New South Wales, Australia.

Guerin, T. F., S. W. L. Kimber, and I. R. Kennedy. 1992. "Efficient One-Step Method for the Extraction of Cyclodiene Pesticides from Aqueous Media and the Analysis of their Metabolites." *Journal of Agricultural and Food Chemistry 40*(11): 2309-2314.

Hatzinger, P. B., and M. Alexander. 1995. "Effect of Aging of Chemicals on Their Biodegradability and Extractability." *Environmental Science and Technology* 29: 537-545.

Howard, P. H., H. C. Sikka, and S. Banerjee. 1981. "Test Methods for Determining the Biodegradation of Organic Chemicals in the Aquatic Environment." In G. Zweig and M. Beroza (Eds.), *Test Protocols for Environmental Fate and Movement of Toxicants*, pp. 150-176. Association of Official and Analytical Chemists, Arlington, VA.

Martens, R. 1977. "Degradation o^{14}C-Endosulfan Under Different Soil Conditions." *Bulletin of Environmental Contamination and Toxicology 17*: 438-446.

McLeod, S. 1975. *Studies on the Wet Oxidation Process for the Determination of Organic C in Soil.* CSIRO Division of Soils, Adelaide, South Australia.

Miles, J. R. W., and P. Moy. 1979. "Degradation of Endosulfan and its Metabolites by a Mixed Culture of Soil Microorganisms." *Bulletin of Environmental Contamination and Toxicology* 23: 13-19.

Monteiro, R. T. R., R. Hirata, M. N. de Andrea, J. M. M. Walder, and F. M. Wiendl. 1989. "Endosulfan-^{14}C Degradation in Soil." *Revisita Brasiliera de Ciencia do Solo 13*: 163-168.

Rhodes, S. H., T. F. Guerin, and P. Peck. 1994. "Composting for the Bioremediation of Contaminated Soil."

Solid Waste in the Pacific. Common Issues—Common Solutions." *6th Annual Conference of the Waste Management Institute New Zealand*, Waste Management Institute of New Zealand, Christchurch.

Spencer, W. F., and M. M. Cliath. 1981. "Evaluating Volatility of Toxicants in Soil and Water." In G. Zweig and M. Beroza (Eds.). *Test Protocols for Environmental Fate and Movement of Toxicants*, pp. 110-121. Association of Official and Analytical Chemists, Arlington, VA.

Van Dyk, L. P., and A. Van Der Linde. 1976. "Persistence of Endosulfan in Soils of the Loskop Dam Irrigation Area." *Agrochemophysica 8*: 31-34.

Anaerobic Biodegradation of the Chlorinated Hydrocarbon Endosulfan

Turlough F. Guerin

ABSTRACT

A study was undertaken to assess the bioremediation potential of indigenous mixed populations of anaerobic microorganisms from selected agricultural soils and freshwater sediments for their ability to biodegrade endosulfan I. Endosulfan I is readily biodegraded by indigenous microorganisms under anaerobic conditions, indicating that the amendment of oxygen-limited soils and sediments with organic cosubstrates may enhance the removal of this compound. The importance of minimizing nonbiological losses has been highlighted as another critical factor in determining anaerobic bioremediation potential.

INTRODUCTION

A number of studies on the aerobic degradation of endosulfan I (the major isomer of the insecticide endosulfan) by microorganisms in liquid cultures have been reported (Miles and Moy 1979; Goebel et al. 1982). There is, however, no reference to the biodegradation of this compound under strictly anaerobic conditions in either soils or liquid culture (Guerin 1993). Studies have been conducted using soils at high moisture contents, where oxygen was not completely limiting. Other studies, conducted under strictly anaerobic conditions in liquid culture, have demonstrated that related cyclodiene compounds may be degraded under these conditions, and these have been reviewed by Guerin (1993). The current paper describes the biodegradation of endosulfan in mixed anaerobic culture by inocula from low-oxygen soils and freshwater sediments. The hypothesis tested in the current research was that endosulfan I can be biodegraded by indigenous microorganisms from low-oxygen soils and sediments under strictly anaerobic conditions.

MATERIALS AND METHODS

Preparation of Soil Inocula, Enrichment, and Routine Incubation Cultures

Soil and sediment samples, and their collection and preparation, have been described in Guerin (1993). An anaerobic enrichment was prepared using endosulfan I and II (isomer ratio 3:1) and the surfactant, Tween 80, to develop a culture potentially capable of degrading endosulfan I at high rates and to compare this with an indigenous population (with little or no previous exposure to endosulfan) (Guerin 1993). The rationale for using the surfactant, Tween 80, was to improve the bioavailability of endosulfan so that it may be more accessible to the growing microorganisms for biodegradation (Rouse et al. 1994). All cultures were kept unshaken and in the dark for 10 days in an incubator at $30\pm1°C$ prior to subculturing. The resulting mixed populations were subcultured 7 times prior to analyzing their degradative capacity in the routine incubation assay.

A medium was prepared for routine (or nonenriched) incubations using the method described by Balch et al. (1979) for cultivating methanogens with the addition of formate and modified by Maule et al. (1987). The pH of the phosphate-buffered medium (100 mM) was adjusted to 7.0 and was purged and degassed with argon for 15 min while the medium was cooled rapidly (Miller and Wolin 1974). All incubations were kept strictly anaerobic during the course of the enrichments, and this was checked by the presence of the redox indicator resazurin (Shelton and Tiedje 1984, Maule et al. 1987).

Incubation Conditions and Pollutant Determination

Serum bottles (125 mL) were used where aliquots of culture were subsampled (Miller and Wolin 1974, Maule et al. 1987). In preliminary time course degradation experiments, butyl rubber stoppers were used to seal the serum bottles and were the same used in the anaerobic studies by Miller and Wolin (1974), Balch et al. (1979), Shelton and Tiedje (1984), Maule et al. (1987), Commandeur et al. (1992), Kohring et al. (1992), and Holliger et al. (1992). In the routine time course experiments, butyl rubber stoppers were replaced with (Teflon™) (polytetrafluoroethylene, or PTFE)-lined stoppers. This approach is the same as that previously employed by Bouwer and McCarty (1983), Grbic-Galic et al. (1990), Van Dort and Bedard (1991), and Sanford and Tiedje (1992). This sealing system was used and tested to determine whether it could retain the relatively insoluble and volatile chlorinated hydrocarbon, endosulfan I, while ensuring the oxygen impermeability of the system. Subsamples were taken from the serum bottles by removing 1- to 2-mL samples from the incubation broth (after homogenizing) using a syringe. Therefore, over the entire period of the trial, there was a decrease in volume of broth of approximately

1 mL at each of the sampling times. Consequently there was an equivalent increase in headspace volume.

Filter-sterilized gas (CO_2:H_2, 80:20) was introduced to the vessels, which were kept unshaken in the dark for 30 days in an incubator at $30\pm1°C$. Subsamples were not taken in the sterile treatments, but rather the entire flask contents were sacrificed because of the problem of adsorption of endosulfan I onto glass. This topic has been described previously (Guerin and Kennedy 1992, Guerin 1993). At various sampling times, the atmospheres were subsampled and analyzed for methane (Shelton and Tiedje 1984). Microbial growth was measured by determining the absorbance of the growing cultures at 700 nm. Controls containing the same media, without added microorganisms, were also analyzed to determine whether the increases in absorbance were due to microbiological activity or to precipitation of media constituents by physical or chemical processes. The culture pH was also determined. Endosulfan I and potential degradation products of endosulfan I were identified by comparing retention times of synthesized standards with the peaks appearing on the treatment chromatograms as a result of endosulfan I biodegradation (Guerin et al. 1992). The microbiological assessment has been described in Guerin (1993).

RESULTS AND DISCUSSION

Endosulfan Biodegradation

Preliminary Studies. In flasks (using butyl rubber seals to restrict oxygen penetration), endosulfan I decreased to very low concentrations after 25 days of incubation in all treatments and controls. No endosulfan I was detected in the treatments containing inoculum from the tailwater drain sediments. Controls containing sterile medium alone, however, indicated a faster initial rate of dissipation than that from inoculated flasks, indicating that there were nonbiological losses. It was apparent that the observed losses were due to absorption of endosulfan I from the medium in the vessels by the unlined butyl rubber seals, or from alkaline hydrolysis, or from a combination of both processes.

The hydrolytic degradation product, endosulfan diol, was detected in all the inoculated treatments. On the other hand, no traces of diol were detected in either of the controls, indicating endosulfan I was quickly removed from the media by absorption onto the butyl rubber stoppers, preventing it from undergoing any hydrolytic reaction. The highest rates of endosulfan diol formation were in the treatments containing the soil inoculum from the highly contaminated endosulfan waste site.

From the preliminary study, no conclusions could be drawn regarding the biodegradation of endosulfan I in anaerobic liquid culture because of the nonbiological processes that were contributing to the loss of endosulfan I. These results stress the importance of using PTFE-lined butyl rubber sealed flasks in biodegradation studies where volatile organics are being studied. When sterile controls were prepared using PTFE-lined butyl rubber, the rate of endosulfan I dissipation was relatively slow at applied concentrations of 1, 2, and 10 μg/mL over the first 10 days of incubation, the period of most active biological growth (Table 1). Thus, provided that incubations in flasks were sealed with inert PTFE, sterile, and buffered at pH 7, relatively little dissipation of endosulfan I was observed.

Biodegradation of Endosulfan in Routine Incubations. In these incubations, butyl rubber seals were replaced with PTFE-lined butyl rubber to reduce non-biological losses of endosulfan I. The inocula used were derived from nonen-riched and enriched mixed cultures (previously described) isolated from a tailwater drain sediment and nonenriched and enriched mixed cultures iso-lated from a chemical sullage (or insecticide waste) pit. Endosulfan I was applied at 2 μg/mL to the routine incubations.

Using sullage pit inocula, not previously enriched by subculturing, led to an almost complete dissipation of endosulfan I after 6 days of incubation. Inoculation with a sample of the tailwater drain sediment (that was unen-riched) also led to an almost total loss of applied endosulfan I after approxi-mately 10 days of incubation. The remaining amounts after 10 days were 6 and 4% of the originally applied endosulfan I in the tailwater drain and sullage pit inoculated incubations, respectively.

In the routine anaerobic incubations containing inocula from enriched sul-lage pit and tailwater drain sediments, the amount of endosulfan I remaining after 10 days incubation was 9 and 6%, respectively. However, the initial rate of dissipation was considerably faster in these incubations, probably a result of the presence of acclimed microorganisms in the starting inoculum. When

TABLE 1. The stability of endosulfan I in sterilized anaerobic media. [a]

Days Incubation	Amount Remaining (%)		
	1 μg/mL	2 μg/mL	10 μg/mL
0	100	100	100
2	98	97	100
4	95	94	98
6	90	96	97
8	80	90	98

[a] The percentage values are for the amounts remaining and are reported as the mean of duplicate analyses. The standard error in these values range between 5 and 10%. The vessels were sealed with PTFE-lined butyl rubber seals.

routine incubations were inoculated with bacteria enriched with Tween 80 (as well as endosulfan I), the remaining endosulfan I concentrations were even lower.

With both the sullage pit and tailwater drain sediment treatments, less than 2% of the originally applied residues remained after 30 days incubation. It is likely that the presence of added surfactant in these treatments increased the bioavailability of the remaining endosulfan residues for microbial decomposition, releasing it from the glass surfaces or from the sediment particles carried over in the inoculum at the beginning of the experiment.

The dissipation of endosulfan I in the nonenriched cultures correlated with an increase in endosulfan diol in the early stages of incubation. At the 10th day, approximately 80% of the applied endosulfan I (2 μg/mL) could be detected as endosulfan diol in the medium of the treatment containing tailwater drain sediment. Endosulfan diol was also formed in the incubations inoculated with endosulfan-enriched microorganisms. This formation of endosulfan diol was greatest when the medium was inoculated with microorganisms from the sullage pit, as compared to tailwater drain sediment.

When applied to the medium containing inocula from an unenriched tailwater drain sediment at the higher concentration of 10 μg/mL, endosulfan I was almost completely degraded by the growing cultures after approximately 15 days of incubation. This was directly comparable to the 1 μg/mL treatment (Table 2). The pattern of almost total dissipation was similar to that observed in

TABLE 2. Decrease in endosulfan I in media inoculated with microorganisms from nonenriched tailwater drain sediment extracts.

Days Incubation	Amount Remaining (%)	
	1 μg/mL	10 μg/mL
0	100	100
2	95	122
4	20	88
6	6	43
8	4	26
15	5	4
30	4	6

[a] The percentage values reported are for the amounts remaining in 2 separate incubations and are the means of duplicate analyses. The standard errors varied between 5 to 10%. The vessels were sealed with PTFE-lined butyl rubber seals.

the routine 2 μg/mL treatments except for the apparent increase in endosulfan I detected in the subsampled aliquots from the culture vessels. This was observed 2 days after the addition of inoculum to the medium. This phenomenon was due to the partitioning of bound endosulfan I from the glass surfaces on the inside of the incubation vessels to the more lipophilic cultures. This concept was introduced and described by Guerin and Kennedy (1992). Endosulfan I was distributed to the glass surfaces in the ungrown (or very young cultures) because 10 μg/mL is a concentration 4 to 5 times greater than its water solubility (Guerin and Kennedy 1992).

In the 10 μg/mL treatment, endosulfan diol production was at a peak at 17 days at 4.3 μg/mL. This decreased to 3 μg/mL at 30 days, indicating that it was further metabolized. Evidence for this was that trace amounts of endosulfan ether, endosulfan hydroxyether, and endosulfan lactone were detected in the incubations, all of which are further degradation products of endosulfan I (Guerin 1993).

The buffered growth medium was effective in maintaining pH stability in that the media retained the initial pH even after the cultures were fully grown. The growing cultures did not increase the pH of the medium more than 0.5 pH units, and the majority of the cultures maintained their original pH values. Furthermore, as demonstrated from the results from the sterile controls, the removal of the small subsamples of broth at each of the sampling times did not increase the amount of endosulfan I partitioning into the gaseous phase in the incubation vessel headspace. This should be borne in mind as a cause of compound loss when conducting similar studies on other volatile organics.

The degradation of endosulfan I was due to the activity of microorganisms and a result of hydrolysis, probably enzymic. In previous studies of microbial degradation with aerobic cultures, much of the observed endosulfan degradation was observed to be due to chemical hydrolysis as an indirect microbiological effect since precautions against these losses were not taken (Miles and Moy 1979, Goebel et al. 1982). Methane was detected in the headspace of all the enrichment cultures as well as those from the routine, nonenrichment incubations. The cells observed in both the enrichment cultures and routine cultures were predominantly gram-negative anaerobic rods and methane-producing bacteria. A description of these organisms is given in Guerin (1993).

CONCLUSIONS

The current study is the first report of endosulfan I biodegradation under strictly anaerobic conditions. However, the preliminary findings did not convincingly show that microorganisms were involved in the degradation, because many of the previously recommended incubation procedures (found in the literature and applied in the preliminary study) caused nonbiological losses of endosulfan I. The routine anaerobic incubations, which minimized nonbiological losses of endosulfan in the assay vessels, demonstrated that mixed cultures (gram-negative anaerobic rods, and methane-producing bacteria) could degrade endosulfan I to the less toxic endosulfan diol at high rates. There was no substantial difference between treatments that were inoculated with enriched or nonenriched cultures, indicating that prior enrichment was not a prerequisite for anaerobic biodegradation of endosulfan I. Endosulfan I could also be readily degraded when applied at 10 μ/mL to the growing cultures, similar to the rates observed at the lower concentrations of 1 and 2 μ/mL.

A number of previous reports have not clearly indicated the importance of the effective sealing of vessels in studies on the anaerobic degradation of organochlorine compounds. Seals for incubation vessels must be both oxygen- and pollutant-impermeable. PTFE-lined butyl rubber provides such a seal because of its resistance to the absorption of volatiles and in preventing volatilization and Bouwer and McCarty (1983) originally reported on the use of Teflon™ for this purpose. Moreover, nonbiological losses have been reduced by including in the anaerobic media a 100-mM phosphate buffer. Thus, losses of endosulfan I due to chemical hydrolysis were kept to a minimum, allowing direct biological degradation to be determined. These findings are therefore of general significance to the design of biodegradation and bioremediation studies.

ACKNOWLEDGMENTS

I. R. Kennedy from the Department of Agricultural Chemistry and Soil Science, University of Sydney (where the study was conducted) is acknowledge for helpful discussions and feedback on early drafts. The ideas of Andrew Maule from PHLS Laboratories, Salisbury, were also appreciated in the design stages of the trial. Funding from the Cotton Research Development Corporation and support from Hoechst (Melbourne and Frankfurt) are gratefully acknowledged.

REFERENCES

Balch, W. E., G. E. Fox, L. J. Magrum, C. R. Woese, and R. S. Wolfe. 1979. "Methanogens: Re-evaluation of a Unique Biological Group." *Microbiological Reviews* 43(2): 260-296.

Bouwer, E. J. and P. L. McCarty. 1983. "Transformations of Halogenated Organic Compounds Under Denitrification Conditions." *Applied and Environmental Microbiology* 45: 1295-1299.

Commandeur, L. C. M., P. C. Slot, J. Gerritse, and J. R. Parsons. 1992. "Reductive Dehalogenation of Polychlorinated Biphenyls by Anaerobic Microbial Consortia Isolated from Dutch Sediments." *Soil Decontamination Using Biological Processes*, pp. 570-577. DECHEMA, Karlsruhe, Germany.

Goebel, H., S. G. Gorbach, W. Knauf, R. H. Rimpau, and H. Huttenbach. 1982. "Properties, Effects, Residues and Analytics of the Insecticide Endosulfan." *Residue Reviews* 83: 1-122.

Grbic-Galic, D., N. Churchman-Eisel, and I. Mrakavic. 1990. "Microbial Transformation of Styrene by Anaerobic Consortia." *Journal of Applied Bacteriology* 69: 247-260.

Guerin, T. F., and I. R. Kennedy. 1992. "Distribution and Dissipation of Endosulfan and Related Cyclodienes in Sterile Aqueous Systems: Implications for Studies on Biodegradation." *Journal of Agricultural and Food Chemistry* 40(11): 2315-2323.

Guerin, T. F., S.W.L. Kimber, I. R. Kennedy. 1992. "Efficient One-Step Method for the Extraction of Cyclodiene Pesticides from Aqueous Media and the Analysis of their Metabolites." *Journal of Agricultural and Food Chemistry* 40(11): 2309-2314.

Guerin, T. F. 1993. "The Relative Significance of Biodegradation and Physico-chemical Dissipation of Endosulfan from Water and Soil." Ph.D Thesis, University of Sydney, New South Wales, Australia.

Holliger, C., G. Schraa, A. T. M. Stams, and A. J. B. Zehnder. 1992. "Enrichment and Properties of an Anaerobic Mixed Culture Reductively Dechlorinating 1,2,3-Trichlorobenzene to 1,3-Dichlorobenzene." *Applied and Environmental Microbiology* 58: 1636-1644.

Kohring, G. W., C. Schirra, S. Schmitt, and F. Giffhorn. 1992. "Anaerobic Degradation of Phenol and Chlorinated Aromatic Compounds by Purple Nonsulfur Bacteria." *Soil Decontamination Using Biological Processes*, pp. 485-490. DECHEMA, Karlsruhe, Germany.

Maule, A., S. Plyte, and A. V. Quirke. 1987. "Dehalogenation of Organochlorine Insecticides by Mixed Anaerobic Microbial Populations." *Pesticide Biochemistry and Physiology* 27: 229-236.

Miles, J. R. W., and P. Moy. 1979. "Degradation of Endosulfan and its Metabolites by a Mixed Culture of Soil Microorganisms." *Bulletin of Environmental Contamination and Toxicology* 23: 13-19.

Miller, T. L., and M. J. Wolin. 1974. "A Serum Bottle Modification of the Hungate Technique for Cultivating Obligate Anaerobes." *Applied and Environmental Microbiology* 27(5): 985-987.

Rouse, J.D., D.A. Sabatini, J.M. Suflita, and J.H. Harwell. 1994. "Influence of Surfactants on Microbial Degradation of Organic Compounds." *Critical Reviews in Environmental Science and Technology* 24(4): 325-370.

Sanford, R. A. and J. M. Tiedje. 1992. "Biodegradation of 2,4-D and Chlorophenols Under Denitrifying Conditions in Soil Column, Microaerobic and Anoxic Enrichments." *Soil Decontamination Using Biological Processes*, pp. 543-548. DECHEMA, Karlsruhe, Germany.

Shelton, D. R., and J. M. Tiedje. 1984. "General Method for Determining Anaerobic Biodegradation Potential." *Applied and Environmental Microbiology* 47: 850-857.

Van Dort, H. M., and D. L. Beddard. 1991. "Reductive Ortho and Meta Dechlorination of a Polychlorinated Biphenyl Congener by Anaerobic Microorganisms." *Applied and Environmental Microbiology* 57: 1576-1578.

Biodegradation of Toxaphene Under Methane Cometabolism and Denitrifying Conditions

*Brian W. Merkley, Lance Hess, Edward A. Shaw,
Roger D. Towe, Richard E. Terry, Jeffrey D. Keith,
Alan L. Mayo, and David G. Tingey*

ABSTRACT

A bench-scale biotreatability study was conducted to investigate the bio-degradation of toxaphene under methane cometabolism and denitrifying conditions. Methanotrophic and denitrifying bacteria were cultured from toxaphene-contaminated soils (from 33,000 mg/kg to 160,000 mg/kg toxaphene). Soils were incubated under various conditions to promote the degradation of toxaphene through denitrification and methane cometabolism. Increases in chloride ion (Cl^-) concentrations and alterations of the toxaphene chromatogram indicated that toxaphene is dechlorinated anaerobically and that methanotrophic bacteria are able to cometabolically dechlorinate toxaphene where methane is the primary substrate.

INTRODUCTION

Background

Toxaphene is a broad-spectrum organochlorine pesticide that was used principally to combat caterpillars, grasshoppers, the cotton boll weevil, and a variety of cattle and sheep skin parasites (Meister 1987). The pesticide is an extremely complex mixture of polychlorinated monoterpenes, with more than 177 components and an overall chlorine content of 67 to 69% (Saleh 1983). The physical and chemical properties of the compound have made it very difficult to determine its fate and transport in the environment (Pepper and Watson 1986). Toxaphene is generally recalcitrant to biometabolism in the environment, with an estimated half-life of 11 to 20 years. Before the United States Environmental Protection Agency (U.S. EPA) finalized an across-the-board ban on toxaphene in 1982, due to its acute toxicity and mutagenic and carcinogenic properties, more than 500,000 tons had been used in the United States (Saleh 1991).

One of the principal producers of toxaphene in the United States was Tenneco Polymers, Inc. (Tenneco) in Fords, New Jersey, where toxaphene was manufactured under the trade name Strobane-T90. Toxaphene was released to the soils through-out the plant during its manufacture, transfer, and storage. Tenneco commissioned a bench-scale study to assess in situ biotreatment as a remediation method for the toxaphene-contaminated soils. Denitrification and methane cometabolism were proposed as potential microbial pathways for the biodegradation of toxaphene.

Denitrifying bacteria are facultative anaerobes that utilize nitrate (NO_3^-) as a terminal electron acceptor. Methanotrophic bacteria, under most conditions, depend on methane (CH_4) as a carbon source (electron donor) and molecular oxygen (O_2) as the respiratory compound (electron acceptor). The enzyme methane monooxygenase (MMO) makes methane available as an electron donor. The soluble form of this enzyme (sMMO) exhibits a low substrate specificity, allowing it to react with (cometabolize) a wide variety of recalcitrant organic compounds generally unavailable as a primary carbon/energy source to microbes (Stirling et al. 1979).

Methane cometabolism (usually sMMO based) has been used to effectively dechlorinate recalcitrant compounds such as trichloroethylene (TCE), dichloroethylene (DCE), and vinyl chloride (Lanzarone and McCarty 1991; Oldenhuis et al. 1989; Roberts et al. 1990). Therefore, an experiment was designed to test the hypothesis that methanotrophic bacteria could cometabolically degrade the toxaphene molecule.

Purpose and Objectives

The purpose of this bench-scale study was to evaluate methane cometabolism and denitrification as mechanisms for treating toxaphene-contaminated soils. Specific objectives were to (1) evaluate the general microbial health of the soils by enumerating aerobic, heterotrophic bacteria; (2) determine the existence of indigenous methanotrophic and denitrifying bacteria in the soils; and (3) evaluate the ability of methanotrophic and denitrifying bacteria to degrade toxaphene.

Site Characterization

Soil samples were obtained from four locations (A, B, C, and D) at the site where elevated levels of toxaphene contamination had been found during site characterization activities. Extensive aromatic hydrocarbons were released throughout area A and toxaphene concentrations were detected in area A soils at concentrations of 120 to 440 mg/kg. Areas B and C have a similar lithology (medium sand) and toxaphene concentrations of 33,000 mg/kg and 41,000 mg/kg, respectively. Soil D was collected from a thin clay layer with a toxaphene concentration of approximately 166,000 mg/kg.

METHODOLOGY

The microbial health of the site soils was evaluated by enumerating aerobic, heterotrophic bacteria. This aspect of the study was designed as a screening

process. An assumption was made that if a viable population of aerobic heterotrophs existed in the soil, then there was a strong likelihood of culturing
indigenous methanotrophic and denitrifying bacteria from the soils. A standard
aerobic plate count was performed for soils A, B, C, and D.

The most probable number (MPN) technique was used to determine the
presence of indigenous denitrifying bacteria in the soils. Several test tubes containing MPN medium were inoculated with various culture dilutions from the
plate count study. MPN medium is a transparent green color, and microbial
activity changes the pH and the color.

Two different techniques were used to identify the presence of methanotrophs
in the soil. These techniques included incubation in batch cultures (120-mL serum
bottles containing methane, a soil inoculum, and a nutrient solution) and incubation of specially prepared agarose agar plates in gastight vessels with methane
as the sole carbon source. The presence of methanotrophs in the batch cultures
initially was determined by turbidity and confirmed by gas chromatography
(GC) analysis of the serum bottle headspace for methane loss over time.

Two batch culture experiments were performed. In the first experiment,
1:10 dilution bottles from the initial population study were used as a culture
source. This suite of treatments is referred to as 1:10 A, 1:10 B, 1:10 C, and 1:10 D.
A 0.1-mL inoculum from the 1:10 dilution bottles was placed into a 120-mL serum
bottle along with 20 mL of a nitrate mineral salts solution (Whittenbury et al.
1970). The serum bottles were capped with Teflon™-lined butyl rubber septa
and then injected with 5.0 mL of methane, for an approximate 5.0% methane/air
final mixture in the headspace. The bottles were inverted to minimize gas loss
through the septa and were placed on a shaker at 1,500 rpm to maximize methane
gas transfer across the liquid/gas interface into the dissolved phase. Controls
were included in the study to demonstrate methane losses through degradation
as opposed to physical gas loss by diffusion through the septa. The batch
cultures were incubated at room temperature (approximately 25°C).

In the second batch culture experiment, a small amount of soil (0.25 g) from
each of the four sample areas was placed into a 120-mL serum bottle along with
20 mL of a CM salts solution that contained a Stienberg C trace elements solution
without copper. This suite of treatments is referred to as treatments A, B, C, and D.
The serum bottles were capped with Teflon™-lined butyl rubber septa and injected
with methane for a final gas mixture of 10 to 20% methane. The bottles were
inverted and incubated in the same manner as the first batch culture experiment.

To confirm the presence of methanotrophic bacteria in the batch cultures,
solution samples (0.1 mL) were collected from treatments A, B, C, and D and
spread on agar plates (15% agarose) prepared with a CM salts solution. Agarose
was chosen as the solid medium because it is nearly 100% free of carbon. The
plates were incubated in gastight pressure vessels in a 20% air and 80% methane
environment and were flushed daily with a fresh gas mixture for 10 min. Replicates of each treatment were incubated in laboratory air as a control.

To further confirm the presence of methanotrophic bacteria in the soils, the
batch culture treatments were analyzed for methane loss over time. Samples
of the headspace (0.1 mL) were injected into a Hewlett Packard 5840A GC with

a [63]Ni electron capture detector (using GC gastight quality syringes), and were measured against prepared standard concentrations.

The ability of methanotrophic and denitrifying bacteria to dechlorinate toxaphene was evaluated by (1) measuring the increase of chloride ion (Cl^-) concentrations in various soil treatments; and (2) analyzing the soils for changes in toxaphene concentrations and changes in the toxaphene chromatogram. If the microbes successfully degraded the toxaphene molecule either directly (anaerobes) or fortuitously (methanotrophs), cleavage of Cl^- ions from the molecule would take place. Dechlorination of the toxaphene molecule will also be evident from alterations in the toxaphene chromatogram. As toxaphene biodegrades there is an increase in the number of compounds having shorter GC retention times. The result is a weathered chromatogram with peaks shifted to the left (Williams and Bidleman 1978).

Soils B, C, and D (50 g) were placed in six separate 1.0-L flasks containing 500 mL of a nitrate mineral salts solution. Air was pumped at a continuous rate into three flasks as a control and a methane/air mixture (from 5 to 30% methane) was pumped at a continuous rate into the other three flasks. The flask solutions were analyzed over time for Cl^-, NO_3^-, and NO_2^- by injecting 5.0-mL filtered samples into a Dionex ion chromatograph.

Soils were analyzed for toxaphene using EPA methods 8080 and 8260 before and after incubation under a varied flow of methane/air (from 0 to 30% methane). Six treatments were prepared by placing 50 g of soil (a composite of soils from areas B, C, and D) and 75 mL of CM salts nutrient solution (± copper) in a 250-mL flask. This experiment was designed, in part, to study the effect of copper-limiting conditions on the methane cometabolism of toxaphene. Four of the treatments were augmented with methanotrophic bacteria cultured earlier in the study. A column of activated carbon was placed in series with treatments one through three and another column was placed in series with treatments four through six. These columns were installed to account for toxaphene losses due to volatilization. The activated carbon was sampled near the inlet port of the columns for analysis at the end of the study.

Toxaphene degradation under anaerobic (denitrifying) conditions was measured by filling biological oxygen demand (BOD) bottles (300 mL) with 200 g of soil (a composite of soils from areas B and C) and 100 mL of nutrient solution. The nitrogen source used in the nutrient solution was either NO_3^- or NH_4Cl. The treatments were amended with glucose and augmented with denitrifying cultures isolated from soil D. The soils were analyzed for toxaphene using EPA methods 8080 and 8260 before and after a 35-day incubation period.

RESULTS AND DISCUSSION

Microbial Activity

Populations of aerobic heterotrophic bacteria ranged from <300 colony-forming units (CFU)/g in soil A to 8×10^5 CFU/g in soil B. A larger bacteria population was expected in the sample A soils due to the low toxaphene concentrations in this area. However, this soil was highly contaminated with aromatic compounds that seemed to suppress microbial growth. There appeared to be an inverse correlation in all of the soils to heterotrophic populations and contaminant concentration (Figure 1). Although populations were relatively low, especially in the soils with higher toxaphene and aromatic compound concentrations (soils A and D), there were sufficient numbers to pursue the remaining study objectives.

Based on color changes in test tubes containing MPN medium, indigenous denitrifying bacteria were identified in soils B, C, and D. No color change was noted in the soil A dilution tubes.

Analysis of the initial results, based on increases in turbidity, indicated the presence of indigenous methanotrophs in the soils from both of the batch culture experiments. However, when samples from the batch cultures were streaked onto agarose plates and incubated in the methane environment, no growth occurred on plates inoculated with soil A. Vigorous growth did occur on the agarose plates inoculated with soils B, C, and D when incubated in the presence of methane. No growth was observed on any of the control plates incubated in laboratory air. Soil A provided an internal control for the experiment due to the nearly sterile conditions of the soil.

When the headspace of each serum bottle was analyzed for methane loss over time, methanotrophic activity was evident in batch cultures inoculated with

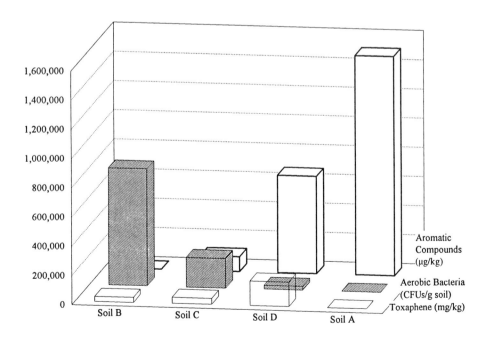

FIGURE 1. Comparison of aerobic bacterial populations and soil contaminant concentrations.

soils B and C (Figure 2). Soil D also demonstrated methanotrophic activity, but only after a prolonged lag phase. This activity was expected considering the higher concentrations of toxaphene in soil D. Methane concentrations remained nearly constant in the batch cultures containing soil A (Figure 2). Analysis of the data indicates that higher concentrations of aromatic compounds are toxic and/or that soil A contained no methanotrophic bacteria.

Chloride Emissions

In the 1.0-L flasks containing soils B, C, and D, no Cl⁻ emissions were measured in the flasks incubated under a continuous flow of methane. However, an increase in Cl⁻ concentrations was observed in the flasks receiving air, especially in the flasks containing soil D. An attempt was made to explain the apparent activity in flasks containing soil D, because aerobic bacteria generally are not able to degrade highly chlorinated compounds.

Three discrete liquid samples were collected from the flasks containing soil D to see if Cl⁻ and other anions were stratified within the flask. One sample location was near the air injection port (an oxygenated region of the flask), another sample was taken from this same location after thoroughly stirring the flask, and the third sample was taken from a stagnant zone in the flask. Analysis of the data suggests that the Cl⁻ increase initially observed was due to anaerobic (denitrification) activity, not aerobic activity (Figure 3). This result is not surprising

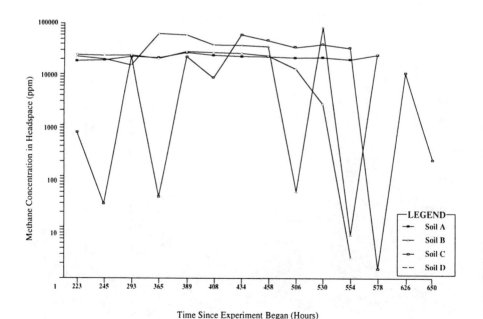

FIGURE 2. Presence of methanotrophic bacteria in soils B, C, and D indicated by GC headspace analysis.

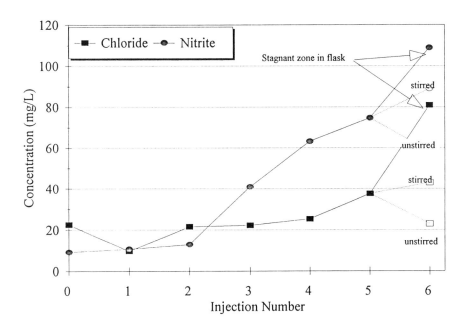

FIGURE 3. Evidence of toxaphene dechlorination through denitrification by Cl⁻ analysis.

considering the presence of compacted clays in soil D and the numerous microsites that are isolated from the influence of oxygen. The flasks were then incubated with a continuous flow of argon to further promote anaerobic conditions. Chloride ion (Cl⁻) and NO_2^- concentrations increased an average of 298% and 176%, respectively, supporting the idea that anaerobiosis was partly responsible for dechlorination of the toxaphene.

Solutions from the second methane batch culture experiment (treatments A, B, C, and D) were subsampled and analyzed at the end of the study for Cl⁻ concentrations (Figure 4). These treatments were analyzed for Cl⁻ because methanotrophic activity was observed in these cultures in the presence of toxaphene. The initial Cl⁻ concentration was not measured in these slurries, but was calculated to be less than 1 mg/L due to the nutrient solution. Because of the small quantities of soil placed into the serum bottles (0.25 g), Cl⁻ contributions from the soils were assumed to be negligible.

Chloride concentrations were an average 244% higher in batch cultures that exhibited methanotrophic activity than in treatments which showed no methanotrophic activity (Figure 4). Strong methanotrophic activity and the largest increases in Cl⁻ concentrations were measured in soil D. Some Cl⁻ concentration increases were observed in the treatments in which little or no methane loss was measured. The increases in Cl⁻ concentrations were not attributed to the degradation of other chlorinated compounds, because of insignificant amounts of these compounds in the soils. Because of the low volumes of soil and relatively large

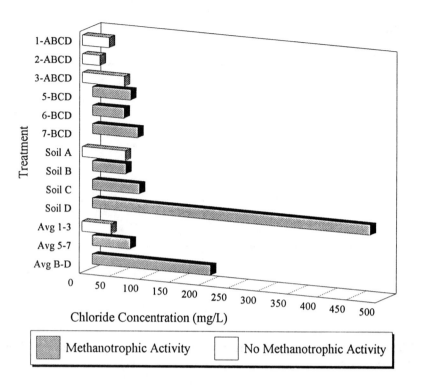

FIGURE 4. Evidence of methane cometabolism of toxaphene indicated by Cl⁻ analysis.

volumes of serum bottle headspace used in the experiment, anaerobiosis is not suspected. It is possible that the measured Cl⁻ increase in treatments which did not show methanotrophic activity may have resulted from some type of aromatic cometabolism of toxaphene. Analysis of the data also suggest that soil A, which had the greatest concentration of aromatic compounds, was more favorable to other types of microbial processes than to methane cometabolism (Figure 4).

Toxaphene Degradation

There was an apparent increase in toxaphene in the soils incubated under a varied flow of methane/air (from 0 to 30% methane) and in the soils treated under anaerobic conditions in BOD bottles of 40,000 mg/kg and 45,000 mg/kg, respectively (Figure 5). This increase in toxaphene was expected because by-products produced from the biodegradation of toxaphene are more soluble than toxaphene. This has resulted in false-positive data indicated from alterations in the toxaphene chromatograms. Soils treated in both anaerobic and methane environments demonstrated a weathered toxaphene fingerprint using EPA method 8080 with an increase in the number of peaks having shorter retention times.

CONCLUSIONS

Although populations of aerobic heterotrophic bacteria were relatively low, especially in soils with higher toxaphene and aromatic compound concentrations, the numbers justified pursuit of the other objectives of the study. Denitrifying bacteria were identified in all of the soils except soil A, and dechlorination of the toxaphene molecule was observed in flasks under denitrifying conditions. Methanotrophic bacteria were cultured from soils B, C, and D in batch cultures, and methanotrophic activity was confirmed through GC analysis of serum bottle headspace gases and by incubating samples on specially prepared solid medium in a 20% methane environment. Additionally, chloride ion (Cl⁻) analysis and changes in the chromatographic fingerprint of toxaphene suggest that toxaphene was dechlorinated by methane cometabolism. Further study is needed to investigate the extent and rate of dechlorination and the toxicity of metabolic by-products.

Analysis of the preliminary results of this study indicate that toxaphene does not appear to inhibit the growth of methanotrophic bacteria and the activity of MMO. Methane uptake in the presence of toxaphene was extremely rigorous in some cases and did not seem to be affected by the presence or absence of copper. It also appears that denitrification may be more closely associated with the enzymatic aromatic cometabolism of toxaphene rather than an anaerobic process. Based on the preliminary data from this study, a field pilot test was implemented

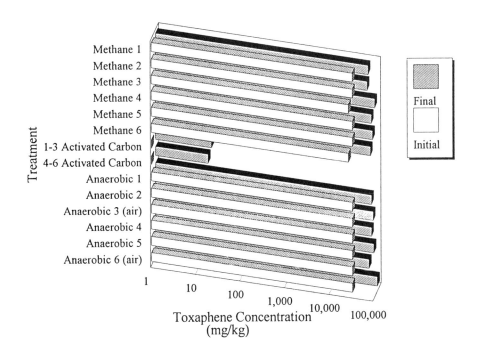

FIGURE 5. Initial vs. final toxaphene concentrations.

to assess the relative effectiveness of in situ denitrification and methane cometabolism. Field pilot study data have been collected and confirm that methane
cometabolism effectively degrades toxaphene. A publication describing the field
pilot study process and results is forthcoming.

REFERENCES

Lanzarone, N. A., and P. L. McCarty. 1991. "Column Studies on Methanotrophic Degradation
of Trichloroethene and 1,2-Dichloroethene." *Ground Water 28*: 910-919.
Meister, R. T. (Ed). 1987. *Farm Chemicals Handbook*. Meister Publishing Co., Willoughby, OH.
Oldenhuis, R., R.L.J.M. Vink, D. B. Janssen, and B. Witholt. 1989. "Degradation of Chlorinated
Aliphatic Hydrocarbons by *Methylosinus trichosporium* OB3b Expressing Soluble Methane
Monooxygenase." *Appl. Environ. Micro. 55*: 2819-2826.
Pepper, I. L., and J. E. Watson. 1986. *Treatment of Pesticide-Containing Soil*. U.S. Environmental
Protection Agency Technical Report, EPA 600/9-87/001, R.S. Kerr Environmental Research
Laboratory, Ada, OK.
Roberts, P. V., G. D. Hopkins, M. M. Douglas, and L. Semprini. 1990. "A Field Evaluation
of In-Situ Biodegradation of Chlorinated Ethenes: Part I, Methodology and Field Site
Characterization." *Ground Water 28*: 591-604.
Saleh, M. A. 1983. "Capillary Gas Chromatography-Electron Impact and Chemical Ionization
Mass Spectrometry of Toxaphene." *J. Agric. Food Chem. 31*: 748-751.
Saleh, M. A. 1991. "Toxaphene: Chemistry, Biochemistry, Toxicity and Environmental Fate."
Reviews of Environ. Contam. Toxicol. 118: 1-85.
Stirling, D. I., J. Colby, and H. Dalton. 1979. "A Comparison of the Substrate and
Electron-Donor Specificities of the Methane Mono-Oxygenases From Three Strains of
Methane-Oxidizing Bacteria." *Biochem. J. 177*: 361-364.
Whittenbury, R., K. C. Phillips, and J. F. Wilkinson. 1970. "Enrichment, Isolation and Some
Properties of Methane-Utilizing Bacteria." *J. Gen. Micro. 61*: 205-218.
Williams, R. R., T. F. Bidleman. 1978. "Toxaphene Degradation in Estuarine Sediments."
J. Agric. Food Chem. 26: 280-282.

Chemical Oxidation and Toxicity Reduction of Pesticide-Contaminated Soils

Christopher M. Miller, Richard L. Valentine,
Jeff L. Stacy, Marc E. Roehl, and Pedro J.J. Alvarez

ABSTRACT

Spills and improper disposal of pesticides by farmers, dealers, and transporters are potentially important sources of groundwater contamination. This paper reports on the chemical oxidation of pesticide-contaminated soils using Fenton's reagent to (1) reduce pesticide levels, and (2) reduce toxicity to the heterotrophic microbial population. Greater than 90% removal of atrazine and pendimethalin was achieved in laboratory batch studies using Fenton's reagent on various contaminated soils. Fenton's treatment also increased heterotrophic microbial activity in seeded contaminated soil (as measured by [UL ^{14}C]-glucose biodegradation to $^{14}CO_2$) to a level similar to that of uncontaminated soil. Microtox® assay after Fenton's treatment showed a significant reduction in toxicity associated with the soil. This study shows that treatment with Fenton's reagent can result in an overall reduction in both pesticide levels and toxicity, which potentially enhances the biodegradability of remaining contaminant and oxidation by-products.

INTRODUCTION

The Iowa Fertilizer and Chemical Association estimates that approximately 90% of agrochemical dealer sites in Iowa have contaminated soil and that 40 to 50% will require remediation. Contamination has been measured at several sites at levels believed to be toxic to indigenous microorganisms, thus reducing the importance of biodegradation as a cleanup alternative. A need exists for an easy, cost-effective method for mitigating spills and reducing the toxicity of pesticide-contaminated soils. Chemical oxidation may not only destroy target compounds, but also reduce toxicity associated with both formulation ingredients and active agents.

The mixture of hydrogen peroxide (H_2O_2) and ferrous iron, commonly referred to as Fenton's reagent, produces hydroxyl radical, ·OH, a powerful and nonspecific oxidizing agent capable of reacting with many organic compounds (Walling and Johnson 1975). Fenton's reagent and hydrogen peroxide have been used to treat pesticide rinsates and contaminated soils (Kelley et al. 1993 and Watts et al. 1990). Chemical treatment can also be an effective pretreatment step for enhanced bioremediation by (1) transforming constituents to by-products that are more readily biodegradable, and (2) reducing overall toxicity to indigenous microorganisms, allowing them to participate in the remediation process (Kearney et al. 1988, Kelley et al. 1993, and Somich et al. 1988, 1990). This paper addresses the effect of Fenton's reagent on pesticide degradation and toxicity reduction of contaminated soils.

EXPERIMENTAL

Three soils with varying properties were obtained from pesticide dealership sites with existing contamination (Table 1). Atrazine and Prowl™ commercial formulations were obtained from a local pesticide dealership. Pendimethalin (active ingredient in Prowl™) was purchased through Chem Service, Inc. Soils were contaminated to desired levels by dissolving pesticide in ethyl acetate, dosing the soil, and then letting the ethyl acetate evaporate.

Fenton's reagent soil treatment experiments were conducted in triplicate 120-mL serum bottles covered with aluminum foil at 20°C. Each reactor consisted of 10 g of soil and 50 mL of solution with varying amounts of hydrogen peroxide (150 to 360 g/kg soil) and ferrous iron (0 to 2 mg Fe(II)/g soil). The remaining atrazine and pendimethalin was measured 48 hours after addition of the appropriate reagents (during which greater than 95% of the hydrogen peroxide decomposed). At this time, the soil and water were allowed to settle. The water phase (leachate) was withdrawn and filtered (0.45 μm), and the remaining soil was air-dried, prior to further testing.

The ability of Fenton's reagent to reduce toxicity of the leachate was evaluated from measurement of BOD_L and COD, and by Microtox® assay. Controls for Fenton's treatment and for pesticide contamination were run. Additionally, the Microtox® assay was used to evaluate air-dried treated soil toxicity. Microbial

TABLE 1. Soil properties.

Parameter	Soil A	Soil B	Soil C
pH	6.7	6.9	7.9
Organic Matter, %	3.4	1.0	0.9
% Sand, Silt, and Clay	12.5, 50.0, 37.5	12.5, 62.5, 25.0	Not determined
Fe, mg/kg	19.0	23.4	12.2
Mn, mg/kg	5.6	10.7	2.8

degradation of [UL ^{14}C]-glucose was measured in soil microcosms to assess the effect of treatment on heterotrophic activity (Parsons & Smith 1989 and Dasappa & Loehr 1991). The microcosms were prepared by mixing 5 g of uncontaminated soil (which served as a microbial seed) and 5 g of treated air-dried soil. [UL ^{14}C]-glucose (2 mL at 1.45 mg/L) was added after 2 days, and $^{14}CO_2$ production was monitored for 10 weeks. All tests were performed on soil A, which was amended with Prowl™ formulation (0 to 72.1 mg/kg soil).

Analysis of atrazine and pendimethalin was by gas chromatography with an electron capture detector. Soil extraction procedures for these compounds used methodology developed by Huang and Pignatello (1990). Hydrogen peroxide was measured by iodometric titration (Kieber & Helz 1986). [UL ^{14}C]-glucose mineralization was measured using liquid scintillation counting. BOD_L and COD were analyzed using standard methods (American Public Health Association 1992).

RESULTS AND DISCUSSION

Fenton's Reagent Treatment of Contaminated Soils

Removal of pendimethalin increased with the amount of hydrogen peroxide added on all three soils, with greater than 99% removal efficiency in soils A and B at the highest peroxide dose of 360 g/kg soil (Table 2). Nevertheless, the maximum removal efficiency of pendimethalin in soil C (23%) was significantly less than in the other two soils. This decrease in removal efficiency is likely attributable to a higher pH value over the course of the reaction (final pH of approximately 7.0) on soil C due to its buffer capacity. Fenton's reagent has been shown to be most effective at a pH between 2 to 4 (Watts et al. 1990).

Similar results were observed for atrazine (Table 3). Its removal efficiency increased with the amount of hydrogen peroxide added on all three soils, with

TABLE 2. Oxidation of pendimethalin in various soils at a fixed ferrous iron dosage of 2 mg Fe(II)/g soil and variable hydrogen peroxide addition. The initial pendimethalin concentration was 50 mg/kg soil and batch reactor with 10 g of soil in 50 mL of solution. The initial pH of the reactors was between 2 and 3.

Peroxide Dose (g peroxide/kg soil)	Removal Efficiency in Soil A	Final pH	Removal Efficiency in Soil B	Final pH	Removal Efficiency in Soil C	Final pH
360	> 99%	3-4	> 99%	3-4	21%	7
250	92%	3-4	97%	3-4	23%	7
150	66%	3-4	90%	3-4	15%	7

greater than 99% in soils A and B and approximately 83% in soil C at a peroxide dose of 250 g/kg soil. The somewhat higher removal of atrazine compared to pendimethalin indicated in soil C may be attributable to the greater solubility of atrazine (20 vs. 0.05 mg/L) (Kearney et al. 1988 and Zimdahl et al. 1984). Solubilization would facilitate oxidation in the aqueous phase where hydroxyl radicals are produced, especially during the initial phase of the reaction when the pH is low enough to enhance their formation. These results suggest that the acid-neutralizing capacity of soils may be particularly important in determining the potential efficacy of Fenton's reagent treatment for immobilized contaminants.

Toxicity Reduction Evaluation

Increasing concentrations of pendimethalin reduced the amount of glucose biodegraded to carbon dioxide in seeded microcosms (Figure 1). Following Fenton's treatment, however, the amount of glucose biodegraded to carbon dioxide was similar to that of uncontaminated soil (Figure 2). Therefore, chemical oxidation of pendimethalin eliminated its inhibitory effect on heterotrophic activity. Microtox® measurement of these soils also showed a significant reduction in relative toxicity. The contaminated soil toxicity was reduced from 150 to 2 toxicity units (TU).

Microtox®, BOD_L, and COD evaluation of soil leachates are summarized in Table 4. Only leachates from Fenton's treated soil exhibited Microtox® toxicity, with leachate from contaminated soil showing greater toxicity than that from uncontaminated soil (13.3 vs. 3.2 TU respectively). Mobilization of toxicity by treatment with Fenton's reagent may be related not only to the formation of water-soluble toxic species and/or the destruction of binding sites on the soil, but also to cosolvency effects and sorption of toxic components to humic material released by Fenton's treatment (Jota & Hassett 1991). This latter possibility is indicated by the large increase in BOD_L (from 26 to 1830 mg/L) and COD (61 to 3,144 mg/L) in the leachate from uncontaminated soils subjected to Fenton's treatment.

TABLE 3. Oxidation of atrazine in various soils at a fixed ferrous iron dosage of 2 mg Fe(II)/g soil and variable hydrogen peroxide addition. The initial atrazine concentration was 50 mg/kg soil and batch reactor with 10 g of soil in 50 mL of solution. The initial pH of the reactors was between 2 and 3.

Peroxide Dose (g peroxide/kg soil)	Removal Efficiency in Soil A	Final pH	Removal Efficiency in Soil B	Final pH	Removal Efficiency in Soil C	Final pH
360	> 99%	3-4	> 99%	3-4	81%	7
250	> 99%	3-4	> 99%	3-4	81%	7
150	81%	3-4	92%	3-4	83%	7

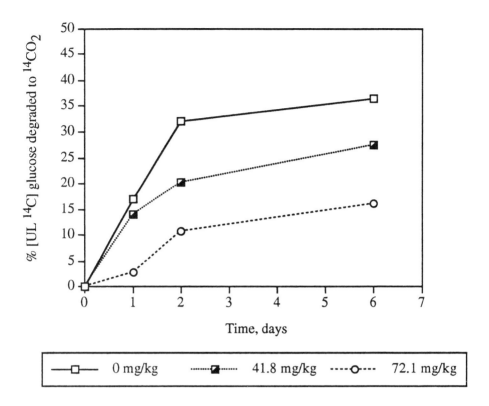

FIGURE 1. Effect of pendimethalin concentration on glucose mineralization in soil A. Each reactor had 10 g of soil and 2 mL of [UL ^{14}C]-glucose (initially 1.45 mg/L). Greater than 98% of the added glucose was metabolized within 6 days. Error bars are not depicted because the standard deviation is smaller than the dimensions of the data symbol.

The release of BOD_L and COD in untreated contaminated soil is likely due to soluble formulation additives or soil organic matter extracted by surfactants present in the formulation. Fenton's treatment greatly increased the BOD_L from 1,450 to 2,128 mg/L and COD from 2,116 to 3,315. The relatively high BOD_L/COD ratio of approximately 0.6 for all Fenton's treated soils indicates the formation of biodegradable by-products. In conclusion, converging lines of evidence showed that Fenton's treatment of pesticide-contaminated soil can decrease toxicity and enhance the potential for subsequent microbial degradation.

ACKNOWLEDGMENTS

This research was supported by the Iowa State Water Resources Research Institute, the National Institute for Environmental Health Sciences, and the EPA Hazardous Substance Research Center for Regions 7 and 8. We would also like to

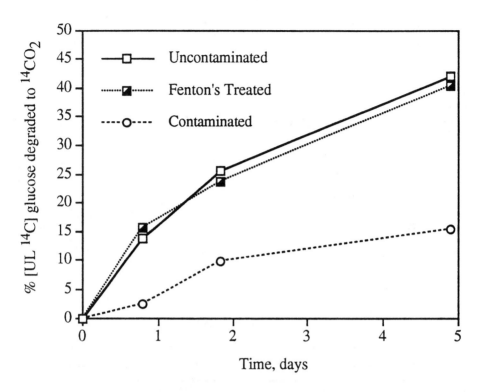

FIGURE 2. Glucose mineralization in Prowl™-contaminated soil after Fenton's treatment. Soil A was treated with 2 mg Fe(II)/g soil and 1,800 g peroxide/kg soil. Pendimethalin concentration was reduced from 72.1 to 5.7 mg/kg. Measurements after 5 days showed no further evolution of $^{14}CO_2$. Error bars are not depicted because the standard deviation is smaller than the dimensions of the data symbol.

TABLE 4. Relative toxicity evaluation of uncontaminated and Prowl™-contaminated soil leachates before and after Fenton's reagent treatment. The Fenton's reagent treatment was 0.4 mg Fe(II)/g soil and 180 g peroxide/kg soil.

Leachate from Soil A	Fenton's Treated	BOD_L mg/L	COD mg/L	BOD_L/COD Ratio	Microtox® Toxicity Units
Uncontaminated	No	26	61	0.43	< 0.5
Uncontaminated	Yes	1,830	3,144	0.58	3.2
Prowl™-contaminated	No	1,450	2,116	0.69	< 0.5
Prowl™-contaminated	Yes	2,128	3,315	0.64	13.3

thank Randy Happel, Dan Freiberg, and the Iowa Fertilizer Chemical Association for providing soils and assistance in obtaining formulated pesticides.

REFERENCES

American Public Health Association, American Water Works Association, and Water Environment Federation. 1992. *Standard Methods for the Examination of Water and Wastewater*, 18th ed.

Dasappa, S. M., and R. C. Loehr. 1991. "Toxicity reduction in contaminated soil bioremediation processes." *Water Research* 9:1121-1130.

Huang, L. Q., and J. J. Pignatello. 1990. "Improved extraction of atrazine and metolachlor in field soil samples." *Journal of the Association of Official Analytical Chemists* 73:443-446.

Jota, M. A., and J. P. Hassett. 1991. "Effects of environmental variables on binding of a PCB congener by dissolved humic substances." *Environmental Toxicology and Chemistry* 10: 483-491.

Kearney, P. C., M. T. Muldoon, C. J. Somich, J. M. Ruth, and D. J. Voaden. 1988. "Biodegradation of ozonated atrazine as a wastewater disposal system." *Journal of Agricultural and Food Chemistry* 36:1301-1306.

Kelley, R. L., W. K. Gauger, and V. J. Srivastava. 1990. "Application of Fenton's reagent as a pretreatment step in biological degradation of polyaromatic hydrocarbons." Paper presented at Gas, Oil, and Environmental Biotechnology III, Chicago, IL. November 8-10.

Kelley, R. L., V. Srivastava, and N. P. Barkley. 1993. "An integrated chemical and biological treatment (CBT) system for site remediation." Paper presented at 19th Annual RREL Hazardous Waste Research Symposium, New Orleans, LA. October 10-12.

Kieber, R. J., and G. R. Helz. 1986. "Two-method verification of hydrogen peroxide determinations in natural waters." *Analytical Chemistry* 58:2312-2315.

Parsons, L. L., and M. S. Smith. 1989. "Microbial utilization of carbon-14-glucose in aerobic vs. anaerobic denitrifying soils." *Soil Science Society of America Journal* 53:1082-1085.

Somich, C. J., P. C. Kearney, M. T. Muldoon, and S. Elsasser. 1988. "Enhanced soil degradation of alachlor by treatment with ultraviolet light and ozone." *Journal of Agriculture and Food Chemistry* 36:1322-1326.

Somich, C. J., M. T. Muldoon, and P. C. Kearney. 1990. "On-site treatment of pesticide waste and rinsate using ozone and biologically active soil." *Environmental Science and Technology* 24:745-749.

Walling, C., and R. A. Johnson. 1975. "Fenton's reagent. VI. Rearrangements during glycol oxidations." *Journal of the American Chemical Society* 97:2405-2407.

Watts, R. J., M. D. Udell, and P. A. Rauch. 1990. "Treatment of pentachlorophenol-contaminated soils using Fenton's reagent." *Hazardous Waste and Hazardous Materials* 7:335-345.

Zimdahl, R. L., P. Catizone, and A. C. Butcher. 1984. "Degradation of pendimethalin in soil." *Weed Science* 32:408-412.

Effect of Compost Addition on Pesticide Degradation in Planted Soils

Michael A. Cole, Xianzhong Liu, and Liu Zhang

ABSTRACT

Samples of soils containing aged herbicide residues were collected from pesticide mixing and loading areas at an agrichemical facility. Herbicide contaminants present in the samples when collected were trifluralin, metolachlor, and pendimethalin. Project goals were to accelerate pesticide degradation and to improve microbial activity and plant growth in the materials. Mixes (50:50, w/w) of contaminated soil and either uncontaminated soil or yard waste compost was planted with herbicide-tolerant sweet corn or left unplanted. After 40 d in the greenhouse, the mixtures were analyzed for pesticides, microbial activity, and plant growth. All mixtures of contaminated soil with compost had higher microbial activity than did mixes of contaminated and uncontaminated soil. Trifluralin was degraded in planted, but not unplanted soil mixes. Metolachlor was not degraded in 100% contaminated soil or unplanted compost mixes, but was degraded in planted compost mixes and soil mixes with or without plants. Pendimethalin degradation was seen in all mixes except unplanted compost mix. The results indicate that bioremediation of herbicide-contaminated soil may be achieved quite rapidly and inexpensively.

INTRODUCTION

Contamination of soil and groundwater at Midwestern U.S. agrichemical retail facilities is evidently common. A survey of 49 locations in Illinois (Krapac et al. 1993) showed that soil contamination with the herbicides alachlor, atrazine, metolachlor, trifluralin, pendimethalin, cyanazine, and metribuzin was very common, whereas contamination with insecticides was much less common. Similar results were found in a survey of Wisconsin agrichemical dealerships (Habecker 1989). Taylor (1993) found detectable pesticides in groundwater samples from wells at agrichemical retail sites, including several compounds for which drinking water standards exist. He suggested that agrichemical facilities are primary sources of groundwater contamination in Illinois. The combination of adverse environmental impacts on the earth's surface and subsurface indicates that remedial activities at these sites would be appropriate.

Felsot and coworkers attempted land-farming of pesticide-contaminated soil from an inactive agrichemical facility and found that pesticide degradation was slow, with detectable parent compounds still present over a year after land application (Felsot et al. 1990; Felsot & Dzantor 1990). However, they also found that herbicides freshly added to agricultural fields were degraded rapidly in comparison to aged materials. Microbial activity (e.g., dehydrogenase levels in soils) increased only slightly in herbicide-contaminated soils amended with corn or soybean stubble (Felsot & Dzantor 1990). The slow degradation of herbicides in the land-farmed material may be attributed to low bioavailability of the herbicides, relatively low microbial activity, or a combination of the two factors.

Considering both technical and legal issues together, more rapid remediation activities that do not require off-site transport or extensive dilution of contaminated materials are attractive, but the costs of many of the currently available methods exceed the property owner's willingness or ability to pay for cleanup. Therefore, we were interested in developing a method by which the contaminated soils could be remediated quickly, inexpensively, and without removing contaminated materials from the facility.

With these factors in mind, a system using herbicide-tolerant plants in combination with compost addition was studied to remediate herbicide-contaminated soils. Plant growth in contaminated soil has been shown to facilitate degradation of several environmental contaminants (Anderson & Coats 1994). Several researchers have found that pesticides degrade quite rapidly in compost (Lemmon & Pylypiw 1992; Michel et al. 1993). We had reported previously that compost in conjunction with planting increased microbial activity and plant growth in soil from one herbicide-contaminated site (Cole et al. 1994) and that addition of compost accelerated herbicide degradation in contaminated material from an agrichemical retail site (Cole et al., submitted for publication).

The purposes of this study were to determine if plant growth, in combination with compost addition, can accelerate herbicide degradation in "aged" soils, in which degradation is often relatively slow (Felsot et al. 1990; Steinberg et al. 1987) and to determine whether or not compost would have beneficial effects on microbial activity and plant growth in herbicide-contaminated soils.

EXPERIMENTAL

Soil and Compost Samples

Soil contaminated with pesticides was obtained from an agrichemical retail site in Illinois, identified as Site 20, and was sampled during the survey described above (Krapac 1993). Sampling sites were near loading areas and other locations where pesticide contamination is relatively common. Contamination was usually restricted to the top 60 cm (depending on site and specific sampling location), and a single composite sample of the contaminated zone was used in these studies. Samples were passed through a 4-mm screen to remove medium to large gravel and used without further processing. The material was primarily road pack containing a mixture of fine gravel, sand, silt, and clay.

Compost derived from yard wastes was obtained from DK Recycling Systems, Inc., Lake Bluff, Illinois. Material that passed through a 6-mm screen was used in both cases. The compost was produced by a thermophilic process, which resulted in a weed-free product. The pesticides studied in this work were not found in the compost, nor in uncontaminated (control) soil.

Control soil was a 50:50 (v/v) mixture of sand and Drummer silty clay loam soil obtained from a local gardener. No pesticides had been applied to the soil for 4 years. Material that passed through a 6-mm screen was used.

Blends containing 0, 1.5, 6, 12.5, 25, or 50% (w/w) contaminated soil and control soil or compost were mixed and transferred into 15-cm-diameter plastic pots for the greenhouse studies (Cole et al. 1994). Four replicate pots were used for each treatment.

Plant Growth Procedures

Four pots of each mixture were planted with 5 seeds of sweet corn (*Zea mays*, cv. 'Golden Beauty') (Cole et al. 1994). Plants were grown in a greenhouse and watered weekly with soluble NPK fertilizer. Four pots were left unplanted, but treated identically to planted pots.

Microbial Activity Analysis

Dehydrogenase activity was determined by reduction of triphenyltetrazolium chloride (TTC) (Cole et al. 1994). Activity is expressed as μmol TTC produced g^{-1} soil 24 h^{-1}.

Pesticide Extraction and Analysis

Extraction and analysis is described in detail elsewhere (Cole et al., submitted for publication). Briefly, a 25-g sample was ground in a Waring blender followed by addition of 10 mL of 1 $\underline{M}$ sodium chloride solution along with sufficient water to make a soil slurry, followed by 50 mL ethyl acetate and 2 mL acetone. The bottle was shaken horizontally on a rotary shaker at 150 rpm for 24 h at 20°C. The ethyl acetate (upper) layer was removed, dehydrated by passage through a column of anhydrous sodium sulfate, and reduced to 1.0-mL final volume in a Kuderna-Danish concentrator. Samples were injected without further purification into a Chrompack CP9000 gas chromatograph (GC) equipped with a nitrogen-phosphorus detector. A 50 m × 0.25 mm (i.d.) column of WCOT fused silica with CP-Sil-8 CB stationary phase (Chrompack, Inc.) was used for all analyses. Extraction efficiency and instrument performance was evaluated by spiking the various matrices with either single pesticides or mixtures.

Because of the complexity of the organic constituents in the extracts, initial identification of pesticides was done by extracting, as described above, and analyzing the extracts by GC-MS procedures.

RESULTS AND DISCUSSION

Plant Growth in Pesticide-Contaminated Soils

There were no significant differences in total corn growth (shoots + roots) in any of the soil mixes (Figure 1), while growth was significantly greater in compost-containing mixes at 25 and 50% compost than it was in any of the soil mixes. Considering all the data as a unit, a 50:50 mixture of contaminated soil and compost was the most satisfactory from the standpoint of maximizing plant growth while minimizing dilution of the contaminated matrix.

Dehydrogenase Activity

In prior work (Cole et al. 1994), we found that fungal populations, as determined by plate count, were significantly higher in all compost mixes when compared with mixes containing uncontaminated soil. Similarly, bacterial populations in compost-containing mixes were significantly greater than populations in soil mixes in most cases. Microbial population size is not always a good indicator of microbial function in soil, because many of the microbes found in soil cannot be isolated and those that grow when plated on a nutrient-rich medium may have been inactive in the soil. Therefore, we used dehydrogenase activity as a broad-spectrum indicator of microbial activity in the mixes. This enzyme has been used by several investigators as an indication of overall heterotrophic activity in soil (Schaffer 1993). Activity in soil mixes was not significantly different

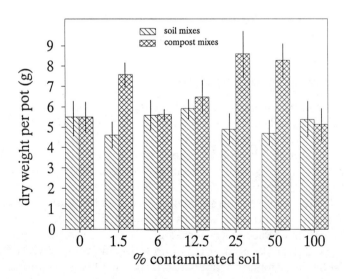

FIGURE 1. Total plant weight (shoots + roots) of corn grown in mixtures of contaminated soil and either uncontaminated soil or compost. Values are the means ± 1 standard deviation of four pots of plants per treatment. Data adapted from Cole et al. (1994), with permission.

among treatments, with a range of values of 16 to 70 µmol TTC reduced g^{-1} soil 24 h^{-1} (Cole et al. 1994). Dehydrogenase activity in compost-amended mixes was 10- to 30-fold higher than seen in soil mixes, with no significant difference between planted and unplanted treatments (Figure 2). Activity in planted or unplanted compost-containing mixes with 25 or 50% contaminated soil was significantly less than what would be expected if no interaction occurred between the contaminated soil and the active organisms in compost. This result indicates that the contaminated soil had significant antimicrobial properties. One possible consequence of inhibition of microbial activity by the contaminated soil is that addition of relatively small amounts of compost or other organic amendments (typically 20 to 40 tons/hectare^{-1}) may not have the beneficial effects seen with addition of large amounts of compost.

Effect of Compost and Planting on Pesticide Degradation

Trifluralin concentrations were significantly reduced only in planted mixes containing either soil or compost (Table 1). Because both plants (Probst & Tepe 1969) and microbes (Zeyer & Kearney 1983) degrade trifluralin, it was not possible to distinguish between plant uptake and metabolism of the herbicide and

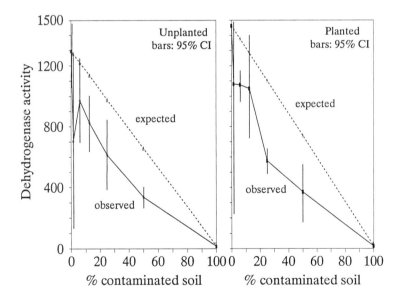

FIGURE 2. Expected and observed dehydrogenase activity in unplanted or planted compost mixes. Solid lines indicate actual data; dotted lines represent expected activity of mixes assuming no interaction between mix components. Values are the means of duplicate analyses of 4 replications per treatment. Vertical bars are the 95% confidence intervals. Data adapted from Cole et al. (1994), with permission.

TABLE 1. Effects of mix composition and planting on pesticide degradation, following 40 d of plant growth. Values are means ± standard deviations of duplicate extractions of four replications per treatment.

Mixture	Treatment	Trifluralin	Metolachlor	Pendimethalin
		mg kg^{-1} soil		
Initial Concentration	None	2.2 ± 0.9	3.0 ± 0.2	11.8 ± 5.1
100% Contamination	Planted	0.80 ± 0.82 (0.27)[a]	3.4 ± 5.0 (0.25)	1.6 ± 0.4 (0.02)
100% Contamination	Not planted	0.48 ± 0.77 (0.77)	0.99 ± 1.4 (0.25)	1.8 ± 0.4 (0.02)
50:50 Soil	Planted	nd[b]	nd	0.5 ± 0.6 (0.01)
50:50 Soil	Not planted	0.52 ± 0.53 (0.07)	0.18 ± 0.16 (<0.001)	1.0 ± 0.2 (0.02)
50:50 Compost	Planted	0.36 ± 0.33 (0.02)	nd	1.5 ± 0.6 (0.02)
50:50 Compost	Not planted	0.44 ± 0.69 (0.08)	2.8 ± 3.4 (0.29)	2.6 ± 3.4 (0.12)

(a) Values in parentheses indicate the probability that the values are less than expected from dilution alone (based on a one-tailed t-test for means of unequal variance).
(b) nd = not detected.

increased microbial degradation in the planted mixes. Metolachlor was not degraded in pots containing only contaminated soil, nor in unplanted 50:50 compost mixes, while degradation to nondetectable levels (0.1 mg kg^{-1}) was achieved in planted mixes of either soil or compost and in unplanted 50:50 soil mixes. With the exception of unplanted compost mixes, pendimethalin was degraded in all treatments. However, degradation to nondetectable levels (0.2 mg kg^{-1}) was not obtained with any treatment. Since there were no significant differences in extractability of freshly added trifluralin, metolachlor, or pendimethalin among treatments (data not shown), degradation is the most likely mechanism for herbicide loss.

The rates of herbicide degradation shown in Table 1 were substantially faster than Felsot et al. (1990) had found for land-farmed soils containing pesticide residues. With their procedure, the contaminated soils were incorporated into the top few centimeters of the field, where there would be relatively little plant root activity. In contrast, the mixtures we used had extensive root growth throughout the matrix, a result that may account for the rapid degradation.

CONCLUSIONS

Several treatments were effective in stimulating degradation of pesticides found in Site 20 material. Since the starting material was obtained from actual contaminated sites in which the pesticides had been in the soil for several years, the results strongly suggest that rapid and effective decontamination of the material can be achieved. Remediation of "aged" soils containing pesticides has been rather difficult because the bioavailability of the compounds tends to decrease with age (Schribner et al. 1992; Steinberg et al. 1987). The fact that herbicides of two different chemical groups (the nitroanilines, trifluralin and pendimethalin and an acetamide, metolachlor) were degraded raises the possibility that the described procedures can be used effectively for a wide variety of pesticide contaminants and mixtures of compounds.

Fungal and bacterial populations in the compost mixes were 2 to 20 times greater than those isolated from soil mixes (Cole et al. 1994), a result which indicates that the microorganisms added with the compost can survive in the soil environment. Increases in microbial activity, as measured by dehydrogenase enzyme activity, indicate that the organisms added in the compost do not simply persist in an inactive state when added to soil, but are metabolically active and may contribute to biodegradation processes. These results are in contrast to some results obtained when exogenous organisms are added to nonsterile soil. In such cases, the organisms may compete poorly against indigenous organisms (Goldstein et al. 1985). The larger populations and activity observed in compost-amended soils strongly suggest that compost can be an effective and inexpensive microbial inoculant. Use of common yard waste compost as an inoculant avoids the potential regulatory problems associated with introduction of laboratory-selected or genetically modified microorganisms. There is extensive information on the safe use of compost as an amendment of field soil; and a substantial literature exists that indicates that compost addition is beneficial to plant growth.

Considering all the data as a group, the most effective treatment for increasing plant growth and microbial activity and accelerating pesticide degradation in materials contaminated with moderate levels of several pesticides is a combination of dilution with either soil or compost in combination with planting. We are currently determining the minimum amount of compost that must be added to achieve significant improvements in microbial activity, plant growth, and pesticide degradation.

ACKNOWLEDGMENTS

The assistance of Warren Goetsch, Illinois Department of Agriculture in obtaining soil samples is gratefully acknowledged. GC-MS analyses were performed by the Illinois Hazardous Waste Research and Information Center.

Funds for this research were provided by Solum Remediation Services, Lake Bluff, Illinois, and by a grant from the Illinois Hazardous Waste Research and Information Center.

REFERENCES

Anderson, T.A., and J.R. Coats (Eds.). 1994. *Bioremediation Through Rhizosphere Technology.* American Chemical Society, Washington, DC.

Cole, M.A., X. Liu, and L. Zhang. 1994. "Plant and microbial establishment in pesticide-contaminated soil amended with compost." In Anderson, T.A. and J.R. Coats (Eds.), *Bioremediation Through Rhizosphere Technology*, pp. 210-222. American Chemical Society, Washington, DC.

Cole, M.A., X. Liu, and L. Zhang. 1994. "Remediation of soil containing trifluralin, metolachlor and pendimethalin by planting and compost addition." Submitted for publication in *Compost Science and Utilization.*

Felsot, A.S., and E.K. Dzantor. 1990. "Enhancing biodegradation for detoxification of herbicide waste in soil." In K. D. Racke and J.R. Coats (Eds.), *Enhanced Biodegradation of Pesticides in the Environment*, pp. 249-268. American Chemical Society, Washington, DC.

Felsot, A., E.K. Dzantor, L. Case and R. Liebl. 1990. *Assessment of problems associated with landfilling or land application of pesticide waste and feasibility of cleanup by microbiological degradation.* Hazardous Waste Resource and Information Center Report HWRIC RR-053. Urbana, IL.

Goldstein, R.M., L.M. Mallory, and M. Alexander. 1985. "Reasons for possible failure of inoculation to enhance biodegradation." *Applied and Environmental Microbiology* 50: 977-983.

Habecker, M.A. 1989. *Environmental Contamination at Wisconsin Pesticide Mixing/Loading Facilities: Case Study, Investigation and Remedial Action Evaluation.* Wisconsin Department of Agriculture, Trade, and Consumer Protection. Madison, WI.

Krapac, I.G., W.R. Roy, C.A. Smyth and M.L. Barnhardt. 1993. "Occurrence and distribution of pesticides in soil at agrichemical facilities in Illinois." In *Agrichemical Facility Site Contamination Study.* Illinois Department of Agriculture, Springfield, IL.

Lemmon, C.R., and H.M. Pylypiw, Jr. 1992. "Degradation of diazinon, chlorpyrifos, isofenphos and pendimethalin in grass and compost." *Bulletin of Environmental Contamination and Toxicology* 48: 409-415.

Michel, F.C., C.A. Reddy, and L.J. Forney. 1993. "Fate of certain lawn care pesticides during yard waste composting." In *Proceedings of Research Symposium, The Composting Council's Fourth National Conference*, pp. 5-8. The Composting Council, Alexandria, VA.

Probst, G.W., and J.B. Tepe. 1969. In Kearney, P.C. and D.D. Kaufman (Eds.), *Degradation of herbicides.* Marcel Dekker, Inc., New York, NY.

Schaffer, A. 1993. "Pesticide effects on enzyme activities in the soil ecosystem." In J.-M. Bollag and G. Stotzky (Eds.), *Soil Biochemistry*, Volume 8, pp. 273-340. Marcel Dekker, Inc., New York, NY.

Schribner, S.L., T.R. Benzing, S. Sun, and S.A. Boyd. 1992. "Desorption and bioavailability of aged simazine residues in soil from a continuous corn field." *Journal of Environmental Quality* 21: 115-120.

Steinberg, S.M., J.J. Pignatello, and B.L. Sawhney. 1987. "Persistence of 1,2-dibromoethane in soils: Entrapment in intraparticle micropores." *Environmental Science and Technology* 21: 1201-1208.

Taylor, A.G. 1993. "The effect of agrichemical use on water quality in Illinois." *Abstracts American Chemical Society Annual Meetings*, Chicago, IL., August, 1993. American Chemical Society, Washington, DC.

Zeyer, J., and P.C. Kearney. 1983. "Microbial dealkylation of trifluralin in pure culture." *Pesticides in Biochemistry and Physiology* 20: 10-18.

Potential for the Biotransformation of Metolachlor in Groundwater

Michaye L. McMaster, Barbara J. Butler, and James F. Barker

ABSTRACT

Metolachlor is a chloroacetanilide herbicide commonly used to control grassy weeds. A groundwater plume of metolachlor has been generated from a point source spill in Cambridge, Ontario. The metolachlor half-life in surface soils ranges from 15 to 50 days, and biotransformation is expected to occur by cometabolism. However, given the restricted nutrient conditions in most aquifer materials, somewhat longer half-lives were expected. The purpose of this research was to evaluate the natural persistence of metolachlor and the potential for in situ bioremediation in the Cambridge groundwater. The potential for metolachlor biotransformation in Cambridge aquifer material and in Canada Forces Base (CFB) Borden aquifer material was investigated in laboratory microcosms. No biotransformation was detected during incubations of up to 360 days, and various nutrient amendment schemes did not promote biotransformation in the aquifer microcosms. In contrast, in comparative microcosms of surface soil with no nutrient amendment, the metolachlor concentration was reduced by 40% after 54 days. These results suggest that the biotransformation rate in the Cambridge aquifer is significantly lower than surface soil rates, which is consistent with the formation of the plume found at the site. The potential for in situ bioremediation of the existing metolachlor-contaminated groundwater at Cambridge appears limited.

INTRODUCTION

The degradation of herbicides in surface or near-surface soils has been studied extensively. Microbial transformation of metolachlor under high nutrient and metolachlor concentrations has been observed in laboratory cultures (Liu et al. 1989, Saxena et al. 1987, Krause et al. 1985). In a soil perfusion experiment performed by Liu et al. (1988), a mixed consortium of microorganisms able to produce CO_2 from metolachlor developed. When applied at suggested field rates, movement of metolachlor into the deep subsurface appears unlikely (Chesters et al. 1989).

Cavalier et al. (1991) investigated the potential for metolachlor degradation in groundwater and estimated a half-life of 1,074 days for 5 µg/L metolachlor at 15°C. This half-life estimate may be excessive because these experiments were conducted with groundwater in the absence of aquifer solids. Many aquifer microorganisms are associated with the solids; only a small part of the biomass is unattached (Holm et al. 1992). Microenvironments suited to microbial activity may develop in a soil-water matrix; these would not be available in a groundwater-only experiment.

At a site in Cambridge, Ontario, point source spills have contaminated the groundwater with metolachlor and other herbicides. The geology of the site consists of an upper sand aquifer, beneath which is a noncontinuous silt and clay aquitard underlain by a lower sand aquifer and a silt till deposit. Metolachlor concentrations in the Cambridge groundwater range up to about 10,000 µg/L, whereas Canadian drinking water guidelines advise that metolachlor concentrations not exceed 50 µg/L (Kent et al. 1991). The Cambridge concentrations substantially exceed those of the study by Cavalier et al. (1991) but are less than levels studied by Liu et al. (1988 and 1989). The goal of this research was to determine if the Cambridge site conditions would promote significant biotransformation, such that bioremediation might be considered at this site.

MATERIALS AND METHODS

Laboratory microcosms were established using aquifer material and groundwater from both CFB Borden and Cambridge, and a Guelph loam surface soil. The surface soil (described by Gallina and Stephenson 1992) plus contaminated groundwater from the Cambridge site served as a comparative control. The Guelph loam soil has a fraction organic carbon content (f_{oc}) of 1.35% (wt/wt) (Allen-King et al. 1995). CFB Borden is a pristine aquifer that has shown herbicide-degrading potential (Agertved et al. 1992). This sandy aquifer has been extensively studied in the past, and served as a reference subsurface material for the present study. CFB Borden aquifer material had an f_{oc} of 0.020%. Cambridge aquifer material was retrieved from ~13 m below ground surface, and groundwater was obtained from the same stratigraphic unit. The Cambridge aquifer material had an f_{oc} of 0.018%.

Microcosms consisted of 160-mL hypovials each containing 20 g of aquifer or surface material and 80 mL of groundwater. Sterile controls were set up as follows: aquifer material or surface soil was autoclaved on 3 consecutive days, then 0.8 mL of 10% sodium azide was added before the groundwater was aseptically added to each hypovial. Series of microcosms were initiated in triplicate, for both sterile control and active samples, and an entire set was sacrificed for analysis at each sampling interval. Results are depicted as the mean value for the 3 replicates, ± the standard deviation. When biotransformation was not observed within a few weeks, nutrient amendments were made to some hypovials (described in the Results section).

Metolachlor was analyzed by gas chromatography (GC) using a flame ionization detector (FID) and a Supelco SPB-5 column (15 m). The operating temperature was 220°C for the column and 275°C for the detector. The estimated method detection limit was 50 μg/L. All samples were extracted for analysis by filling a 60-mL hypovial with microcosm liquid, replacing 1.5 mL with 1.0 mL of high-performance liquid chromatography (HPLC)-grade hexane, then agitating the sealed sample vial for 20 min at 400 rpm. The CO_2 content of microcosm headspace gas was monitored using a Fisher/Hamilton Model 29 Gas Partitioner.

RESULTS

Borden Microcosms

Three concentrations (100, 500, and 1,000 μg/L) of metolachlor were tested with Borden aquifer material. After 48 days, no biotransformation had been observed (Figure 1), so the possibility of nutrient limitation(s) was investigated. Two sets of sample vials were amended with 2 mL of N- and P-containing mineral salts medium (Modified Bushnell Haas medium, Mueller et al. 1991). After 56 more days, there was no decline in metolachlor concentrations. To investigate the possibility of carbon limitation, 0.5 mL of a 10% glucose solution was added to 2 of every 3 replicates of each remaining sample set. Microcosm headspace gas was then monitored for CO_2 production as an indicator of microbial activity. CO_2 levels were elevated after 8 days in active, carbon-amended microcosms, but there was no significant reduction in metolachlor concentrations, and 32 days after carbon amendment there was still no evidence of metolachlor biotransformation (Figure 1).

Cambridge Microcosms

Cambridge microcosms amended with Modified Bushnell Haas medium (NA) or without amendment (UA) were prepared. Metolachlor levels of 10,000 μg/L and 250 μg/L simulated concentrations found at the site. Figures 2 and 3 show results for the low and high levels of metolachlor in Cambridge aquifer material, respectively. The large deviations in results from the early 10,000 μg/L samples were later corrected by modifying the extraction protocol to allow 25 min of shaking.

After 91 days, active microcosm metolachlor concentrations were not significantly different from sterile controls. A mixed carbon source (1 mL of a 2% sucrose and 0.4% yeast extract solution, plus 100 μL of ethanol) was then added to 2 of every 3 hypovials per set, and headspace CO_2 analyzed. After 8 days only small amounts of CO_2 were detected in carbon-amended microcosms. After an additional 49 days, significant microbial activity had taken place in carbon-amended microcosms, and the presence of a black precipitate (presumed to be metal sulfide and indicative of microbial H_2S production) was noted in several

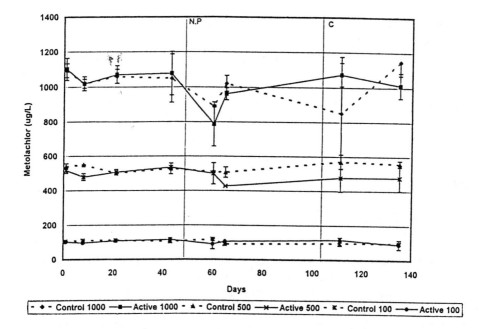

FIGURE 1. CFB Borden microcosm results. At Day 48 Modified Bushnell Haas was added to the next two sample sets. At Day 103 carbon source was added to 2 out of 3 replicates in the remaining sample sets. Points are the average of triplicate samples with ± standard deviation shown.

of the active samples. However, there was no significant difference in the metolachlor levels in sterile control and active microcosms.

After 360 days, no biotransformation of metolachlor was apparent and sterile controls in fact displayed lower metolachlor levels than the active samples. This finding cannot be explained readily. Perhaps some abiotic transformation mechanism was impaired by the microbial growth in active microcosms.

Surface Soil Microcosms

Figure 4 shows results for the comparative control, surface soil microcosms. No amendments were made to these microcosms. Allen-King et al. (1995) demonstrated the biotransformation of the herbicide glufosinate-ammonium in the surface soil. The f_{oc} of this soil is much greater than that of the two aquifer materials. After 54 days, the concentration of metolachlor had decreased by 40% in active microcosms, a much larger change than in the sterile controls (Figure 4). Elevated levels of CO_2 were detected in the active microcosms relative to the controls. Further biotransformation had occurred by day 154, although the decrease in the slope of the plot suggests that the system was becoming limited in some way.

DISCUSSION

The CFB Borden aquifer is considered a relatively pristine aquifer with very homogeneous properties. Biotransformation of metolachlor was not supported in microcosms of Borden material, even when nutrient-amended. Similarly, metolachlor biotransformation was not observed in Cambridge aquifer material, after ~1 year of incubation. The lack of biotransformation in Cambridge microcosms implies that naturally occurring in situ biotransformation at the Cambridge site is exceedingly slow. The extent of the metolachlor plume found at the site is consistent with this view, and this is also consistent with the metolachlor half-life of 1,074 days in groundwater determined by Cavalier et al. (1991). The metolachlor biotransformation detected in the surface soil microcosms indicated that no inherent factor(s) in the groundwater collected from the Cambridge site was restricting metolachlor biotransformation, because these microcosms contained the groundwater. Either some factor(s) other than inorganic N, inorganic P, and/or readily utilizable organic matter was required for biotransformation, or the requisite microorganisms are very rare in the aquifer materials.

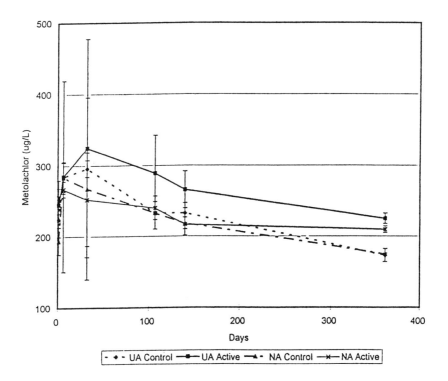

FIGURE 2.   Cambridge microcosm results: 250 µg/L concentration.   UA, unamended; NA, amended with Modified Bushnell Haas medium. Points are the average of triplicate samples with ± standard deviation shown.

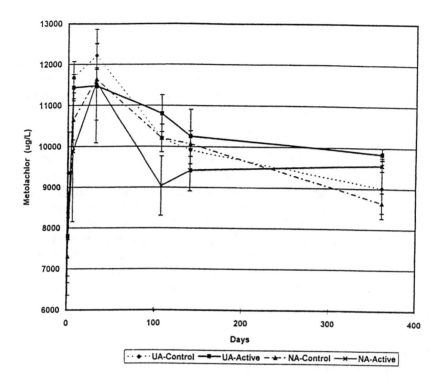

FIGURE 3. Cambridge microcosm results: 10,000 µg/L concentration. UA, unamended; NA, amended with Modified Bushnell Haas medium. Points are the average of triplicate samples with ± standard deviation shown.

Point source metolachlor spills may cause groundwater contamination due to the compound solubility (530 mg/L) (Kent et al. 1991) and mobility. Our results suggest that if an opportunity for metolachlor attenuation does not occur during transport through the surface soil and the shallow unsaturated zone, then the potential for attenuation via biotransformation in aquifer zones appears poor. Techniques such as nutrient injections into the aquifer appear unlikely to improve the biotransformation potential. Further work is being conducted to investigate differences between the Guelph loam soil and the aquifer materials. Alternative remedial options for the Cambridge site are being considered, because at this time bioremediation is not considered feasible.

ACKNOWLEDGMENTS

This work has been supported by Ciba Canada and by a Natural Science and Engineering Research Council Industrial Oriented Research Grant to J. F. Barker.

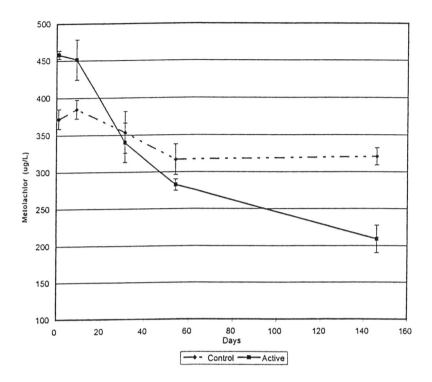

FIGURE 4. Surface soil microcosm results, using Guelph loam and Cambridge groundwater. Points are the average of triplicate samples with ± standard deviation shown.

REFERENCES

Agertved, J., K. Rugge, and J. F. Barker. 1992. "Transformation of the Herbicides MCPP and Atrazine under Natural Aquifer Conditions." *Groundwater* 30(1):500-506.

Allen-King, R. M., B. J. Butler, and B. Reichert. 1995. "Fate of the Herbicide Glufosinate-ammonium in the Sandy, Low-organic Carbon Aquifer at CFB Borden, Ontario." *Journal of Contaminant Hydrology* (in press).

Cavalier, T. C., T. L. Lavy, and J. D. Mattice. 1991. "Persistence of Selected Pesticides in Groundwater Samples." *Groundwater* 29(2):225-231.

Chesters, G., G. V. Simsiman, J. Levy, B. J. Alhajjar, R. N. Fathulla, and J. M. Harkin. 1989. "Environmental Fate of Alachlor and Metolachlor." *Reviews of Environmental Contamination and Toxicology* 110:1-74.

Gallina, M. A., and G. R. Stephenson. 1992. "Dissipation of [^{14}C]Glufosinate Ammonium in Two Ontario Soils." *Journal of Agricultural and Food Chemistry* 40(1):165-169.

Holm, P. E., P. H. Nielsen, H.-J. Albrechtsen, and T. H. Christensen. 1992. "Importance of Unattached Bacteria and Bacteria Attached to Sediment in Determining Potentials for Degradation of Xenobiotic Organic Contaminants in an Aerobic Aquifer." *Applied and Environmental Microbiology* 58(9):3020-3026.

Kent, R. A., B. D. Pauli, D. M. Trotter, and J. Gareau. 1991. *Canadian Water Quality Guidelines for Metolachlor*. Scientific Series No. 184. Environment Canada, Ottawa, Ontario.

Krause, A., W. G. Hancock, R. D. Minard, A. J. Freyer, R. C. Honeycutt, H. M. LeBaron, D. L. Paulson, S-Y. Liu, and J-M. Bollag. 1985. "Microbial Transformation of the Herbicide Metolachlor by a Soil Actinomycete." *Journal of Agricultural and Food Chemistry* 33(4):584-589.

Liu, S-Y., R. Zhang, and J-M. Bollag. 1988. "Biodegradation of Metolachlor in a Soil Perfusion Experiment." *Biology and Fertility of Soils* 5:276-281.

Liu, S-Y., Z. Zheng, R. Zhang, and J-M. Bollag. 1989. "Sorption and Metabolism of Metolachlor by a Bacterial Community." *Applied and Environmental Microbiology* 55(3):733-740.

Mueller, J. G., S. E. Lantz, B. O. Blattman, and P. J. Chapman. 1991. "Bench-scale Evaluation of Alternative Biological Treatment Processes for the Remediation of Pentachlorophenol- and Creosote-contaminated Materials: Solid Phase Remediation." *Environmental Science and Technology* 25(6):1045-1055.

Saxena, A., R. Zhang, and J-M. Bollag. 1987. "Microorganisms Capable of Metabolizing the Herbicide Metolachlor." *Applied and Environmental Microbiology* 53(2):390-396.

Composting of Herbicide-Contaminated Soil

Maureen A. Dooley, Karen Taylor,
and Bobby Allen

ABSTRACT

Bench-scale testing was conducted to evaluate the feasibility of using composting to remediate the herbicide dicamba, its breakdown product 3,6-dichlorosalicylic acid (3,6-DCSA), and other site-specific contaminants such as benzonitrile and benzoic acid, in soil (predominantly clay) found in a landfill in Georgia. Compost microcosms were prepared by mixing soil with shredded tree waste to improve permeability to air and water and with manure to provide additional moisture retention and trace minerals. Test compost microcosms were prepared with different percentages of bulking agent and using two concentrations of dicamba. A control compost was prepared to measure abiotic losses. The results indicate that the herbicide dicamba and other organic compounds such as 3,6-DCSA, benzonitrile, and benzoic acid can be biodegraded successfully, even in soil with extremely low permeability. Biodegradation rates were dependent upon bulking agent ratios, where the higher amounts of bulking agent provided an environment with better air and water permeability and, therefore, exhibited faster rates of biodegradation. Treatment criteria established for this site could be achieved in 21 to 28 days in the compost with a high proportion (41%) of bulking agent.

INTRODUCTION

A treatability study was conducted by ABB Environmental Services (ABB-ES) to evaluate the feasibility of using bioremediation to reduce the concentration of the target compounds dicamba (3,6-dichloro-2-methoxybenzoic acid), 3,6-dichlorosalicylic acid (DCSA), benzonitrile, and benzoic acid in soil located at a landfill (the Site), in Walker County, Georgia. The treatment goals established for this site were 25 mg/kg combined dicamba/3,6-DCSA and 25 mg/kg combined benzonitrile/benzoic acid. Data generated from this study were used to prepare conceptual design and cost estimates for full-scale biological treatment.

The Site is located approximately 4 miles north of LaFayette, Georgia, in Walker County. The former landfill area comprises approximately 5 acres (2 ha). The area contained buried drums and waste residue from the manufacture of dicamba, benzonitrile, and styrene-butadiene rubber. Wastes and soils designated for remediation were removed from the landfill and placed into lined cells adjacent to the site for containment.

Technical Approach

Dicamba is used as a herbicide for control of several broadleaf and grassy weeds (Krueger et al. 1989). Biodegradation studies have demonstrated that dicamba could be mineralized and that a major metabolite formed was 3,6-DCSA (Smith 1973). Biodegradation studies conducted by Krueger et al. (1989) suggested that dicamba was susceptible to aerobic biodegradation, but an acclimated culture may be required before extensive biodegradation would occur. In that study 3,6-DCSA did not accumulate. Other site-specific contaminants included benzonitrile and benzoic acid which are both readily degraded under aerobic conditions (Harper 1977 and Rubin et al. 1982).

Testing was designed to initially demonstrate that indigenous microorganisms capable of metabolizing the target compounds were present in site soil, but primarily to measure biodegradation under simulated full-scale treatment conditions to evaluate the feasibility of using bioremediation to reduce target compounds to below the cleanup standards. Because of the high percentage of clay in the landfill soil, a composting treatment system was evaluated. The bulking agents were used to increase permeability and moisture retention in soil as well as to provide additional bacteria and inorganic and organic nutrients.

TREATABILITY TESTING PROCEDURES AND RESULTS

Soil Characterization

Soil samples (grab) were collected from six different locations at the site and composited. The soil was predominantly clay and was broken up manually to create a homogeneous mixture for testing. Results from target compound analysis (Modified EPA Method 3550/8150) of the composite indicated that dicamba/3,6-DCSA, benzoic acid, and benzonitrile were not detected (<20 mg/kg) and 3,5-DCSA was measured at 300 mg/kg. This compound was not designated as one of the target compounds; however, the concentration is notable because it was higher than the other target compounds and will be an available carbon source for the bacteria. Because additional soil was not readily available for testing, pure compounds and soil leachate from the site (containing dicamba and 3,5-DCSA) were mixed into the composite soil for testing.

Aerobic bacteria able to use dicamba, 3,6-DCSA, benzonitrile, and benzoic acid as sole sources of carbon were isolated in liquid culture from the site soils. Liquid culture that were composed of mineral salts medium and only one of the individual target compounds as a source of carbon were maintained by regularly transferring a portion of the culture into fresh medium. The results from the isolation indicated that indigenous populations were present in the soil that could metabolize each of the target compounds.

Biodegradation Testing

Compost Preparation. Three microcosms were prepared initially using 4.5% wood chips and 6.3% cow manure (by weight). The dicamba concentration was varied from approximately 750 to 2,800 mg/kg. The compost test conditions are summarized in Table 1. An additional compost microcosm was prepared using a higher ratio of shredded tree waste (35%) 14 days later because there was concern that 10% bulking agent was not sufficient to allow

TABLE 1. Compost microcosm test conditions.

| | Starting Conditions of Compost | | | |
	Low Dicamba	High Dicamba	Killed Control	Low Dicamba
Soil	89%	89%	89%	58.8%
Wood chips (by weight)	4.5%	4.5%	4.5%	35%
Manure (by weight)	6.3%	6.3%	6.3%	6%
Lime	0	0	0	0.2%
Bacteria[a]	added	added	added	added
Dry Weight	75%	73%	74%	72%
Orthophosphate	150 mg/kg	150 mg/kg	150 mg/kg	188 mg/kg
Nitrogen (NH_4 as N)	375 mg/kg	300 mg/kg	350 mg/kg	400 mg/kg
pH	7	6.5	6.5	7.3
Mercuric chloride	NA[b]	NA	2%	NA
Dicamba	813 mg/kg	2,766 mg/kg	659 mg/kg	674 mg/kg
3,6-DCSA	612 mg/kg	654 mg/kg	476 mg/kg	398 mg/kg
3,5-DCSA	549 mg/kg	602 mg/kg	701 mg/kg	744 mg/kg
Benzonitrile	756 mg/kg	570 mg/kg	853 mg/kg	467 mg/kg
Benzoic acid	852 mg/kg	889 mg/kg	967 mg/kg	1,168 mg/kg

(a) Dicamba-degrading bacteria isolated from the site were added to soil.
(b) NA = not analyzed.

adequate oxygen transfer through the compost. The killed control was prepared using 10% bulk and was amended with the biocide mercuric chloride.

Mineral nutrients (300 mg/kg N-NH$_4$ and 44 mg/kg orthophosphate) and target compound-degrading bacteria isolated from the site soil were added to each of the live microcosms. The mineral nutrients were added in the form of dry fertilizer and manure, which was also a source of nitrogen and phosphate. Aeration was provided by mixing the soil three times a week, and moisture and nutrient levels were maintained throughout the study.

Radiotracer Study. A radiotracer study was conducted in parallel with the solid-phase biodegradation test to measure dicamba mineralization. A portion of each compost that was prepared for treatability testing was amended with approximately 10^6 cpm of ^{14}C-dicamba (uniformly labeled) and placed in a 1,000-mL biometer flask. The soil conditions were maintained in the same manner as in the biodegradation study. The ^{14}CO$_2$ was trapped in 1N NaOH found in the side arm of the biometer flask. When analyzed, the NaOH was removed from the side arm and replaced with fresh NaOH. A known portion of the NaOH (0.5 mL) was added to the scintillation cocktail and analyzed for ^{14}C on a liquid scintillation counter. The results were reported in counts per minute (cpm).

Chemical Analysis Results. Duplicate samples were removed from each of the test composts and analyzed for the target compounds over the course of the study. The results from the analyses, which are presented as an average of the duplicates, showed that benzonitrile and benzoic acid were biodegraded within 7 days in each of the live compost microcosms (Figures 1 and 2). Abiotic losses of benzonitrile were observed in the killed control, where approximately 50% of the benzonitrile had volatilized.

DCSA biodegradation occurred at different rates in the live compost microcosms and the most rapid rate of 3,5- and 3,6-DCSA biodegradation occurred in the microcosms with 41% bulking agent (Figures 3 and 4). There was a lag period of at least 7 days was observed before biodegradation began and the final concentrations of 3,5- and 3,6-DCSA were 18 and 14 mg/kg, respectively, after 39 days.

A longer lag period of approximately 21 days was observed before DCSA biodegradation began in microcosms with 10% bulking agent. DCSA biodegradation was only slightly slower in the presence of high levels of dicamba. The combined DCSA concentration at day 53 was approximately 80 mg/kg for both 10% composts. There were no significant losses of 3,6-DCSA observed in the killed control over the testing period; however, the results from the analysis of 3,5-DCSA were more variable and reduction of 3,5-DCSA was seen between days 42 and 53. The fluctuations in 3,5-DCSA may have been due to non-homogeneous distribution of 3,5-DCSA in the control compost soil.

Dicamba biodegradation of greater than 98% was observed in each of the live compost microcosms; however, the rates of biodegradation were very

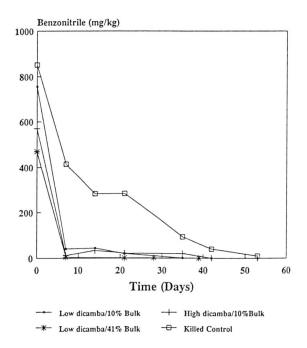

FIGURE 1. Benzonitrile biodegradation in compost microcosms.

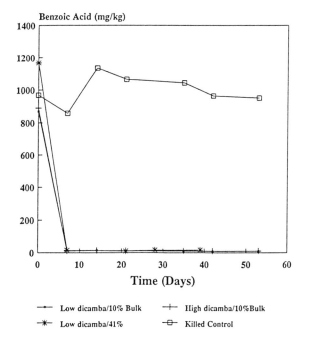

FIGURE 2. Benzoic acid biodegradation in compost microcosms.

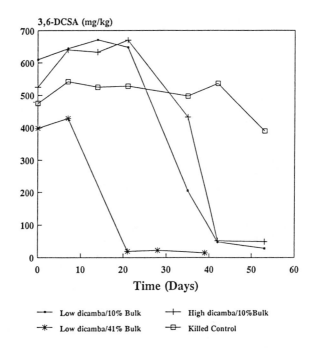

FIGURE 3. 3,6-dichlorosalicylic acid biodegradation in compost microcosms.

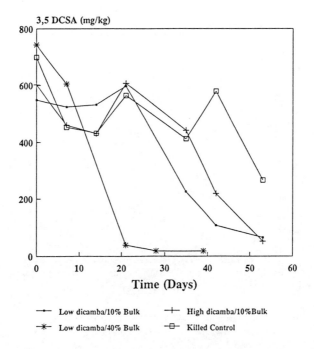

FIGURE 4. 3,5-dichlorosalicylic acid biodegradation in compost microcosms.

different. Dicamba removal occurred most rapidly in the compost with 41% bulking agent (Figure 5) and no dicamba was detected above 0.5 mg/kg after 21 days. A lag period of approximately 35 days was observed before dicamba removal was observed in the 10% composts. The lag period was not any longer in the compost with the high concentration of dicamba. Dicamba levels were reduced to less than 0.5 mg/kg after 42 days (low-concentration dicamba) and 53 days (high-concentration dicamba). No significant losses of dicamba were observed in the killed control compost. These results are similar to the DCSA data where the most rapid degradation occurred in the 41% bulking agent compost.

Mineralization Results. The greatest extent of dicamba mineralization (27%) occurred in the compost with 41% bulking agent (Figure 6). Most of the dicamba mineralization occurred by day 28, which corresponds to the dicamba biodegradation data obtained from the gas chromatographic (GC) measurements. The $^{14}CO_2$ evolution continued at a slower rate over the remainder of the test period. Because the GC data show that the dicamba was completely removed in the compost, it appears that a large percentage of the dicamba either was converted into biomass or was transformed to other nonvolatile compounds.

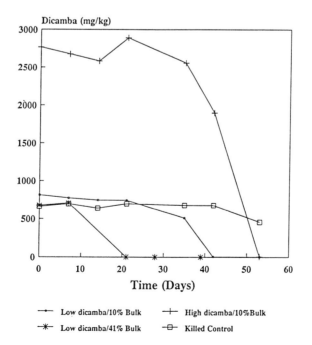

FIGURE 5. Dicamba biodegradation in compost microcosms.

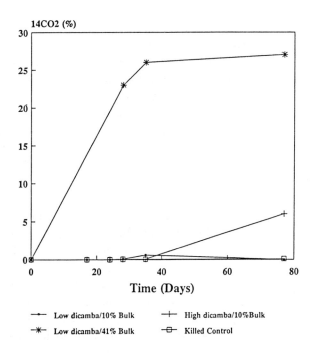

FIGURE 6. ^{14}C-dicamba mineralization results.

Limited mineralization was demonstrated in the other test composts, which showed only 3% mineralization in the low dicamba/10% bulking agent compost and 6% mineralization in the high dicamba/10% bulking agent microcosm. Also, a longer period of time was required before mineralization began in these microcosms, which indicates that the use of 41% bulking agent created a more favorable condition for dicamba biodegradation.

Bacteria Analysis Results. Results from bacteria analysis of composts indicate that there was an increase in both the total and dicamba-degrading population and the largest increase was observed in the 41% compost. The number of dicamba-degrading bacteria increased from 3.7×10^4 CFU/g to $>1.0 \times 10^6$ CFU/g in all live test composts (65 days). Although dicamba-degrading bacteria were isolated from the site soil, additional bacteria may have been introduced through the addition of the manure and tree waste. The total number of bacteria increased in each of the live microcosms from 10^6 CFU/g to 10^7 CFU/g in the 10% compost and 8×10^7 in the 41% compost.

CONCLUSIONS

The results from the treatability study demonstrate that compost treatment can be used to reduce the concentrations of target compounds to below the cleanup levels of 25 mg/kg (dicamba/3,6-DCSA) and 25 mg/kg (benzonitrile/benzoic acid) established for this site. Dicamba mineralization was demonstrated without the transitory formation or accumulation of 3,5- or 3,6-DCSA. The higher quantity of bulking agent (41%) in the clay soil resulted in a shorter lag time and a more rapid rate and greater extent of target compound reduction. Also, a greater extent of dicamba mineralization was observed in the 41% compost.

A large amount of bulking agent is required to achieve rapid biodegradation rates for target compounds in the landfill soil. Pilot-scale testing is recommended to confirm the results from bench-scale testing. Based on the results of this study, 41% bulking agent is recommended however, less bulking agent may be required at sites with soil with lower clay content.

REFERENCES

Harper, D. 1977. *Biochem. J.* 167(3): 685-692.

Krueger, J. P., R. G. Butz, Y. H. Atallah and D. J. Cork. 1989. "Isolation and Identification of Microorganisms for the Degradation of Dicamba." *J. Agric. Food Chem.* 37(2): 534-538.

Rubin, H. E., R. V. Subba-Rao and M. Alexander. 1982. "Rates of Mineralization of Trace Concentrations of Aromatic Compounds in Lake Water and Sewage Samples." *Appl. Environ. Microbiol.* 43(5): 1133-1138.

Smith, A. E. 1973. "Transformations of Dicamba in Regina Heavy Clay." *J. Agric. Food Chem.* 21: 708-710.

U.S. EPA. 1986. *Test Methods for Evaluating Solid Waste-Physical/Chemical Methods.* SW-846 3rd ed. Office of Solid Waste, U.S. EPA, Washington, DC.

Biodegradation of Carbazole and Related Compounds

Toshio Omori

ABSTRACT

Pseudomonas sp. CA10, which can grow on carbazole (CAR) as the sole source of carbon and nitrogen; *Staphylococcus auricularis* DBF63, which can grow on dibenzofuran (DBF) or fluorene (FN) as the sole source of carbon and energy; and *Rhodococcus* sp. SY1, which can grow on dibenzothiophene (DBT) as a sole source of sulfur, were isolated. DBF-grown cells of *S. auricularis* DBF63 formed oxidation products of dibenzo-*p*-dioxin (DD). The possibility of degradation of tetrachlorodibenzo-*p*-dioxin is discussed. Based on the detection of the metabolic intermediates of CAR, DBF, FN and DD, the degradation pathways of these compounds are proposed.

INTRODUCTION

To cope with the acid rain caused by the release of sulfur dioxide from organic sulfur of fossil fuels during combustion, it is very important to obtain bacteria that are capable of efficiently removing the sulfur from organic sulfur compounds. *Rhodococcus* sp. SY1, which can grow on dibenzothiophene (DBT) as a sole source of sulfur; *Pseudomonas* sp. CA10, which can grow on carbazole (CAR) as the sole source of carbon and nitrogen; and *Staphylococcus auricularis* DBF63, which can grow on dibenzofuran (DBF) or fluorene (FN) as the sole source of carbon and energy were isolated. However, carbazole, fluorene, dibenzofuran, and dibenzothiophene are xenobiotics and structurally related recalcitrant compounds found in fossil fuels. Xenobiotic compounds such as carbazole, fluorene, and dibenzofuran have been found to contaminate both soil and groundwater.

Contamination of the environment poses a potentially serious problem since these xenobiotic compounds have been reported to show toxic, mutagenic, and carcinogenic activities, even when present in low concentrations. Microbial degradation of such pollutants is one of the most important factors in reducing their concentrations.

EXPERIMENTAL PROCEDURES AND MATERIALS

Organisms

Rhodococcus sp. strain SY1 was previously described in references as *Corynebacterium* sp. strain SY1.

Growth Conditions

All cultivation was performed for 3 or 4 days at 30°C with rotary shaking with use of media and growth conditions as described previously (Omori 1992; Monna 1993; Ouchiyama 1993).

Analytical Methods

The concentrations of CAR in the culture medium were assayed by high-performance liquid chromatography (HPLC). An Inertsil ODS2 column was used with an acetonitrile-water-0.05% trifluoroacetic acid mobile phase. Gas chromatograms were made on Hitachi 163 gas chromatograph by using a glass column with silicon OV-17. Mass spectra and proton and carbon magnetic resonance spectra were determined as described previously (Omori 1992; Monna 1993; Ouchiyama 1993). Ultraviolet (UV) spectra were determined by using methanol as a solvent in a Shimazu UV 1 60A spectrophotometer. The optical density (OD) of the cell suspensions was measured at 610 nm.

RESULTS AND DISCUSSION

Biodegradation of Carbazole
by *Pseudomonas* sp. CA 10

A bacterial strain CA10, that assimilates carbazole (CAR) as the sole source of carbon and nitrogen, was isolated from activated sludge samples and identified as *Pseudomonas* sp. Anthranilic acid (AN) and catechol (CAT) were identified as the main metabolites of CAR by HPLC and gas chromatography-mass spectrometry (GC-MS). An initial oxidation product, 2'-aminobiphenyl-2,3-diol, and a *meta*-cleavage product, 2-hydroxy-6-oxo-6-(2'-aminophenyl)-hexa-2,4-dienoic acid, were tentatively identified in the culture broth of CAR by GC-MS.

When AN was used for a substrate in culture by these strains, CAT and a small amount of *cis,cis*-muconate was detected by HPLC. This conversion suggested the existence of an *ortho*-cleavage pathway. Based on these results, a CAR degradation pathway was proposed (Figure 1). The effects of CAR concentrations on the degradation of CAR and the formation of metabolites were investigated. At a CAR concentration of 17 mM, the consumption rate of substrate was slow compared with rates of low substrate concentrations, and 1.4 mM of AN and 0.1 mM of CAT were accumulated in the culture broth.

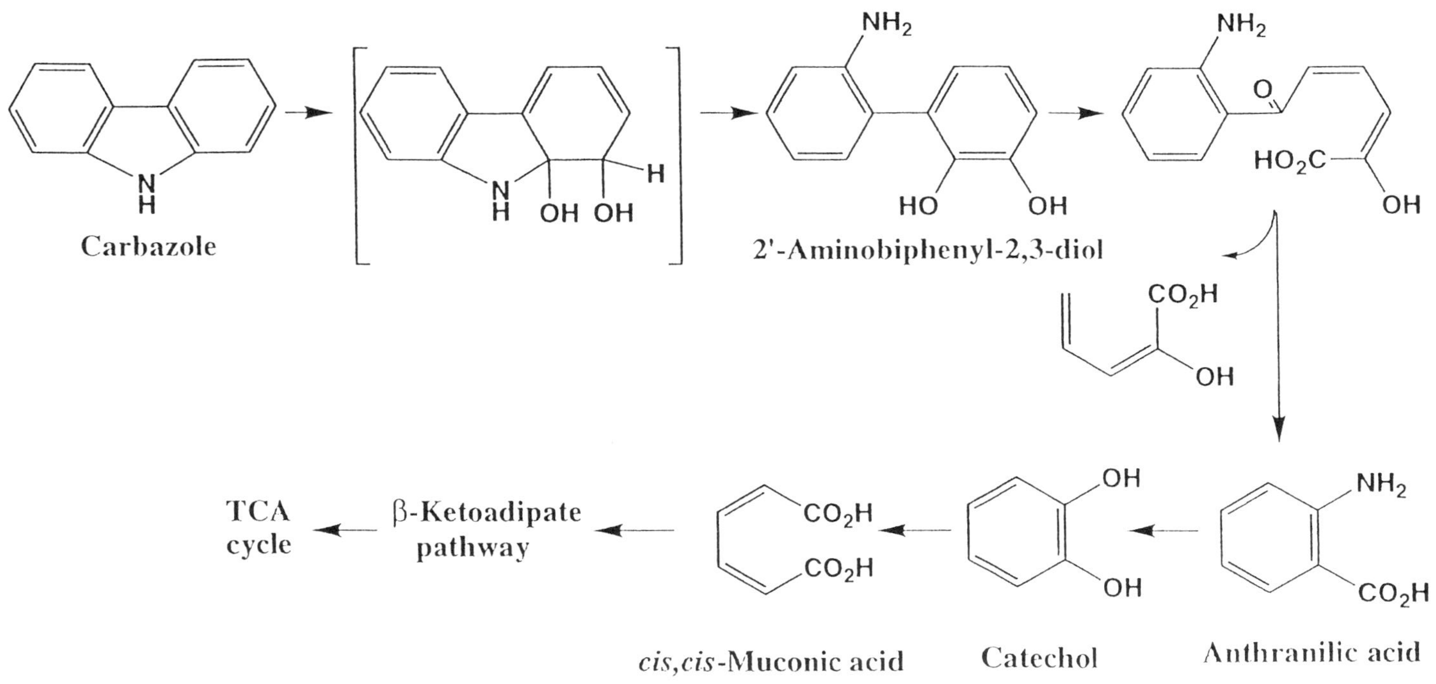

FIGURE 1. Proposed degradation pathway of carbazole.

Biodegradation of Dibenzofuran, Fluorene, and Dibenzo-*p*-dioxin by *Staphylococcus auricularis* DBF63

Staphylococcus auricularis DBF63, which can grow readily on dibenzofuran (DBF) or fluorene (FN) as the sole source of carbon and energy, was isolated. The cooxidation of various polycyclic aromatic compounds was also investigated. Dibenzothiophene, carbazole, and naphthalene were cooxidized in the presence of succinate or DBF as a carbon source, but no reaction or growth was observed with dibenzothiophene-sulfone, phenazine, *o,o'*-biphenol, 2,3-benzofuran, or DD. Salicylic acid and gentisic acid accumulated in the culture broth of this strain when DBF was supplied as a growth substrate. Also, the formation of 9-fluorenol, 9-fluorenone, 4-hydroxy-9-fluorenone, and 1-hydroxy-9-fluorenone was demonstrated, and accumulation of 1,9a-dihydroxy-1-hydro-9-fluorenone was observed when this strain grew on FN.

On the basis of these results, the degradation pathways of DBF and FN were proposed (Figure 2). The analogous oxidation product of dibenzo-*p*-dioxin such as 1,9a-dihydroxy-1-hydro-9-dibenzo-*p*-dioxin were obtained by incubation with DBF-grown *S. auricularis* DBF63 cells. Because the product 1-hydroxy-1,9a-dihydroxy-DD is chemically unstable, and with possible spontaneous conversion into a catechol-type compound, DBF-grown DBF cells appear to have high levels of metapyrocatechase activity for various catechols, further degradation of DD could be possible under suitable conditions (Figure 2). By adapting strain DBF63 to tetrachlorodibenzo-*p*-dioxin or genetically changing the dioxygenase which attacks the angular position of DBF, the microbial degradation of tetrachloro-dibenzo-*p*-dioxin is highly expected (Figure 3).

Desulfurization of Dibenzothiophene by *Rhodococcus* sp. Strain SY1

Dibenzothiophene has been used for as a model compound for biodesulfurization of organic sulfur found in fossil fuels. Strain SY1, identified as a *Rhodococcus* sp. was isolated on the basis of its ability to utilize dibenzothiophene (DBT) as a sole source of sulfur. Strain SY1 can utilize a wide range of organic and inorganic sulfur compounds, such as DBT-sulfone, dimethyl sulfide, dimethyl sulfoxide, dimethyl sulfone, CS_2, FeS_2, and even elemental sulfur. Strain SY1 metabolized DBT to sequential products of dibenzothiophene-5-oxide, DBT-sulfone, and 2-hydroxybiphenyl. The later was subsequently nitrated to produce at least two different hydroxynitrobiphenyls during cultivation. Resting cells of SY1 desulfurized toluenesulfonic acid and released sulfite anion. On the basis of these results, a new DBT degradation pathway was proposed (Figure 4). Figure 5 shows the growth of SY1, the decrease of DBT, and the formation of 2-hydroxybiphenyl and sulfate. DBT was exhausted after 65 h, followed by the stationary growth phase. As SY1 could grow to an OD at 560 nm of 0.8 when enough DBT was supplied, the relatively low OD observed was probably due to the shortage of DBT as a sulfur source. Sulfate release was observed, but in a small amount, indicating that the sulfate released from DBT was effectively

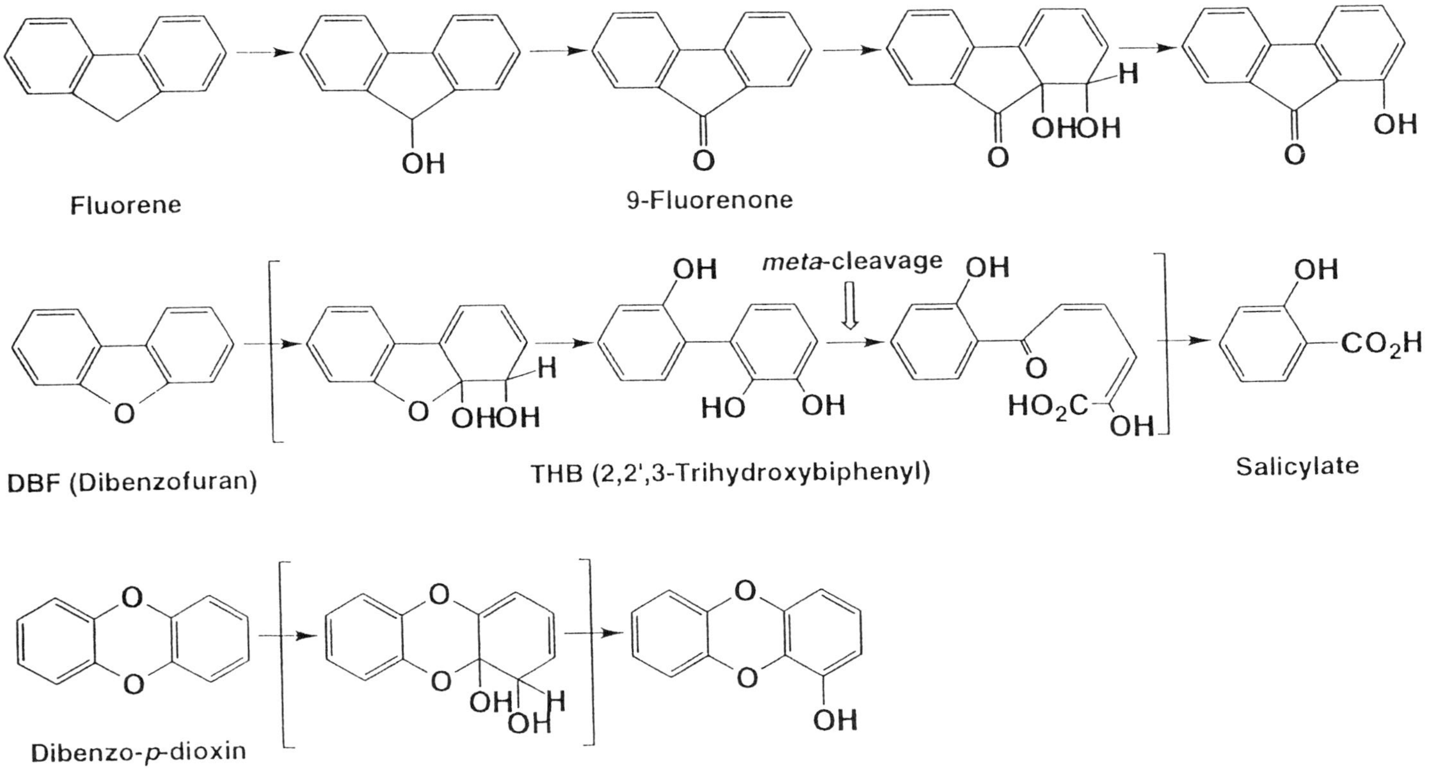

FIGURE 2. Degradation pathways of dibenzofuran and fluorene, and cooxidation of dibenzo-*p*-dioxin by *Staphylococcus auricularis*.

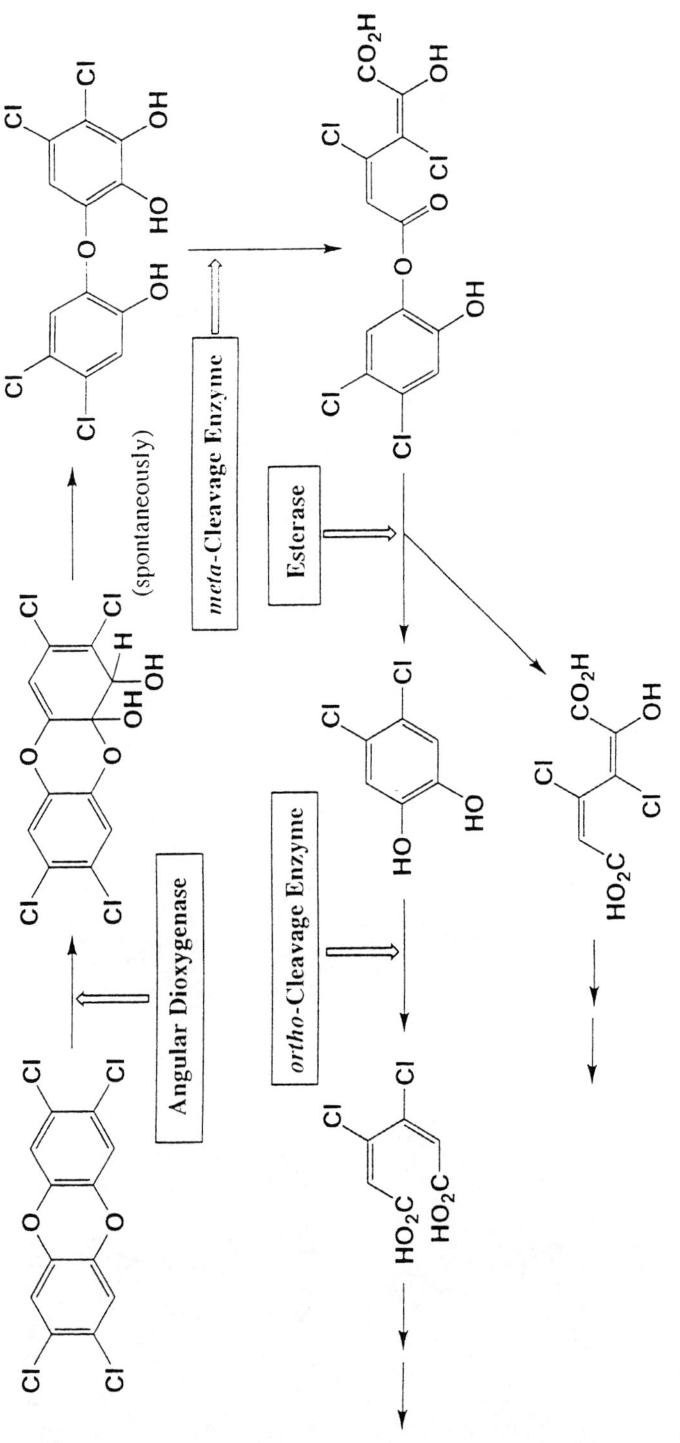

FIGURE 3. Strategy for degradation of tetrachlorodibenzo-*p*-dioxin.

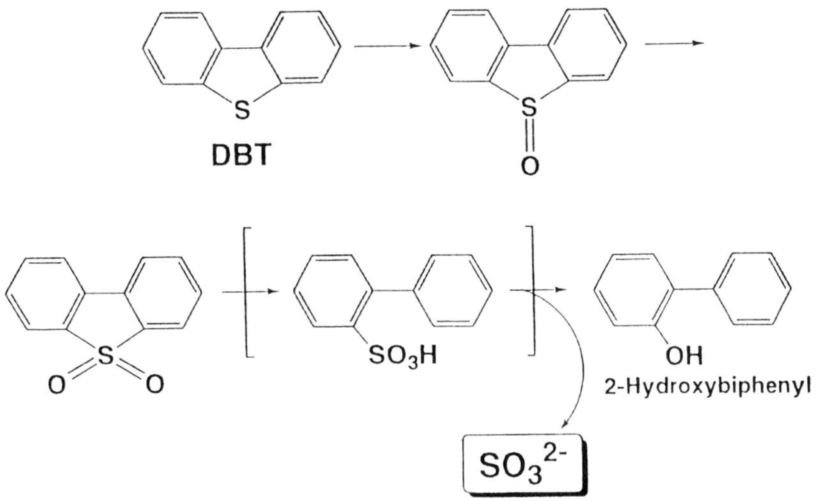

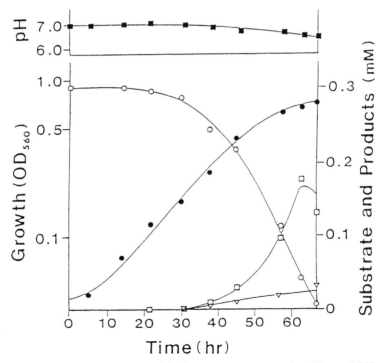

FIGURE 4. Desulfurization pathway of dibenzothiophene by *Rhodococcus* sp.

Growth (●) of *Rhodococcus* sp. strain SY1 on DBT (○) as a sole source of sulfur, formation of 2HBP (□) and sulfate (▽), and pH change during growth (■).

FIGURE 5. Growth curve of *Rhodococcus* sp. strain SY1 grown on dibenzo-thiophene.

utilized by SY1; 2-hydroxybiphenyl was formed at much smaller rate than stoichiometrically estimated from observed decrease of DBT. This may be due to its further conversion of nitrate products and accumulation of DBT-5-oxide and DBT sulfone.

REFERENCES

Monna, L., T. Omori, and T. Kodama. 1993."Microbial Degradation of Dibenzofuran, Fluorene, and Dibenzo-*p*-Dioxin by *Staphylococcus auricularis* DBF63." *Appl. Environ. Microbiol. 59*, 285-289.

Omori, T., L. Monna, Y. Saiki, and T. Kodama. 1 992."Desulfurization of Dibenzothiophene by *Corynebacterium* sp. Strain SY1." *Appl. Environ. Microbiol. 58*, 91 1-915.

Ouchiyama, N., Y. Zhang, T. Omori, T. Kodama. 1993."Biodegradation of Carbazole by *Pseudomonas* sp. CA06 and CA10." *Biosci. Biotech. Biochem. 57*, 455-460.

Effect of Heavy Metals on *ortho*-Nitrophenol Biodegradation by *Pseudomonas aeruginosa* B$_8$

Yana I. Topalova and Krasimira D. Petrova

ABSTRACT —————————————————————————

The *ortho*-nitrophenol (ONP)-degrading strain of *Pseudomonas aeruginosa* B$_8$, treated by the "high selective pressure" method toward high xenobiotic concentrations, shows wide substrate specificity toward phenol, ONP, 2,4-dinitrophenol (2,4-DNP), 2,5-DNP, 2,6-DNP, and pentachlorophenol (PCP) (Topalova et al. 1993). The ONP biodegradational activity of strain B$_8$, and biomass accumulation when ONP in a concentration of 0.29 mMol was the sole carbon and energy source, have been investigated. The influence of different heavy metals (Cu^{2+}, Fe^{2+}, Fe^{3+}, Pb^{2+}, and the combination Fe^{3+}/Pb^{2+}) in a concentration of 40 mg/L on ONP biodegradation, biomass accumulation, and oxygenase activities by whole cells and in cell extracts was studied comparatively. Cu^{2+} and Pb^{2+} inhibited biodegradational activity (32 to 97%), and Fe^{2+} and Fe^{3+} showed a stimulatory effect (30 to 60%), but the combination Fe^{3+}/Pb^{2+} eliminated the Pb^{2+} inhibition.

INTRODUCTION

Nitroaromatic xenobiotics and heavy metals are listed as U.S. Environmental Protection Agency priority pollutants (U.S. EPA, 1979). The biodegradation of nitroaromatics has been investigated by many authors (Zeyer and Kearney 1984, Zeyer et al. 1986, Spain and Gibson 1988, 1991). But there is little information about the biodegradational process in the presence of heavy metals. In the United States, 37% of sites polluted with organic compounds were found also to contain inorganic pollutants, such as heavy metals (Springael et al. 1993).

Heavy metals as waste components potentially affect the rate and mechanisms of the biodegradational processes (Said and Lewis 1991). The effect of heavy metals on nitrophenol biodegradation is of special interest. In this report, we describe the effect of Cu^{2+}, Fe^{2+}, Fe^{3+}, Pb^{2+}, and Fe^{3+}/Pb^{2+} on the ONP degradation by *Pseudomonas aeruginosa* B$_8$ in whole cells and on oxygenase activities in cell extracts.

EXPERIMENTAL PROCEDURES AND MATERIALS

Isolation and Cultivation of Bacteria

The investigated microbial culture was isolated from crude oil-contaminated soil in the region of Sofia (Bulgaria). The bacterial culture was treated by the "high selective pressure" method toward increasing concentrations of phenol, ONP, 2,4-DNP, 2,5-DNP, 2,6-DNP, and PCP, which are used as a sole carbon and energy source (Topalova et al. 1994). The isolate designed as B_8 was identified as *Pseudomonas aeruginosa* B_8 by virtue of its physiological and biochemical characteristics according to *Bergey's Manual* (Murray 1986).

The cultivation was realized in batch culture in flasks (500 cm^3) on a rotary shaker at 220 rpm and a temperature of 28°C. The mineral medium described by Furukawa and Chakrabarty (1983) was modified by addition of 20% v/v Tris-SO$_4$ buffer pH 7.5, reducing the concentration of CaCl$_2$·H$_2$O (0.01 g/L), and eliminating of KH$_2$PO$_4$. *Ortho*-nitrophenol was added in a concentration of 0.29 mMol. The metallic cations (40 mg/L), Cu^{2+} (0.629 mMol), Fe^{2+} (0.716 mMol), Fe^{3+} (0.716 mMol), Pb^{2+} (0.487 mMol), and Fe^{3+}/Pb^{2+} (0.358 mMol + 0.243 mMol) were added as CuSO$_4$·5H$_2$O, FeSO$_4$·7H$_2$O, FeCl$_3$, Pb(CH$_3$COO)$_2$, and FeCl$_3$+ Pb(CH$_3$COO)$_2$, respectively.

Analytical Procedures

The biomass accumulation was measured either by following the increasing Abs 600 nm, or by determination of protein according to Lowry for whole cells after extraction with trichloracetic acid (Herbert et al. 1971). The calibration curve was constructed with defined dilutions of a cell suspension (Hanson and Phillips 1981). ONP in cell suspension was determined by high-performance liquid chromatography (HPLC). The apparatus consisted of a U6K injector, two pumps (model 510), a solvent programmer (model 680), Automatic Gradient Controller (data model 745), and absorbance detector (model 484), operating at 254 nm (all components from Waters Associates Inc., Mifford, Massachusetts). The separations were made on 25 × 4.5 mkm, C18 column (Knauer, Germany), using 70% methanol:30% water in 0.015% trifluoroacetic acid (TFA).

Enzyme Assays

The biomass for determining enzyme activities was harvested in the late log phase (at 120 h). The crude cell extracts were prepared by sonic disruption of the bacteria (Feist and Hegeman 1969) modified as described previously (Topalova et al. 1994). ONP-oxygenase (ONPO) activity was determined after the method of Zeyer et al. (1986) by measuring the decrease in ONP at 410 nm. Catechol 1,2-dioxygenase (C12DO) activity was determined after the method of Nakazawa and Nakazawa (1970). The accumulation of the ring-fission product *cis-cis* muconic acid was measured spectrophotometrically at 260 nm. Catechol 2,3-dioxygenase (C23DO) activity was measured as described by Nozaki (1970). Phenol hydroxylase (PHH) activity was assessed according to Neujahr and Gaal (1973). The reaction was followed up by measuring the decrease of absorption at 340 nm. The enzyme investigation was carried out in a Shimadzu spectrophotometer.

All chemicals used were of analytical grade. The chemicals were supplied by Merck (Germany) or Fluka (Germany).

RESULTS

Effects of Cu^{2+}, Fe^{2+}, Fe^{3+}, Pb^{2+}, and Fe^{3+}/Pb^{2+} on Biomass Accumulation

The results are presented in Figure 1. The biomass accumulation by B_8 without heavy metal ions, equal to 200 mg/L protein (at 48 h) and 810 mg/L protein (at 120 h), was used as a control. Cu^{2+} decreased 97% biomass accumulation at 48 h and 120 h of cultivation. The presence of Fe^{2+} stimulated 60% biomass production at 48 h, but decreased this indicator by 19% at 120 h. Fe^{3+} stimulated both early and late biomass accumulation, respectively, by 29% and 32%. The early biomass accumulation was stimulated (51%) by $Pb(CH_3COO)_2$, but biomass production was inhibited by Pb^{2+} at 120 h. The combination Fe^{3+}/Pb^{2+} slightly influenced biomass production.

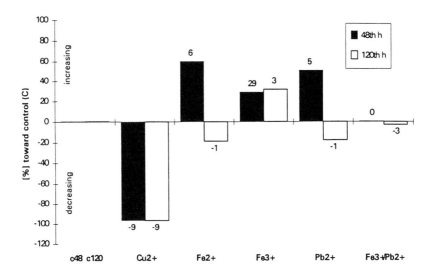

FIGURE 1. Effects of heavy metals on biomass accumulation by *Pseudomonas aeruginosa* B_8 at 48 and 120 h from cultivation; ONP (0.29 mMol) was added as a sole carbon and energy source, C as a control was used biomass accumulation without metal ions (C at 48 h = 200 mg/L protein, C at 120 h = 810 mg/L protein).

Effects of Cu^{2+}, Fe^{2+}, Fe^{3+}, Pb^{2+}, and Fe^{3+}/Pb^{2+} on ONP Biodegradation

The results are presented in Figure 2. The degraded ONP without heavy metals in the medium was used as a control (0.07 mMol at 48 h and 0.21 mMol at 120 h). Cu^{2+} and Pb^{2+} caused a decrease in degraded ONP. Fe^{2+} weakly stimulated only the early ONP biodegradation. The effect of Fe^{3+} on the whole biodegradation process was activating. The combination Fe^{3+}/Pb^{2+} slightly depressed the early ONP biodegradation, but the inhibitory effect was eliminated at 120 h.

Effects of Fe^{3+}, Pb^{2+}, and Fe^{3+}/Pb^{2+} on Oxygenase Activities by Whole Cells

Oxygenase activities when strain B_8 was cultivated in medium with ONP (0.29 mMol) as a sole carbon and energy source and in the presence of metal ions are presented in Figure 3. As a control were used the oxygenase activities in the medium without heavy metals (for ONPO, 0.470 µMol/min·mg; PHH, $7.0 \cdot 10^{-3}$ µMol/min·mg; C12DO, 0.160 µMol/min·mg; C23DO, 0.052 µMol/min· mg protein). Fe^{3+} caused a decrease in ONPO activity by 39%, Pb^{2+} by 7%, but Fe^{3+}/Pb^{2+} increased ONPO activity by 44%. The PHH activity was stimulated by the presence of Fe^{3+}, Pb^{2+}, and Fe^{3+}/Pb^{2+} by 108%, 59%, and 66%, respectively. Fe^{3+} increased C12DO activity by 19%; Pb^{2+} and Fe^{3+}/Pb^{2+} decreased it by 45% and 25%, respectively. C23DO activity was stimulated by Fe^{3+} and Fe^{3+}/Pb^{2+} by 82% and 149%, respectively, but Pb^{2+} inhibited this activity.

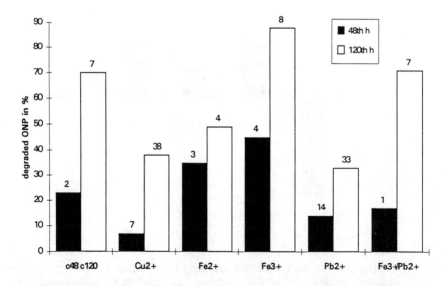

FIGURE 2. Effects of heavy metals on the degraded ONP by *Pseudomonas aeruginosa* B_8; ONP (0.29 mMol) was added as a sole carbon and energy source. The degraded ONP without heavy metals was used as a control (C at 48 h = 0.07 mMol, C at 120 h = 0.21 mMol).

Effects of Fe^{3+}, Pb^{2+}, and Fe^{3+}/Pb^{2+} on Oxygenase Activities in Cell Extracts

Oxygenase activities, at the initiation of reaction in cell extract when Fe^{3+}, Pb^{2+}, or Fe^{3+}/Pb^{2+} was added, were investigated. The results are presented in Figure 4. The oxidative potential in cell extract was slightly influenced by the investigated metal ions. The ONPO activity was not significantly (<3%) affected by addition of Fe^{3+}, Pb^{2+} or Fe^{3+}/Pb^{2+}. The addition of Fe^{3+} stimulated PHH activity 38%, but Pb^{2+} and Fe^{3+}/Pb^{2+} did not influence PHH. Fe^{3+} stimulated C12DO activity by 44% toward the control, Pb^{2+} inhibited C12DO 53%, but a combination of Fe^{3+}/Pb^{2+} had no apparent effect, a 6% inhibition. Pb^{2+} inhibited C23DO activity by 44%, but only 2 to 5% stimulation of this activity was obtained by addition of Fe^{3+} or Fe^{3+}/Pb^{2+}.

DISCUSSION

In this study the effect of different heavy metals on the ONP biodegradation and oxygenase activities by *Pseudomonas aeruginosa* B_8 was examined. In whole cells, the effect of metal ions on the total ONP-biodegradational potential was assessed. Biomass production, biodegraded ONP, and oxygenase activities were used as biological indicators. Cu^{2+} inhibited both biomass accumulation and ONP biodegradation. The inhibitory effect of Cu^{2+} on the oxidative microbial

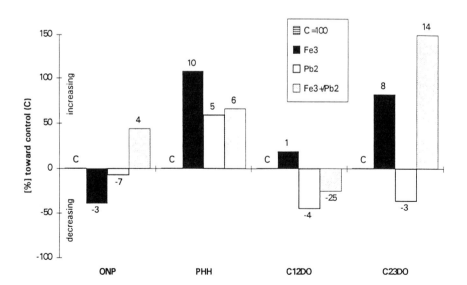

FIGURE 3. Effects of Fe^{3+}, Pb^{2+}, and Fe^{3+}/Pb^{2+} on oxygenase activities by whole cells; as a control (C = 100%) was used the enzyme activity without metal ions for ONPO = 0.470 μkMol/min·mg protein, PHH = $7.0 \cdot 10^{-3}$, C12DO = 0.160, C23DO = 0.052 μkMol/min·mg protein.

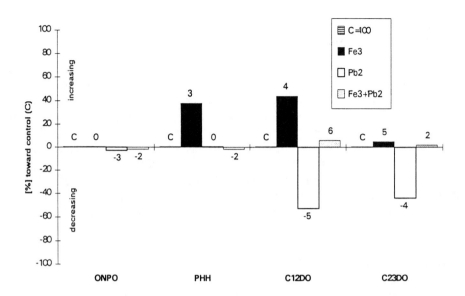

FIGURE 4. Effects of Fe^{3+}, Pb^{2+}, and Fe^{3+}/Pb^{2+} on oxygenase activities in cell extracts; as a control (C = 100%) was used the enzyme activity without metal ions for ONPO = 0.470 µkMol/min·mg protein, PHH = $7.0 \cdot 10^{-3}$, C12DO = 0.160, C23DO = 0.052 µkMol/min·mg protein.

potential was reported by other authors (Neujahr and Gaal 1973, Zeyer et al. 1986, Gurujeyalakshmi and Oriel 1989). In our study, Fe^{2+} stimulated only the early ONP biodegradation (at 48 h) but the later effect (at 120 h) was inhibitory. This fact was in agreement with the literature data about the influence of Fe^{2+} on the PHH and ONPO activities, measured at the late log phase (Zeyer et al. 1986, Gurujeyalakshmi and Oriel 1989). Fe^{3+} had an activatory effect on PHH, C12DO, and C23DO, both in whole cells and in cell extracts and on the total biodegradational process. Only ONPO activity was not stimulated by Fe^{3+} in both systems. It is not farfetched to speculate that Fe^{3+} is potentially able to change some biodegradational mechanisms.

The stimulatory effect of $Pb(CH_3COO)_2$ on early biomass production was based on acetate as an additional carbon and energy source. At same time, Pb^{2+} inhibited ONP biodegradation, which corresponded to different degree (7 to 45%) of decrease in oxygenase activities in whole cells. As an exception, Pb^{2+} considerably activated PHH activity in both in whole cells and in cell extracts. This fact, as well as the fact that Fe^{3+} activated all investigated oxygenase activities, explains why Fe^{3+}/Pb^{2+} eliminated the inhibitory effect of Pb^{2+}. The combination Fe^{3+}/Pb^{2+} and acetate (from $Pb(CH_3COO)_2$) favorably influenced ONP biodegradation and help to overcome the Pb^{2+} inhibition.

These data may be used to regulate nitrophenol biodetoxification by highly specialized microbial cultures in the presence of heavy metal pollution. Besides that, heavy metals play an important ecological role, and their regulating effect on the biodegradational process in situ cannot be ignored. Further experiments will be aimed at expanding and elucidating the above-discussed regulating possibilities.

ACKNOWLEDGMENTS

We thank I. Ivanov for discussing HPLC analysis. This work has been supported by the National Science Fund at the Ministry of Education, Science and Technologies, Bulgaria.

REFERENCES

Feist C. F., and G. D. Hegeman. 1969. "Phenol and Benzoate Metabolism by *Pseudomonas putida*: Regulation of Tangential Pathways." *Journal of Bacteriology* 100: 869-877.

Furukawa, R., and S. J. Chakrabarty. 1983. "Common Induction and Regulation of Biphenol, Xylene, Toluene and Salicylate in *Pseudomonas paucimobilis*." *Journal of Bacteriology* 154: 1356-1363.

Gurujeyalakshmi, G., and P. Oriel. 1989. "Isolation of Phenol-Degrading *Bacillus stearothermophilus* and Partial Characterization of the Phenol Hydroxylase." *Applied and Environmental Microbiology* 55: 500-502.

Hanson, R. S., and J. A. Phillips. 1981. "Chemical Composition." In P. Gerhardt, R.G.E. Murray, R. N. Costilow, E. W. Nester, W. A. Wood, N. R. Krieg, and G. B. Phillips (Eds.), *Manual of Methods for General Bacteriology*, pp.358-359. American Society of Microbiologists, Washington, DC.

Herbert, D., P. J. Phillips, and J. R. Strange. 1971. "Determination of Protein." In J. R. Norris, and D. W. Ribbons (Eds.), *Methods in Microbiology*, Vol. 5B, pp. 243-265. Academic Press, London and New York.

Murray, R.G.E. (Eds.). 1986. *Bergey's Manual of Systematic Bacteriology*. Vol. 2, pp. 140-190. Williams & Wilkins, Baltimore, MD.

Nakazawa, T., and A. Nakazawa. 1970. "Pyrocatechase (*Pseudomonas*)." In H. Tabor and C. W. Tabor (Eds.), *Methods Enzymology*, Vol. 17A, pp. 518-522.

Neujahr, H. Y., and T. Gaal. 1973. "Phenol Hydroxylase from Yeast." *European Journal of Biochemistry* 35: 385-400.

Nozaki, M. 1970. "Metapyrocatechase (*Pseudomonas*)." In H. Tabor and C. W. Tabor (Eds.), *Methods Enzymology*, Vol. 17A, pp. 522-525.

Said, W. A., and D. L. Lewis. 1991. "Quantitative Assessment of the Effects of Metals on Microbial Degradation of Organic Chemicals." *Applied and Environmental Microbiology* 57: 1498-1503.

Spain, J. C., and D. T. Gibson. 1988. "Oxidation of Substituted Phenols by *Pseudomonas putida* F1 and *Pseudomonas sp.* strain JS6." *Applied and Environmental Microbiology* 54: 1399-1404.

Spain, J. C., and D. T. Gibson. 1991. "Pathway for Biodegradation of *p*-Nitrophenol in a *Moraxella sp.*" *Applied and Environmental Microbiology* 57: 812-819.

Springael, D., L. Diels, L. Hooyberghs, S. Kreps, and M. Mergeay. 1993. "Construction and Characterization of Heavy Metal-Resistant Haloaromatic-Degrading *Alcaligenes eutrophus* Strains." *Applied and Environmental Microbiology* 59: 334-339.

Topalova, Y., R. Dimkov, and T. Donev. 1993. "Biodegradation of Nitrophenols by Fresh and Lyophilized Microbial Preparations." In S. Neychev, Ts. Angelova and S. Dimov (Eds.),

Proceedings of Eighth Congress of the Microbiologists in Bulgaria, pp. 255-260. International Centre of Scientists, Black Sea Resort "St. Constantine," Varna, Bulgaria.

Topalova, Y., R. Dimkov, and R. Manolov. 1994. "Influence of the Concentration of Aryl-Containing Xenobiotics on the Activity of Oxygenase Enzymes." *Biotechnology and Biotechnology Equipment 1*: 62-68.

U.S. Environmental Protection Agency. 1979. *Water-Related Environmental Fate of 120 Priority Pollutants*. EPA-440/4-79-029. Washington, DC.

Zeyer, J., and C. Kearney. 1984. "Degradation of *o*-Nitrophenol and *m*-Nitrophenol by *Pseudomonas putida*." *Journal of Agricultural and Food Chemistry 32*: 238-242.

Zeyer, J., H. P. Kocher, and K. N. Timmis. 1986. "Influence of para-Substituents on the Oxidative Metabolism of *o*-Nitrophenols by *Pseudomonas putida B2*." *Applied and Environmental Microbiology 52*: 334-339.

Two-Stage Biotransformation of 2,4,6-Trinitrotoluene

Sarah L. VanderLoop, Makram T. Suidan,
Moustafa A. Moteleb, and Stephen W. Maloney

ABSTRACT

An anaerobic fluidized-bed granular activated carbon (GAC) bioreactor in series with an activated sludge reactor was used to treat 2,4,6-trinitrotoluene (TNT). A wastewater solution of 100 mg/L 2,4,6-TNT, as well as carbonate buffer and nutrient solutions, were fed to the anaerobic reactor. Ethanol and ammonia were supplied for microbial growth at 540 mg/L and 20 mg/L, respectively. The mixed anaerobic culture completely transformed the TNT to unknown products. Attempts to detect these products by gas chromatography (GC) and high-performance liquid chromatography (HPLC) were unsuccessful. The effluent from this reactor was treated further in an activated sludge system. The TNT transformation products were partially oxidized in the aerobic stage, resulting in recovery of 91% of the total influent nitrogen as ammonia, nitrate, and biomass. This corresponds to recovery of 82% of influent TNT nitrogen.

INTRODUCTION

TNT is the explosive most widely used by the military, but current technologies for treating munitions production wastewater are costly and produce a secondary waste, spent GAC. Bioremediation holds promise as a more effective and less costly treatment strategy. Direct oxidation of TNT is difficult due to the electron-withdrawing nitro constituents. Partial reduction of TNT, followed by oxidation, may be the most successful route to mineralization. For example, acidogenic and sulfidogenic isolates have successfully reduced TNT to triaminotoluene (TAT) (Preuss et al. 1993). Strictly anaerobic conditions were required for converting the intermediate diaminonitrotoluene to TAT, but further conversion of TAT with the sulfidogenic isolate was reported under only aerobic conditions. Another sulfate-reducing species of the genus *Desulfovibrio* used TNT as the sole nitrogen source under nitrogen-limiting conditions, reducing it to toluene, which is readily mineralized aerobically (Boopathy and Kulpa 1992).

Although anaerobic transformation of TNT is widely reported, the partially reduced products are rarely mineralized. The rationale behind a two-stage system is, then, to initiate reduction anaerobically in the first stage, these products being more amenable to oxidation or possible mineralization in the aerobic, second stage. This research evaluated the effectiveness of an anaerobic fluidized-bed granular activated carbon (AFBGAC) bioreactor in series with an activated sludge unit in treating 2,4,6-TNT. A similar system was successful in stoichiometric reduction of 2,4-dinitrotoluene to 2,4-diaminotoluene (2,4-DAT) anaerobically, followed by aerobic mineralization of 2,4-DAT (Berchtold et al. 1995). The AFBGAC bioreactor has also been proven effective in treating toxic and inhibitory waste such as coal gasification wastewater and wastewater containing chlorinated hydrocarbons and phenols (Suidan et al. 1983, 1991; Flora et al. 1993). The GAC serves as a superior microbial attachment surface, as well as a buffer against shock loads or buildup of toxic intermediates. In this study, the effluent from the first stage, potentially rich in reduced compounds, is fed to the activated sludge for further treatment.

MATERIALS AND METHODS

The 10-L AFBGAC reactor was charged with 1.0 kg of 16 × 20 U.S. mesh F400 GAC (Calgon Corporation, Pittsburgh, Pennsylvania). The recycle flowrate was controlled to maintain a bed expansion of 30%. TNT, buffer, and nutrient solutions were supplied continuously through the recycle lines, resulting in an unexpanded, empty-bed hydraulic retention time of 0.34 day. The anaerobic reactor was seeded with one bed-volume of effluent from an identical system used to treat 2,4-DNT. A water jacket maintained the temperature at 35°C and the pH was held at 7.2.

The activated sludge unit consisted of a 17.5-L plexiglas tank with an upflow clarifier. It was seeded with mixed liquor from a municipal wastewater treatment plant in Cincinnati, Ohio, and the pH was maintained at 8.0 to optimize for nitrifiers. The sludge age was maintained at 15 days.

The reactors were monitored daily for feed and gas flowrates, temperature, and effluent pH. Mixed liquor volatile suspended solids in the aerobic reactor and in the clarifier were determined weekly. Aqueous samples were withdrawn twice weekly from each reactor to analyze for ammonia, nitrate, and 2,4,6-TNT. The volatile suspended solids were measured according to Standard Methods 209C and D. Ammonia concentrations were determined using a model 13-620-505 ammonia ion selective electrode (Fisher Scientific, Pittsburgh, Pennsylvania) and an Orion model 215284-A01 ATC probe. Standard U.S. EPA method 353.1 was adapted for nitrate/nitrite analysis which involved a copper-catalyzed reduction of the nitrate to nitrite with hydrazine sulfate. A color reagent (H_3PO_4, sulfanilamide, and N-1-naphthylethylenediamine dihydrochloride) was added and its absorbance was measured at 520 nm. This method is described in detail in Wrenn et al. (1994). TNT analyses were performed on a Hewlett Packard series 1050 liquid chromatograph equipped with a variable-wavelength UV detector.

A 15-cm reverse-phase AccuBond ODS II column was used (J&W Scientific, Folsom, California). The TNT was eluted isocratically with a 65% water, 35% methanol solution at 1.5 mL/min and was detected by absorbance at 230 nm.

Efforts to detect TNT transformation products were unsuccessful, although several detection methods were attempted. A Hewlett Packard 5890 gas chromatograph equipped with flame ionization detectors (FID) and a DB-1 capillary column (J&W Scientific, Folsom, California) was used for GC analyses. Nitrogen was the carrier gas and the injection and detector temperatures were maintained at 250°C. The oven temperature for all runs started at 120°C, was increased at 30°C per minute to 200°C, and was maintained at 200°C for an additional 40 minutes. The DB-1 column, used routinely for nitrotoluene, aminotoluene, and aminonitrotoluene analyses, requires injection of organic solutions. Analyte extractions into ether and phenol were attempted. Because extraction efficiency can be pH dependent, as is the case for aminotoluenes, extraction into each solvent was performed at acidic, neutral, and basic pH. If the unknown products were sufficiently polar, very little would enter the organic phase. The resulting flat chromatograms indicate that either this was the case or the analytes were retained irreversibly on the column.

Further attempts to detect the products involved direct aqueous sample injection into a Hewlett Packard series 1050 high-performance liquid chromatograph equipped with a variable-wavelength UV detector. A 15-cm reverse-phase Accu-Bond ODS II column was used (J&W Scientific, Folsom, California). Diode array scans of effluent samples showed slight absorbance at 240 and 284 nm. These detector wavelengths were used in the analyses. Mobile-phase compositions of 60% methanol, 40% water, and 100% methanol were used at each wavelength, at a flowrate of 1.5 mL/min. Standards of several common TNT transformation products, including 2-amino-4,6-dinitrotoluene, 4-amino-2,6-dinitrotoluene, 2,4-diamino-6-nitrotoluene, and 2,4,6-triaminotoluene, were injected onto a C18 HPLC column and identified at 240 nm using these methods, while samples from both reactors showed no peaks throughout the study. If the products were very polar, it is possible that they would be irreversibly retained by any exposed silica in the column. The colored compound was, in fact, isolated on a silica gel column, but it did not elute even with methylene chloride.

Efforts were made to detect azo-dyes or possible diazo products using a Waters HPLC unit equipped with a model 990 photodiode array detector. A mobile-phase gradient from 100% water to 100% methanol was used to elute potential products from a Waters 15 cm Nova-Pak C18 column, although, again, nothing was detected.

RESULTS AND DISCUSSION

The GAC used in the anaerobic reactor was preloaded with 100 g of TNT prior to system startup. Startup of the system allowed a significant period (150 days) for biomass growth and acclimation as well as further sorption of organics onto the GAC. At this point, based on isotherm studies for a variety of compounds, it was judged that the GAC was fully loaded, or more accurately, that loading and bioregeneration of the GAC had reached some equilibrium. TNT did not break through at any point in this study, even after over 500 days of operation.

The TNT was completely transformed to unknown products in the anaerobic reactor. Samples of GAC from the anaerobic reactor, extracted with methanol and methylene chloride, also showed no traces of TNT. A brownish tint to the filtered effluent earlier in system operation indicated possible formation of an aromatic amine or an azo-dye, but attempts to identify this compound were not successful and the color has since dissipated. Efforts to detect the TNT transformation products, as outlined in Materials and Methods, were unsuccessful.

The effluent from the first stage, as well as additional nutrient and buffer solutions, were fed directly to the activated sludge reactor. Nitrogen data indicate that the activated sludge was able to oxidize a significant fraction of the partially reduced TNT products, although mineralization was not complete. The inorganic nitrogen content of the final effluent was measured as ammonia and nitrate. Volatile suspended solids (VSS) in the reactor and in the clarifier were measured to determine the mass of wasted sludge and biomass losses over the weir. A conservative 12% nitrogen content of biomass was used in calculating the biomass contribution to the nitrogen balance.

Weekly averages of total nitrogen values in the system influent and in the effluent from each reactor, tabulated from the sources mentioned above, are plotted against day of operation in Figure 1. Total nitrogen is given in mg/day to account for dilution effects of additional nutrient and buffer solutions to the

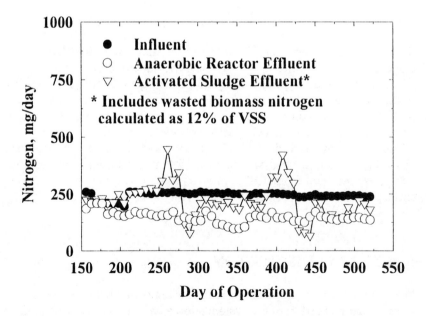

FIGURE 1. Total nitrogen balance on each stage.

second stage and also for minor variations in feed flowrates. Influent nitrogen to the system was composed of approximately half ammonia and half TNT. Only ammonia-N was detected in the effluent from the anaerobic reactor. The remaining nitrogen was bound in the unknown TNT biotransformation products. This nitrogen was partially recovered in the final effluent from the activated sludge reactor, however, as ammonia, nitrate, and wasted biomass. The clarifier failed on two occasions on approximately days 250 and 400 of operation. This led to significant biomass losses over the weir, giving high biomass nitrogen values, followed by periods when no sludge was wasted to allow recovery. The resulting sludge was depleted in nitrifiers and was not as effective in oxidizing the ammonia, leading to temporarily high ammonia and low nitrate values.

Average values for each of the measured nitrogen sources in the system influent and in the effluent from each reactor are given in Table 1 as nitrogen in mg/day. These average values cover a period of almost 400 days of system operation. Data from the initial 150-day acclimation phase of operation is not included in the table. Because growth rates are relatively low for anaerobes, the nitrogen content of anaerobic organisms was neglected. The total measured nitrogen content from the listed sources was summed for the readers' convenience. Large standard deviations in final effluent nitrogen values arose from the aforementioned clarifier failures. It is suspected that some ammonia was lost to stripping during these periods. However, an average of 91% of the total influent nitrogen was recovered in the final effluent from the activated sludge. This corresponds to recovery of 82% of the initial TNT nitrogen.

CONCLUSIONS

The mixed anaerobic culture suspended in the anaerobic fluidized-bed GAC reactor was able to completely transform 110 mg/L 2,4,6-TNT to unknown

TABLE 1. Average contributions to the total nitrogen balance.

Parameter (mg/day)	Influent	Anaerobic Reactor Effluent	Final Effluent
TNT-N	123.6 (15.6)	ND	ND
NH_3-N	121.1 (8.0)	148.7 (27.6)	40.6 (4.0)
NO_3-N	0	ND	104.6 (68.4)
biomass-N	0	NM	77.0 (54.8)
Total-N	244.7 (14.7)	148 (27.6)	222.3 (76.0)

Standard deviations are noted in parentheses.
ND = Not detected.
NM = Not measured.

products. Efforts to identify these products are ongoing. This material was partially mineralized in the second-stage activated sludge reactor, as evidenced by the recovery of 91% of the total influent nitrogen as ammonia, nitrate, and biomass. This corresponds to the release of approximately 82% of the initial TNT nitrogen.

ACKNOWLEDGMENT

This research was supported by the U.S. Army Construction Engineering Research Laboratory, Champaign, Illinois.

REFERENCES

American Public Health Assoc., American Water Works Assoc., and Water Environment Federation. 1992. *Standard Methods for the Evaluation of Water and Wastewater*, 18th ed.

Berchtold, S. R., S. L. VanderLoop, M. T. Suidan, and S. W. Maloney. 1995. "Treatment of 2,4-dinitrotoluene using a two-stage system: Fluidized-bed anaerobic GAC reactors and aerobic activated sludge reactors." Accepted for publication in *Water Env. Res.*

Boopathy, R., and C. F. Kulpa. 1992. "Trinitrotoluene (TNT) as a sole nitrogen source for a sulfate-reducing bacterium *Desulfovibrio* sp. (B strain) isolated from an anaerobic digester." *Current Microbiol.* 25: 235-241.

Flora, J.R.V., M. T. Suidan, A. M. Wuellner, and T. K. Boyer. 1993. "Anaerobic treatment of a simulated high-strength industrial wastewater containing chlorophenols." *Water Env. Res.* 26(1): 21-31.

Preuss, A., J. Fimpel, and G. Diekert. 1993. "Anaerobic transformation of 2,4,6-trinitrotoluene (TNT)." *Arch. Microbiol.* 159: 345-353.

Suidan, M. T., G. L. Siekerka, S. W. Kao, and J. T. Pfeffer. 1983. "Anaerobic filters for the treatment of coal gasification wastewater." *Biotechnol. Bioeng.* 55: 1581.

Suidan, M. T., A. M. Wuellner, and T. K. Boyer. 1991. "Anaerobic treatment of a high strength industrial waste bearing inhibitory concentrations of 1,1,1-trichloroethane." *Water Sci. Technol.* 23: 1385.

Wrenn, B. A., J. R. Haines, A. D. Venosa, M. Kadkhodyan, and M. T. Suidan. 1994. "Effects of nitrogen source on crude oil biodegradation." *J. Ind. Microbiol.* 13: 279-286.

Aerobic Biotransformation and Mineralization of 2,4,6-Trinitrotoluene

Bum-Han Bae, Robin L. Autenrieth, and James S. Bonner

ABSTRACT

Respirometric mineralization studies of 2,4,6-trinitrotoluene (TNT) were conducted with microorganisms isolated from a site contaminated with munitions waste in Illinois. Nine aerobic bacterial species were isolated under a carbon- and nitrogen-limited condition and tentatively identified as: one *Pseudomonas* species; one *Enterobacter* species; and seven *Alcaligenes* species. Experiments were performed using each of the nine organisms individually and with a consortium of all nine bacterial species. The aerobic microorganisms were cultured in a sterile nutrient solution with glucose and 20 mg/L TNT. Mineralization was determined using uniformly ring-labeled ^{14}C-TNT in a respirometer that trapped the evolved CO_2. Biodegradation behavior was characterized based on oxygen consumption, distribution of ^{14}C activity, and high-performance liquid chromatography (HPLC) analysis of TNT and its transformation products. While the microbial consortium did not conclusively mineralize TNT, one isolate, *Enterobacter* sp. (I5), mineralized 3% of the TNT within 10 days of incubation. Although the extent of mineralization was low, it was reproducible. Highly polar intermediates of TNT were formed, but azoxy-compounds were not detected either in the medium or in the acetonitrile extracts of the biomass. The distribution of ^{14}C activity showed 3.2% as evolved CO_2, 13.4% in the biomass, and 80.7% in the liquid phase. The highest TNT transformation rate (first-order reaction coefficient of 6.23/day) was observed with one of the *Alcaligenes* sp. I15, which generated polar compounds. Efforts are currently being focused on optimizing mineralization and understanding consortium dynamics.

INTRODUCTION

TNT is a highly oxidized nitroaromatic compound that is toxic and difficult to biodegrade. It is recalcitrant to biological degradation due to the presence of highly oxidized nitro groups on the aromatic ring that inhibit electrophilic attack by oxygenase enzymes (Bruhn et al. 1987). The nitro groups can be either reduced to amino groups or eliminated from the aromatic ring. Formation of highly reactive aromatic amines subsequently can be polymerized, which is undesirable (Schackmann and Müller 1991). The denitrification reaction, which generally leads to mineralization, is rarely observed for TNT (Kaplan 1990).

Reductive biotransformation of TNT has been reported with gram-positive as well as gram-negative organisms from soil, sediment, and sewage, among others. Aerobically, the formation of azoxy compounds as a result of abiotic coupling reactions of unstable hydroxylamino compounds poses significant restrictions for bioremediation efforts (Carpenter et al. 1978). The elimination of a nitro group from TNT was reported by Naumova et al. (1986). Funk et al. (1993) reported the formation of methyl-phloroglucinol (MPG) and *p*-cresol after a stepwise reduction of nitro groups in TNT with an anaerobic culture. Fungal degradation of TNT has been concentrated on a white-rot fungus, *Phanerochaete chrysosporium* to mineralize TNT with its nonspecific peroxidase system. With this fungal species, 18.4 %, and 19.6% of initial TNT was mineralized in 90 days in soil and liquid cultures, respectively (Fernando et al. 1990). However, significant mineralization of TNT was not observed in other fungal systems (Parrish 1977), activated sludge (Carpenter et al. 1978), or bacterial isolates (Traxler et al. 1975).

The objectives of this study were to determine if TNT would be mineralized by any of nine aerobic isolates and to evaluate conditions for aerobic mineralization. Mineralization of TNT was determined respirometrically for an aerobic bacterial culture, *Enterobacter* sp. I5. Less than 1.0% of ^{14}C activity in CO_2 trap was detected in other cultures. The azoxy compounds were detected in the acetonitrile extracts of suspended solids in the culture medium pH at 7.0 but not at 6.0.

MATERIALS AND METHODS

Respirometry Study

Nine aerobic bacterial species were isolated from a munitions waste-contaminated site (Bae et al. 1994). These species were one *Enterobacter* species, one *Pseudomonas* species, and seven *Alcaligenes* species. After enrichment in Oxoid nutrient broth for 7 days, the biomass from each pure culture reactor was harvested by centrifugation (7,500 g for 15 min) and suspended in 50 mM phosphate buffer. This was repeated three times. The total suspended solids (TSS) concentration of cell suspension was determined before use.

A respirometer equipped with an electronic data acquisition system (N-Con system) was used. In a 500-mL reactor, 100 mL of sterilized nutrient solution was spiked with 85 mg/L NH_4NO_3, 20 mg/L TNT. Glucose (1.0 g/L) was added as a supplemental carbon source throughout the experiments. The internal CO_2 trap containing sodium hydroxide (NaOH) was replenished every 5 days. Oxygen consumption was monitored and pure oxygen was supplied automatically. Incubation temperature was 25°C. The reactors were covered with aluminum foil to prevent photodegradation of TNT. A control reactor with no inoculum was operated under the same conditions every time. At the end of the experiments, the reactor medium was acidified by adding 0.5 mL of concentrated H_2SO_4. To identify intermediates of TNT, batch reactors without ^{14}C TNT were operated separately in a 250-mL Erlenmeyer flask with the same procedure as the respirometer study except for the TNT concentration (30 to 75 mg/L).

Chemicals and Analysis

A symmetrical 2,4,6-TNT (purity 99.0%) was purchased from Chem Service Inc. (West Chester, Pennsylvania). Reported intermediates of TNT were supplied by Dr. Spanggord (SRI International, California). Available standard solutions for HPLC calibration were purchased either from AccuStandard (New Haven, Connecticut) or from Polyscience (Niles, Illinois). Radioactively labeled TNT (^{14}C-U-ring labeled; specific activity 21.58 mCi/mmol; purity >98%) was purchased from Chemsyn Science Laboratories (Lenexa, Kansas). The purity of ^{14}C TNT was reconfirmed prior to the experiment by HPLC analysis.

TNT and its biotransformation products were analyzed using HPLC. The HPLC analytical conditions were reported previously (Bae et al. 1994). The activity of ^{14}C was counted using a Beckman Liquid Scintillator (LS3801) and Ready Gel cocktail (Fullerton, California). Duplicate 1.0-mL samples were collected from the respirometer reactors. After centrifugation for 15 min at 12,800 g, the supernatant was collected for HPLC analysis and LS counting. The biomass pellets were washed with 1.0 mL phosphate buffer and suspended in 1.0 mL deionized water before analysis.

RESULTS

Duplicate mineralization studies were performed for all the nine species at pH 7.0. After 10 days of incubation, no significant amount of ^{14}C activity in CO_2 trap was detected (less than 1.0% as counting per minute). The pH was lowered to 6.0 and the same experiment was performed on 6 of the pure isolates and a consortium of all 9 species. The total ^{14}C activity added in each reactor was $5.4*10^6$ disintegrations per minute (DPM). After 10 days of incubation, significant [14]CO_2 activity (3.2% of the initial ^{14}C activity) was detected in the reactor CO_2 trap inoculated with *Enterobacter* sp. I5 compared to less than the 1.0% of [14]CO_2 activity detected in the other reactors (Table 1). The average count in the NaOH fraction (870 DPM/mL) was significantly higher than the background count (36 DPM). A methylene chloride extraction of the NaOH and a precipitation of $Na^{14}CO_3$ with $BaCl_2$ were performed. Neither of these fractions was significantly higher than the background count. The complete mass balance for the *Enterobacter* sp. I5 reactor is presented in Table 2. The dissolved oxygen concentration in the reactors was 6.2 ± 0.5 mg/L at the end of experiments. HPLC analysis of 10-day-old supernatant of the sp. I5 reactor showed that most of the intermediates were polar, whereas nonpolar intermediates dominated in the other reactors.

TABLE 1. ^{14}C activity distribution after 10 days of incubation at pH 6.0.

Microbial Species	Initial Biomass (mg/L)	As $^{14}CO_2$	In Biomass	In Solution	Total
Enterobacter sp. I5	284	3.6	13.7	83.3	100.6
Alcaligenes sp. I6	452	0.8	4.6	91.2	96.6
Alcaligenes sp. I10	420	0.9	13.6	87.3	101.8
Alcaligenes sp. I15	562	1.0	27.0	63.2	91.2
Alcaligenes sp. I19	380	0.9	6.9	91.2	99.0
Alcaligenes sp. I22	525	0.8	6.7	92.0	99.5
Mix of nine	516	1.0	5.3	92.3	98.6
Control	—	0.7	—	101.8	102.5

Degradation of TNT by *Enterobacter* sp. occurred rapidly with the increase in the intermediates: 2-amino-4,6-dinitrotoluene (2amDNT), 2-amino-4,6-dinitrotoluene (4amDNT), and 4-hydroxylamino-dinitrotoluene (4OH) (Figure 1). The concentration of 4amDNT (11.8 mg/L) was much higher than that of 2amDNT (2.0 mg/L). More than 85% of the initial TNT (75 mg/L) was transformed within 5 days, and the quantities of diaminonitrotoluenes were negligible (less than 0.5 mg/L) after 5 days. However, the area of unknown peaks (around 3.61 min of peak retention time) increased more than twice. Also, small peaks at the elution times of 2,4- and 2,6-dinitrotoluene (DNTs), and 2- and 4-nitrotoluene (MNTs) were detected consistently (less than 1.0 mg/L as DNTs and MNTs) in replicate experiments. The peak area of 4OH, a precursor of azoxy compounds by an abiotic coupling reaction, remained high despite the aerobic culture condition.

One *Alcaligenes* sp. (I15) was the fastest TNT transformer among the nine isolated bacterial species. Batch experiments were performed with this species

TABLE 2. Mass balance for 2,4,6-trinitrotoluene in an *Enterobacter* sp. I5 reactor.

Phase	% Activity
As $^{14}CO_2$	3.2 ± 0.1% as DPM
In biomass	13.4 ± 0.6% as DPM
In solution	80.7 ± 4.2% as DPM
Total Recovery	97.3% as DPM

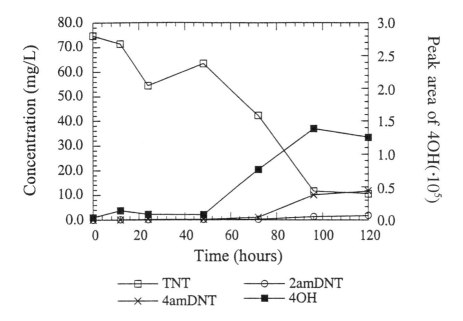

FIGURE 1. Transformation of TNT by *Enterobacter* sp. I5 at pH 6.0.

at pH 6.0 and 7.0. The concentrations of TNT and liquid-phase intermediates and is presented in Figure 2. The rates of TNT removal at either pH were comparable. The concentrations of major reduction products were negligible except for the concentration of 4amDNT, which was 1.8 mg/L at 48 hours. However, a large amount of 4OH was generated at pH 6.0 than at pH 7.0. Analysis of acetonitrile extracts of the solid phase of both reactors revealed azoxy compounds at pH 7.0, but not at 6.0.

DISCUSSION

The major obstacles observed in the biodegradation of TNT were the formation of azoxy compounds as a result of abiotic coupling reactions and the cleavage of the benzene ring. Funk et al. (1993) reported that TNT intermediates were polymerized at an alkaline pH, regardless of the presence of oxygen. In their culture, the reduction of pH from 7.0 to 6.0 significantly inhibited removal of TNT, but the increase of pH from 7.0 to 8.0 slightly improved degradation of TNT. They suggested an optimum pH range of 6.5 to 7.0 for the removal of TNT. In our study, however, TNT removal was not affected by the reduction of pH. Analysis of intermediates revealed that pH reduction stabilized reactive hydroxyl-amino compounds, which otherwise polymerize to azoxy compounds and concentrate in the biomass. Transformation of hydroxylamino compounds is a binary reaction. These compounds can be either biologically reduced to amino

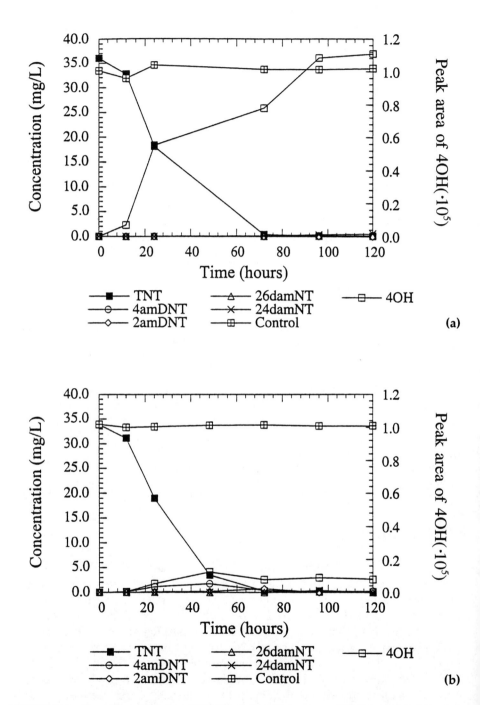

FIGURE 2. Transformation of TNT by *Alcaligenes* sp. I15 (a) at pH 6.0, and (b) at pH 7.0.

congeners or chemically polymerized to azoxy compounds. If the polymerization reaction can be made less favorable by changing the culture conditions, the amount of 4OH transformation into amino congeners will be increased. Because the reduction step is an energy-demanding reaction (McCormick et al. 1976), the rate can be increased by supplying an additional energy source such as glucose.

The presence of peaks at the elution time of DNTs and MNTs is interesting. If these compounds are truly MNTs and DNTs, the peaks may suggest (1) the reduction of nitro groups followed by deamination, and/or, (2) the oxygenolytic liberation of a nitro group and corresponding formation of phenol (Kaplan 1990). In a separate experiment with different nitrogen sources (data not shown), we observed that the presence of nitrite in the liquid medium significantly inhibited transformation of TNT, whereas TNT was transformed rapidly in the presence of nitrogen as NH_4^+ or in the absence of any nitrogen source. Therefore, the TNT might serve as a nitrogen source in the form of nitrate for *Enterobacter* sp. I5 because nitrite can be quickly oxidized into nitrate in aerobic conditions.

The results provide evidence for benzene ring cleavage of TNT intermediates with an aerobic bacterial culture leading to eventual mineralization. The formation of polar compounds other than diamino congeners also was observed. Approximately 13% of the ^{14}C activity was detected in the biomass, but the identification of the labeled compounds was not determined. The obvious question arises whether these are adsorbed intermediates or this carbon has been assimilated by these organisms. However, the majority of ^{14}C activity remained in the liquid phase. Considering the low initial TSS for I5 (284 mg/L), optimization of mineralization should be further investigated. Currently, fractionation of ^{14}C activity in the biomass, identification of intermediates, and reaction dynamics during treatment of TNT are under investigation.

ACKNOWLEDGMENTS

The authors would like to thank Dr. R. Spanggord at SRI International (California) for providing standard compounds of TNT intermediates. Also, the assistance of Muruganandam Sivaraju is appreciated. This research was supported by the U.S. Army Research Office URI Grant No. DAAL03-92-G-0171.

REFERENCES

Bae, B.-H., R. L. Autenrieth, and J. S. Bonner. 1994. "Analysis and Separation of TNT and its Biotic and Abiotic Transformation Products." In: D. W. Tedder (Ed.), *Emerging Technologies in Hazardous Waste Management VI, Proceedings of American Chemical Society Symposium*. Atlanta, GA. pp. 1157-1160.
Bruhn, C., L. Lenke, and H. J. Knackmuss. 1987. "Nitrosubstituted Aromatic Compounds as Nitrogen Source for Bacteria." *Appl. Environ. Microbiol.* 53: 208-210.

Carpenter, D. F., N. G. McCormick, J. H. Cornell, and A. M. Kaplan. 1978. "Microbial Transformation of 14C-Labeled 2,4,6-Trinitrotoluene in an Activated-Sludge System." *Appl. Environ. Microbiol.* 35: 949-954.

Fernando, T., J. A. Bumpus, and S. D. Aust. 1990. "Biodegradation of TNT (2,4,6-trinitrotoluene) by *Phanerochaete chrysosporium*." *Appl. Environ. Microbiol.* 56: 1666-1671.

Funk, S. B., D. J. Roberts, D. L. Crawford, and R. L. Crawford. 1993. "Initial-Phase Optimization for Bioremediation of Munition Compound-Contaminated Soils." *Appl. Environ. Microbiol.* 59: 2171-2177.

Kaplan, D. L. 1990. "Biotransformation Pathways of Hazardous Energetic Organo-Nitro Compounds." In: D. Kamely, A. Chakrabarty, and G. Omenn (Eds.), *Biotechnology and Biodegradation*. Gulf Publishing Co., Houston, TX. pp. 155-182.

McCormick, N. G., F. E. Feeherry, and H. S. Levinson. 1976. "Microbial Transformation of 2,4,6-Trinitrotoluene and Other Nitroaromatic Compounds." *Appl. Environ. Microbiol.* 31: 949-958.

Naumova, R. P., S. Selivanovskaya, and F. A. Mingatina. 1988. "The Possibility of 2,4,6-Trinitrotoluene Deep Destruction by Bacteria." *Mikrobiologiia (USSR)* 157:222.

Parrish, F. W. 1977. "Fungal Transformation of 2,4-Dinitrotoluene and 2,4,6-Trinitrotoluene." *Appl. and Environ. Microbiol.* 34: 232-233.

Schackmann, A., and R. Müller. 1991. "Reduction of Nitroaromatic Compounds by Different *Pseudomonas* species under Aerobic Conditions." *Appl. Microbiol. Biotechnol.* 34: 809-813.

Traxler, R. W., E. Wood, and J. M. Delaney. 1975. "Bacterial Degradation of Alpha-TNT." *Dev. Ind. Microbiol.* 6: 71-76.

Biotransformation of Trinitrotoluene (TNT) by *Streptomyces* Species

Stephen B. Funk, Maria B. Pasti-Grigsby,
Elise C. Felicione, and Don L. Crawford

ABSTRACT

Composting has been proposed as one process for use in the bioremediation of 2,4,6 trinitrotoluene (TNT)-contaminated soils. However, the biotransformations of TNT that occur during composting, and the specific compost microorganisms involved in TNT metabolism, are not well understood. Both mesophilic and thermophilic actinomycetes are important participants in the biodegradation of organic matter, and possibly TNT, in composts. Here we report on the biotransformation of TNT by *Streptomyces* species growing aerobically in a liquid medium supplemented with 10 to 100 mg/L of TNT. *Streptomyces* spp. are able to completely remove TNT from the culture medium within 24 hours. As has been observed with other bacteria, these streptomycetes transform TNT first by reducing the 4-nitro and 2-nitro groups to the corresponding amino group; reducing TNT first to 4-amino-2,6-dinitrotoluene and then 2,4-diamino-6-nitrotoluene. These intermediates are transitory and are themselves removed from the medium within 7 days.

INTRODUCTION

TNT represents a major environmental hazard because it contaminates soil and groundwater throughout the United States and in other countries (Kaplan 1992). TNT and its metabolites have been shown to be toxic as well as mutagenic (Won et al. 1976; Tan et al. 1992). Biotransformation of TNT has been reported in many studies, under both anaerobic and aerobic conditions (for a review see Crawford 1995; Kulpa and Boopathy 1995; Michels and Gottschalk 1995; Preuss and Rieger 1995; Stahl and Aust 1995; Spain 1995). The initial step of transformation, either anaerobic or aerobic, seems to be reduction of a nitro side group. TNT is transformed to aromatic amines, leading to the formation of diaminonitrotoluene. Reduction of the third nitro group, forming triaminotoluene, has normally been observed only under strictly anaerobic conditions (Preuss et al. 1993). Furthermore, under aerobic conditions azoxy compounds or polymers can form, which are a condensation of hydroxylaminotoluene with nitrosotoluene (Michels and Gottschalk 1994; Bumpus and Tatarko 1994).

The current method for remediating TNT from soil is incineration, an expensive technology costing $\geq \$800/ton$ (Funk et al. 1992). Composting explosives has been proposed as a bioremediation technology. Kaplan and Kaplan (1982) showed transformation of ^{14}C-TNT under simulated composting conditions: no $^{14}CO_2$ was detected, but a significant percentage of the ^{14}C-labeled products was bound to the humic fraction. Most of the labeled carbon was found in solvent extracts from the compost and identified as mono- and diaminonitrotoluene. Isbister et al. (1984) showed a rapid decrease in TNT concentration (initial 10,000 mg/kg), but there was no evidence of ring cleavage and only minor amounts of the normal TNT reduction products were found in the compost extract. The ^{14}C label was associated with the residual humus-like compost material, indicating binding or polymerization of the biologically transformed TNT intermediates into insoluble macromolecules.

The initial TNT concentration seems to affect the mineralization rate by the white-rot fungus *Phanerochaete chrysosporium*. Fernando et al. (1990) reported 35% mineralization of the added ^{14}C-TNT (6 µM initial concentration) by the lignin-degrading fungus in 18 days. At initial concentrations of 44 µM, only 20% of the added ^{14}C-TNT was recovered as $^{14}CO_2$ after 90 days. Spiker et al. (1992) recovered only 10% of the counts as $^{14}CO_2$ from the initial 22 µM ^{14}C-TNT using the same fungus. At 66 µM (a typical concentration found at contaminated sites), TNT mineralization by *P. chrysosporium* was completely inhibited.

Bioremediation of nitroaromatic compound-contaminated soils has been demonstrated by using anaerobic soil slurry systems (Kaake et al. 1992, Funk et al. 1994), as well as aerobic composting (U.S. Army 1991, Kaplan & Kaplan 1982). While *Clostridium* species appear to play an important role in nitroaromatic compound degradation using anaerobic systems (Regan & Crawford 1994), little is known about the type of compost microorganisms involved during composting. *Streptomyces* are actinomycetes known to be important decomposers in composts. *Streptomyces* spp. degrade some xenobiotic compounds; for example, they have been used to decolorize azo dyes (Pasti-Grigsby et al. 1992). Because actinomycetes are common in soils and composts, we have examined a number of these organisms for their ability to tolerate and transform TNT.

Streptomyces chromofuscus A11 (ATCC 55184), also described elsewhere (Pasti & Crawford 1991), was used in optimization experiments to study TNT degradation in liquid culture flasks, as well as in a 10-L glass jar fermentation vessel. This organism was also used as an inoculum in a bench-scale composting experiment which biotransformed and mineralized ^{14}C-TNT. In addition, a number of *Streptomyces* species that were isolated from TNT-contaminated soil were found to proliferate well on solid media containing TNT. These isolates were also tested in liquid media supplemented with TNT, to assess their TNT biotransformation capabilities. This is the first report of TNT biotransformation by streptomycetes.

MATERIALS AND METHODS

Microorganisms

Streptomyces spp. strains A2 through A20 were isolated from the digestive tract of Kenyan termites (Pasti & Belli 1985). *Streptomyces* spp. strains ESA 1 through ESA 16 were isolated in our laboratory from soil contaminated with TNT from an ordnance works site near Weldon Spring, Missouri (Pasti-Grigsby, unpublished). *Streptomyces* sp. SR10, lignocellulose-decomposing *Streptomyces badius* 252 (ATCC 39117), and *Streptomyces viridosporus* T7A (ATCC 39115) were all strains isolated previously in our laboratory (Pasti & Crawford 1991).

Substrates

TNT and ^{14}C U-ring-labeled TNT (0.2595 μCi/mg) were synthesized in our laboratory by Dr. Stefan Goszczynski, University of Idaho, Moscow; 4-amino-2,6-dinitrotoluene (4A26DNT) and 2,4-diamino-6-nitrotoluene (24DA6NT) were obtained from Dr. Ronald Spanggord (SRI International, Menlo Park, California). The wood chip/horse manure compost was provided by Dr. Ronald Crawford (University of Idaho, Moscow).

Culture Conditions

Solid Media. To test their tolerance to TNT, the actinomycetes were grown on plates containing 25, 50, 75, and 100 mg/L of TNT. Two media were tested: BHI (Bacto Brain Heart Infusion; 15 g/L agar, Difco Laboratories, Detroit, Michigan) and YMA (per liter of deionized H_2O: 4 g yeast extract, Difco; 10 g malt extract, Difco; 5 g bacto peptone, Difco; 10 g glucose; 15 g agar, Difco). In both media, the pH was adjusted to 7.00. A TNT stock solution (10 mg/mL dissolved in methanol) was added to the media after autoclaving. Plates were inoculated and incubated at 37°C for 1 week.

Shake-Flask Culture. A liquid medium optimization experiment for TNT transformation was carried out using *S. chromofuscus* A11. The final TNT concentration was 50 mg/L. Medium 1 contained: 2.5 g of yeast extract, 1 g each of asparagine, glutamate and proline, 5 g of glucose in a nitrogen-free mineral salts solution (5.03 g of Na_2HPO_4, 1.98 g of KH_2PO_4, 0.20 g of $MgSO_4 7H_2O$, 0.2 g of NaCl, 0.05 g of $CaCl_2 2H_2O$, plus 1 mL of trace elements solution [Pridhman and Gottlieb 1948] per liter of deionized water). Medium 2 contained: 1 g of NH_4NO_3, 0.4 g of KH_2PO_4, 0.2 g of lactose, 0.2 g of yeast extract, and 0.67 g of yeast nitrogen base per liter of deionized water. Medium 3 contained: 3 g of yeast extract, 3 g of malt extract, 5 g of bacto tryptone, 10 g of glucose and 2.68 g of NH_4Cl. Based on the findings of this test, a follow-up experiment was performed using all the isolates that tolerated ≥50 mg/L TNT on YMA plates. The actinomycetes were grown in 125-mL cotton plugged flasks containing 50 mL of medium 1 at 37°C, with shaking at 300 rpm. Uninoculated controls were incubated as well. The TNT was added (final concentration 50 mg/L) after autoclaving during the inoculation step. In a similar experiment, TNT was added (10 to 100 mg/L) to the flask after *S. chromofuscus* A11 was grown overnight.

Fermentor Culture. Batch cultivation was carried out in a 10-L jar fermentor (Model BIOFLO III, New Brunswick Scientific, Edison, New Jersey) with a total working volume of 8 L. The agitation speed was 700 rpm. The temperature was maintained at 37°C. Foaming was manually controlled with Antifoam A Emulsion (Sigma Chemical Company, St. Louis, Missouri), while the pH was uncontrolled. Medium 1 was used to batch-cultivate *S. chromofuscus* A11. An 18-hour-old, 800-mL culture grown in medium 1 was used as the inoculum for the fermentor. After 24 hours the culture broth was gently drained, leaving a heavy mycelial mat covering a polyurethane mesh, which was wrapped around the inside wall of the fermentation vessel. Fresh sterile medium 1 (8 L) containing TNT (final concentration 50 mg/L) was used to refill the jar. After 24 hours incubation, another equal dose of TNT was added.

Bench-Scale Composting. Composting processes were carried out with both nonsterilized compost and sterilized (methylene chloride treated, dried, and autoclaved two times for 90 min) compost inoculated with 0.5 mL of a 10% glycerol spore suspension (equivalent to 2.65×10^9 CFU) of *S. chromofuscus* A11. Each treatment was performed in triplicate. In addition, one uninoculated sterile compost-containing flask was sealed and stored at 4°C. Ten grams of compost material were placed in 250-mL rubber-stopped flasks, and ^{14}C-TNT was added to a final concentration of 175 mg/kg dry weight (1.01×10^6 dpm). The compost was 60% saturated with 20 mL of 0.1M phosphate buffer (pH 7.0). The flasks were set in a cabinet at 35°C and flushed from 2 to 4 hours a day with a slow stream of filtered (Bacterial Air Vent Filter, Gelman Sciences, Ann Arbor, Michigan) CO_2-free, humidified air. The air was stripped of CO_2 and other volatiles and humidified by sparging through 5 consecutive flasks. The first flask contained 2 L of water; the second and third, 2 L of 1N NaOH; the fourth, 2 L of H_2SO_4; and the fifth, 2 L of pre-warmed water at 35°C. The flasks containing compost were agitated manually twice a week. The airstream exiting each flask and was sparged first through two 1N NaOH traps (5 mL each, for trapping CO_2) followed by one 1N H_2SO_4 trap (5 mL, for trapping volatile amines).

Sampling and Sample Preparation

Shake-Flask Culture. Samples (0.5 to 1 mL) were taken periodically over the 5-day growth period. The supernatant samples were centrifuged (10,000 G for 10 min) and 0.45-mL aliquots were treated with trichloroacetic acid (1 M TCA final concentration) and incubated on ice for 30 min before being centrifuged again: a 0.4-mL aliquot of the treated sample with TCA was then diluted with 0.2 mL of 30% acetonitrile and stored at 4°C until analyzed.

Fermentor Culture. Samples (5 mL) were taken periodically over the 24-hour incubation period. The supernatant samples were treated as described above for the shake-flask culture. After 3 days the test was stopped, the jar was drained, and the supernatant was separated from the mycelium and concentrated by a spiral-wound hollow fiber system (Amicon, Beverly, Massachusetts). Both fractions (supernatant and mycelium) were kept frozen until they were analyzed. A portion of the cellular fraction was centrifuged (5,000 G for 15 min), and the 50-mL pellet was resuspended in 100 mL of deionized water. This suspension was extracted with 50 mL of methylene chloride. The organic fraction was completely evaporated by using a rotary evaporator, and the residue was resuspended in 5 mL of acetonitrile. Aliquots were diluted ten-fold with 11mM phosphate buffer (pH 4) and kept refrigerated until analysis.

Bench-Scale Composting. All traps (NaOH and H_2SO_4) were replaced every 7 to 10 days. At the end of the experiment, the traps were replaced and 10 mL of 1N HCl was added to the compost. The compost was aerated for an additional 1 h. Then, 2.5-mL aliquots of the trapping solution were diluted with 2 mL of water and mixed with 15 mL of Ready Safe (Beckman Instruments Inc., Fullerton, California) scintillation cocktail. After 6 weeks the compost material was extracted with 75 mL of acetonitrile (shaking at 37°C overnight). The extract was filtered through 3MM (Whatman, Maidstone, England) filter paper and evaporated down to 20 to 30 mL. A sample (2.5% volume) was mixed with 15 mL Ready Safe scintillation cocktail. An additional 1-mL aliquot was filtered through a 0.45-mm-pore-size Acrodisc filter (Gelman Sciences, Ann Arbor, Michigan) and 0.2 mL was diluted with 0.8 mL of 20% acetonitrile and placed in a glass vial. The extracted compost and filter paper were air dried, and a 10% sample (wt/wt; 0.7 to 1.0 g) was treated with 10 mL of benzethonium hydroxide (1M in methanol; Sigma Chemical Co., St. Louis, Missouri) at 55°C overnight. Then, 0.25 mL was added to 15 mL Ready Safe scintillation cocktail.

Analysis

Culture supernatants and the extractions were analyzed using high-performance liquid chromatography (HPLC, Hewlett-Packard model 1090 Series II). TNT and the reduced intermediates, 4-amino-2,6-dinitrotoluene (4A26DNT) and 2,4-diamino-6-nitrotoluene (24DA6NT), were monitored by a modified version of a previously published method (Kaake et al. 1992), which uses a C18 reverse-phase column and an 11mM phosphate buffer/acetonitrile (10:90) gradient solvent system. Analytical standards for TNT, 4A26DNT, and 24DA6NT were used to quantitate samples.

A liquid scintillation counter (Beckman Instruments model LS-7000) was used to determine the ^{14}C content of the compost samples. Sample counts per minute (cpm) values were converted to disintegrations per minute (dpm) using an equation derived from a set of ^{14}C quenched standards. Acetonitrile extractions were also run on a preparative column and fractions were collected and counted.

RESULTS

Cultures were initially grown on solid media to test their tolerance to TNT. All actinomycete isolates were able to grow on BHI plates supplemented with up to 100 mg/L TNT. Isolates grown on YMA plates exhibited a variety of growth densities on varying concentrations of TNT (Table 1). The isolates that were able to tolerate at least 50 mg/L TNT on the YMA plates were selected for liquid culture experiments. Shake flask cultures of *S. chromofuscus* A11 grew and biotransformed TNT the most efficiently in the well-balanced medium 1. No identifiable intermediates were detected. Liquid medium 2 gave the poorest results.

Previously grown cultures of *S. chromofuscus* A11 were able to transform TNT (10 to 100 mg/L) in 24 h. Table 2 shows the extent of binding of TNT and partially reduced metabolites to *S. chromofuscus* cells grown in liquid media and the percent recoveries. Cultures receiving 10 to 50 mg/L of TNT did not have any identifiable breakdown products in the supernatant or bound to cells after 13 days of incubation. However, cultures that received 60 to 100 mg/L of TNT-bound TNT and/or reduced intermediates that were extractable with methylene chloride. When cultures were inoculated in the presence of 50 mg of TNT per liter, growth was slowed and biotransformation rates decreased. *S. chromofuscus* A11 took 72 hours to grow and to subsequently remove TNT and the two reduced intermediates from the supernatant. No cellular-bound compounds were extractable after 100 hours of incubation. Only four of the other actinomycetes examined, ESA 1, 2, 5, and 12, were able to flourish and transform TNT to reduced or unknown intermediates in the supernatant after 100 hours.

S. chromofuscus A11 was able to completely remove TNT from the fermentor supernatant in less than 1 hour. Figure 1 shows the time course analysis of contaminant concentration in the 8-L fermentation vessel. At 24 hours an additional dose of TNT (50 mg/L final concentration) was added, and at 72 hours the supernatant and cellular fraction were analyzed, revealing no identifiable compounds in the supernatant. However, the extracted cells were found to have 8.2 and 8.1 mg of 4A26DNT and 24DA6NT, respectively, which calculates to 1.6% and 1.8% of the parent compound recovered as 4A26DNT and 24DA6NT, respectively. It was also observed that the pH, which was not controlled, rose from neutral to pH 9.0 over the 3-day incubation after the TNT was added.

For the composting experiment, the mineralization of ^{14}C-TNT was calculated by the amount of ^{14}C captured in the culture traps. Figure 2 shows a time course analysis of the average daily amount of mineralization for sterilized and non-sterilized compost inoculated with *S. chromofuscus* A11. The total amount of ^{14}C trapped (e.g., as CO_2 and volatile organics) for the sterile and nonsterile compost treatment was 1,600 ±770 and 38,000 ±7,800 dpm, respectively. The remaining labeled carbon was associated with the acetonitrile extract and the compost material itself. Table 3 shows where the label was found and the percent of label recovered. The refrigerated sterile compost control was treated the same

TABLE 1. Tolerance to TNT by various selected *Streptomyces* spp. grown on yeast extract/malt extract agar. +++, very dense growth; ++, dense growth; +, moderate growth; -, minimal growth; 0, no growth.

Streptomyces spp.	TNT-Supplemented YMA Plates				
	0 mg/L	25 mg/L	50 mg/L	75 mg/L	100 mg/L
ESA 1	+++	+++	++	+	+
ESA 2	+++	+	+	0	0
ESA 3	+++	++	+	0	0
ESA 4	+++	+	0	0	0
ESA 5	+++	+++	+	+	0
ESA 6	+++	+++	0	0	0
ESA 7	+++	++	0	0	0
ESA 8	+++	+	0	0	0
ESA 9	+++	+++	+	0	0
ESA 10	+++	+++	0	0	0
ESA 11	+++	+++	++	0	0
ESA 12	+++	+++	++	+	+
ESA 13	+++	++	+++	0	0
ESA 14	0	0	0	0	0
ESA 15	+++	+	0	0	0
ESA 16	0	0	0	0	0
A2	+++	+	0	0	0
A3	+++	+	0	0	0
A4	+++	+	0	0	0
A6	+++	++	–	–	–
A7	+++	++	0	0	0
A8	+++	++	0	0	0
A10	0	0	0	0	0
A11	+++	+++	++	–	0
A12	+++	+++	++	0	0
A13	+++	+++	++	–	–
A14	+++	+++	++	0	0
A20	0	0	0	0	0
T7A	+++	++	0	0	0
SR10	+++	++	0	0	0
S.252	+++	++	0	0	0

TABLE 2. Recovery of TNT, 4A26DNT, 24DA6NT bound to *Streptomyces chromofuscus* A11 cells grown in a liquid medium and sacrificed after 13 days incubation.

Recovery (μM)	Initial TNT Concentration (μM)					
	220 (50 mg/L)	264 (60 mg/L)	308 (70 mg/L)	352 (80 mg/L)	396 (90 mg/L)	440 (100 mg/L)
as TNT	0	0	0	0	204	227
as 4A26DNT	0	82	11	148	45	137
as 24DA6NT	0	66	237	191	89	58
Total	0	148 (56%)	248 (81%)	339 (96%)	338 (85%)	422 (96%)

as the experimental cultures, but without air sparging or trapping. The aceto-nitrile extracts were analyzed by HPLC, and no identifiable compounds for either treatment were detected. The control did, however, have detectable amounts of TNT in the extract (63% of the total added ^{14}C-TNT was recovered as ^{14}C-TNT).

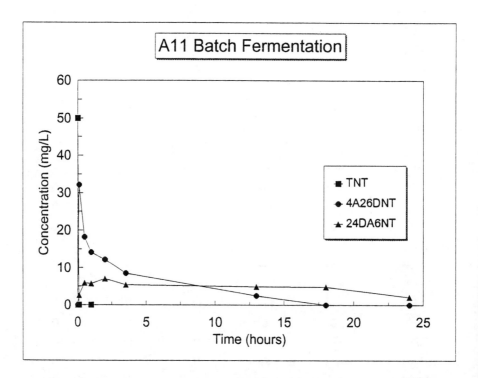

FIGURE 1. Time course analysis of TNT degradation using *Streptomyces chromofuscus* A11 during an 8-L batch fermentation.

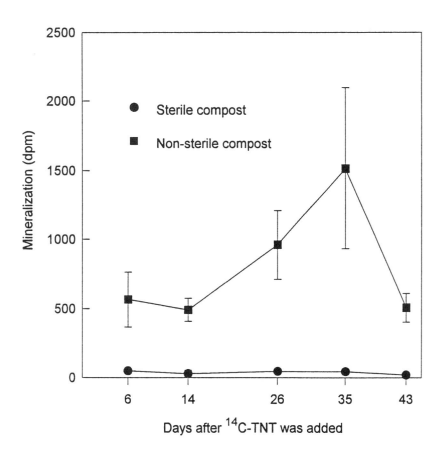

Average daily mineralization

y-axis: Mineralization (dpm)

legend:
● Sterile compost
■ Non-sterile compost

x-axis: Days after ^{14}C-TNT was added

FIGURE 2. Time course analysis of the average daily amount of ^{14}C-CO$_2$ captured in a trapping train (dpm ± s.d.), indicating mineralization.

The label associated with extractions of both sterile and nonsterile compost plus inoculum was more polar than any of the known breakdown products of TNT because it eluted from the reverse-phase HPLC column early during the run.

DISCUSSION

The richer solid medium (BHI) was able to culture all of the isolates we examined even at high concentrations of TNT. YMA solid medium was then used because it is selective for actinomycetes at higher concentrations of TNT. Many of the actinomycetes isolated from TNT-contaminated soil were able to tolerate only low concentrations of TNT on solid media. As the TNT concentration was

TABLE 3. ^{14}C-TNT composted for 43 days using sterile and nonsterile compost inoculated with *Streptomyces chromofuscus* A11. An uninoculated, refrigerated control flask of sterilized compost was also maintained. Percentages are calculated from the total counts added to each flask (1.01×10^6 dpm).

Sample	Trapped[a] (dpm)	%	Extracted (dpm)	%	Bound (dpm)	%	Recovery (%)
Sterilized compost	1,600	0.16	494,000	49.1	480,000	47.6	96.8
Nonsterile compost	38,000	3.8	187,000	18.5	395,000	39.2	61.5
Control	—	—	928,000	92	67,000	6.7	98.7

(a) Nearly all of the ^{14}C-label that was trapped was found in the NaOH traps.

increased, growth was inhibited. A select few, however, were able to proliferate at TNT concentrations of up to 100 mg/L. Several of the ESA isolates are still the objects of further study, while *S. chromofuscus* A11 remains our model actinomycete for the study of TNT degradation.

S. chromofuscus A11 was able to completely degrade up to 50 mg/L of initial TNT in liquid media; from 50 to 80 mg/L, there was an accumulation of degradation products, and above 80 mg/L, degradation of TNT was inhibited. The mechanism used in TNT biodegradation by *S. chromofuscus* A11 involves a reductive pathway with transient formation of 4A26DNT and 24DA6NT. However, we have not shown that TNT is used as a carbon and energy source, only that it is degraded by cometabolism. ^{14}C-TNT liquid culture experiments are planned in order to determine if any labeled carbon is assimilated into cell biomass. *S. chromofuscus* A11 biodegraded TNT in flasks as well as in the fermentation run.

During the fermentation of *S. chromofuscus* A11 in the presence of TNT, the culture became brownish in color, which we initially suspected might be due to the formation of azoxy polymers. However, we did not detect the dimeric compounds 4,4',6,6'-tetranitro-2,2'-azoxytoluene or 2,2',6,6'-tetranitro-4,4'-azoxytoluene normally seen when the hydroxylamine intermediates of TNT polymerize. The observed rise in pH may be due to cleavage of the amino side groups from the reduced intermediates, although no ammonia analyses were performed to support this hypothesis.

The addition of *S. chromofuscus* A11 to the compost did not enhance TNT degradation, possibly because conditions did not allow *S. chromofuscus* A11 to proliferate throughout the compost material (possibly residual methylene chloride in the sterilized compost). The indigenous microbial population present in the compost transformed and mineralized some TNT. However, data were insufficient to determine a biodegradation pathway using this compost system. There was insignificant recovery of labeled carbon in the acid traps, indicating volatile organic amines were not produced during TNT transformation. The low label recovery from the nonsterile compost treatment may have been due to several factors. First, there may have been some undetected loss of CO_2 because the first and second NaOH traps often had similar counts. In comparison, the sterile treatment had most of the counts associated only with the first NaOH trap. The NaOH traps from the nonsterile treatment may have become saturated with CO_2 (from TNT or another carbon source) and thus lost the ability to trap all of the $^{14}CO_2$. This could be prevented by changing the traps more often or adding more traps in line. Second, the ^{14}C-carbon may have been tightly bound to the humus or may have existed in a polymerized state and was not released by our extraction methods. The solid compost material that persisted was not countable and may have had some of the label associated with it. The ^{14}C-label may have been tightly bound to this humic-like material. HPLC analyses of the extract did not show the presence of any of the normal TNT biotransformation products, possibly indicating an alternative degradation pathway to the anaerobic systems. The refrigerated control had only 63% of the labeled recovered as TNT. The remaining label was mostly extractable, but not in an identifiable form (i.e. polymerized or bound with other extractable material originating from the compost).

ACKNOWLEDGMENTS

We would like to thank Deanna Moser for her help with the fermentor setup. We would also like to thank Dr. Andrzej Pasczcynski for his helpful insight throughout all of the experiments. This research was supported in part by U.S. EPA grant no. CR 820804-02-0, NSF grant no. OSR-9350539, and by the Idaho Agriculture Experiment Station.

REFERENCES

Bumpus J. A., and M. Tatarko. 1994. "Biodegradation of 2,4,6-Trinitrotoluene by *Phanerochaete chrysosporium*: Identification of Initial Degradation Products and the Discovery of a TNT Metabolite that Inhibits Lignin Peroxidases." *Current Microbiology* 28:185-190.

Crawford, R. L. 1995. "Anaerobic Biodegradation of TNT (2,4,6-Trinitrotoluene and Dinoseb (2-*sec*-butyl-4,6-dinitrophenol)." In Spain (Ed.), *Biodegradation of Nitroaromatic Compounds*. In press.

Fernando, T., J. A. Bumpus, and S. D. Aust. 1990. "Biodegradation of TNT (2,4,6-Trinitrotoluene) by *Phanerochaete chrysosporium*." *Applied and Environmental Microbiology* 56:1666-1671.

Funk, S. B., D. L. Crawford, R. L. Crawford, G. Mead, and W. Davis-Hoover. 1994. "Full Scale Anaerobic Bioremediation of Trinitrotoluene (TNT) Contaminated Soil: A U.S. EPA SITE Program Demonstration." *Applied Biochemistry and Biotechnology* (in press).

Funk, S. B., D. J. Roberts, D. L. Crawford, and R. L. Crawford. 1992. "Bioremediation of Munition Compound-Contaminated Soils." *Applied and Environmental Microbiology* 59:2171-2177.

Isbister, J. D., G. L. Anspach, J. F. Kitchens, and R. C. Doyle. 1984. "Composting for Decontamination of Soils Containing Explosives." *Microbiologies* 7:47-73.

Kaake, R. H., D. J. Roberts, T. O. Stevens, R. L. Crawford, and D. L. Crawford. 1992. "Bioremediation of Soils Contaminated with the Herbicide 2-sec-Butyl-4,6-Dinitrophenol (Dinoseb)." *Applied and Environmental Microbiology* 58(5):1683-1689.

Kaplan, D. L., and A. M. Kaplan. 1982. "Thermophilic Biotransformations of 2,4,6-Trinitrotoluene Under Simulated Composting Conditions." *Applied and Environmental Microbiology* 44:757-760.

Kaplan, D. L. 1992. "Biological Degradation of Explosives and Chemical Agents." *Current Opinion in Biotechnology* 3:253-260.

Kulpa, C. F., and R. Boopathy. 1995. "Anaerobic Metabolism of Nitro-Compounds by Sulfate Reducing and Methanogenic Bacteria." In Spain (Ed.), *Biodegradation of Nitroaromatic Compounds*. In press.

Michels, J., and G. Gottschalk. 1994. "Inhibition of the Lignin Peroxidase of *Phanerochaete chrysosporium* by Hydroxylamino-Dinitrotoluene, an Early Intermediate in the Degradation of 2,4,6-Trinitrotoluene." *Applied and Environmental Microbiology* 60:187-194.

Michels, J., and G. Gottschalk. 1995. "Pathways of TNT Degradation by *Phanerochaete chrysosporium*." In Spain (Ed.), *Biodegradation of Nitroaromatic Compounds*. In press.

Pasti, M. B., and M. L. Belli. 1985. "Cellulolytic Activity of Actinomycetes Isolated from Termites (Termitidae) Gut." *FEMS Microbiology Letters* 26:107-112.

Pasti, M. B., and D. L. Crawford. 1991. "Relationships Between the Abilities of Streptomycetes to Decolorize Three Anthron-Type Dyes and to Degrade Lignocellulose." *Canadian Journal of Microbiology* 37: 902-907.

Pasti-Grigsby, M. B., A. Paszczynski, S. Goszczynski, D. L. Crawford, and R. L. Crawford. 1992. "Influence of Aromatic Substitution Patterns on Azo Dye Degradability by *Streptomyces* spp. and *Phanerochaete chrysosporium*." *Applied and Environmental Microbiology* 58: 3605-3613.

Preuss, A., J. Fimpel, and G. Diekert. 1993. "Anaerobic Transformation of 2,4,6-Trinitrotoluene (TNT)." *Archives in Microbiology* 159:345-353.

Preuss, A., and P. G. Rieger. 1995. "Anaerobic Transformation of Trinitrotoluene and Other Nitroaromatic Compounds." In Spain (Ed.), *Biodegradation of Nitroaromatic Compounds*. In press.

Pridhman and Gottlieb. 1948. "The Utilization of Carbon Compounds by Some Actinomycetes as an Aid for Species Determination." *Journal of Bacteriology* 56:107-114.

Regan, K. M., and R. L. Crawford. 1994. "Characterization of *Clostridium bifermentans* and its Biotransformation of 2,4,6-Trinitrotoluene (TNT) and 1,3,5-Triaza-1,3,5-Trinitrocyclohexane (RDX)." *Biotechnology Letters* 16(10):1081-1086.

Spain, J. C. 1995. "Bacterial Degradation of Nitroaromatic Compounds Under Aerobic Conditions." In Spain (Ed.), *Biodegradation of Nitroaromatic Compounds*. In press.

Spiker, J. K., D. L. Crawford, and R. L. Crawford. 1992. "Influence of 2,4,6-Trinitrotoluene (TNT) Concentration on the Degradation of TNT in Explosive-Contaminated Soils by the White Rot Fungus *Phanerochaete chrysosporium*." *Applied and Environmental Microbiology* 58:3199-3202.

Stahl, J. C., and S. Aust. 1995. "Biodegradation of 2,4,6-Trinitrotoluene by the White Rot Fungus *Phanerochaete chrysosporium*." In Spain (Ed.), *Biodegradation of Nitroaromatic Compounds*. In press.

Tan, E. L., C. H. Ho, W. H. Griest, and R. L. Tyndall. 1992. "Mutagenicity of Trinitrotoluene and its Metabolites Formed During Composting." *Journal of Toxicology* 36:165-175.

U.S. Army Corps of Engineers. November 1991. *Optimization of Composting for Explosives Contaminated Soil*. Final report, contract DAAA15-88-D-0010.

Won, W. D., DiSalvo, L. H., and J. Ng. 1976. "Toxicity and Mutagenicity of 2,4,6-Trinitrotoluene and its Microbial Metabolites." *Applied and Environmental Microbiology* 31:576-580.

Biodegradability of Selected Highly Energetic Pollutants Under Aerobic Conditions

Alison M. Jones, Charles W. Greer, Guy Ampleman,
Sonia Thiboutot, Jacques Lavigne, and Jalal Hawari

ABSTRACT

Laboratory studies were conducted to assess the biodegradation potential of four energetic compounds — 1,3,5-trinitro-1,3,5-triazacyclohexane (RDX), 2,4,6-trinitrotoluene (TNT), glycidyl azide polymer (GAP), and nitrocellulose (NC) — under aerobic conditions. Soils contaminated with energetic compounds were screened for microorganisms able to degrade these compounds. *Rhodococcus* sp. strain A was enriched and purified from RDX-contaminated soil by its ability to use RDX as sole source of nitrogen. Using [^{14}C]-labeled RDX, its ability to mineralize RDX to CO_2 in pure culture under aerobic conditions was verified. Bioaugmentation of RDX-contaminated soil with this bacterium enhanced the rate and extent of biodegradation. Although TNT is a more toxic and recalcitrant pollutant, concentration-dependent [^{14}C]TNT mineralization activity was detected in TNT-contaminated soil. A *Pseudomonas* sp., enriched from this soil, was able to extensively transform TNT to partially reduced amine metabolites in axenic liquid culture. In contrast to these isolates, which utilized RDX and TNT as sources of nitrogen, soil consortia developed from soils exhibiting [^{14}C]GAP and [^{14}C]NC mineralization activities used GAP and NC as sole carbon sources for growth.

INTRODUCTION

Soils and groundwaters at many military sites and industrial production plants are contaminated with energetic chemicals derived from the manufacture and handling of explosive munitions. Energetic compounds pose serious health and ecological hazards due to their mutagenicity, toxicity, and relative recalcitrance in the environment (Walker & Kaplan 1992). Conventional decontamination technologies, such as incineration, are expensive and suboptimal. Alternative remediation technologies for energetic compound-contaminated environments currently are being investigated. In situ bioremediation may be a viable treatment option due to its potential effectiveness without releasing new hazardous pollutants into the environment, its cost-effectiveness, its adaptability to specific compounds and environments, and its acceptance to the general public (Walker & Kaplan 1992). However, due to the xenobiotic nature and toxicity of organo-nitro compounds, especially polynitroaromatics, only limited catabolic potential is found in native microbial communities (Dickel & Knackmuss 1991).

Recent studies by Spain (e.g., Haigler et al. 1994, Spanggord et al. 1991) suggest that aerobic bacteria may play a more important role in degrading organonitro compounds than was previously thought. Identification of novel aerobic bacteria that can degrade recalcitrant energetic compounds such as RDX, TNT, GAP, and NC, and characterization of the catabolic pathways involved, are imperative to developing in situ bioremediation strategies for these hazardous xenobiotics. Moreover, a detailed knowledge of the environmental conditions required to foster optimum rates of biodegradation may lead to the exploitation of these microorganisms for biorestoration of energetic compound-contaminated soils.

The present study assessed the aerobic biodegradation potential of four highly energetic compounds by microorganisms indigenous to contaminated soils.

MATERIALS AND METHODS

Soil samples were collected at three contaminated sites from depths of 15 to 45 cm. Two sites were chosen for specific contamination by RDX, TNT, and NC. The third site, a disposal site for energetic compounds treated by "open-pit burning," was suspected of being contaminated with various energetic compounds. Soils were extracted and analyzed for the extent of RDX and TNT contamination by a high-performance liquid chromatography (HPLC) technique (U.S. EPA 1990). To confirm contamination of soils with GAP and NC, a gel permeation chromatography technique is under development. The ^{14}C-labeled energetic compounds for use in ^{14}C-mineralization studies were synthesized according to Ampleman et al. (in press).

To isolate bacteria from contaminated soil capable of degrading the energetic compounds, enrichment cultures were prepared using a minimal salts medium (MSM; Greer et al. 1990) containing the energetic compound (100 mg/L) as either sole nitrogen (glucose as carbon source) or carbon (ammonium sulfate as nitrogen source) source. To confirm that the energetic compounds had been degraded, ^{14}C-substrate mineralization studies were conducted, and the percentage of ^{14}C-substrate evolved as $^{14}CO_2$ was monitored by liquid scintillation spectrometry. Laboratory-scale microcosms enriched with ^{14}C-substrates were used to assess spontaneous mineralization in contaminated soils, and to study bioaugmentation of RDX-contaminated soil with an actively degrading bacterium. In the latter microcosms, an aqueous suspension of washed cells was added to the surface of soil to yield an initial inoculum density of 10^8 colony-forming units (CFU)/g soil. Uninoculated soils acted as negative controls. All incubations were in the dark at 25 to 27°C.

Chemical analyses were conducted on culture supernatants (0.22 µm) as follows: RDX and its potential metabolites were monitored by reverse-phase HPLC with UV (254 nm) detection, TNT and its metabolites were analyzed following benzene extraction using gas chromatography/mass spectroscopy (GC/MS), and nitrite and ammonium ion concentrations were estimated colorimetrically (Hanson & Phillips 1981). Authentic compounds were used as standards.

RESULTS AND DISCUSSION

Soil contaminated with RDX (9640 mg/kg) yielded a microbial consortium with the ability to mineralize RDX when provided as the sole nitrogen source for growth. Following incubation for 4 days, approximately 40% of the radiolabel was evolved as $^{14}CO_2$, suggesting that the use of energetic compounds, which are typically rich in nitrogen, as the sole nitrogen source in enrichment cultures may favor the selection of microbial consortia capable of metabolizing these compounds by extracting the nitrogen. *Rhodococcus* sp. strain A was isolated and purified from this consortium. With glucose as a carbon source, this bacterium was able to mineralize 35% of the [^{14}C]RDX to $^{14}CO_2$ following aerobic incubation for 48 h. To characterize its RDX biodegradation pathway, *Rhodococcus* sp. strain A was grown in MSM with 1.4 mM glucose and 50 mg/L RDX as sole nitrogen source. RDX was rapidly metabolized; within 40 h of incubation, RDX was depleted in the culture medium (Figure 1).

To date only trace levels of nitrite (detected during the initial phase of growth) and ammonium (<0.5 mg/L detected over the growth cycle) ions have

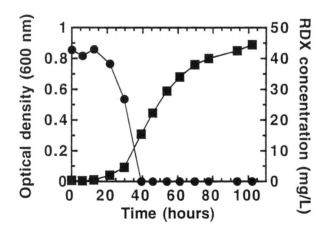

FIGURE 1. Metabolism of RDX by *Rhodococcus* sp. strain A during growth in defined medium with RDX as sole nitrogen source. Increase in cell density (■) was determined spectrophotometrically. Disappearance of RDX (●) was monitored by HPLC.

been identified as metabolites of RDX biodegradation by *Rhodococcus* sp. strain A. This suggests that under nitrogen-limiting conditions, the nitrite- and ammonium-nitrogen liberated from RDX metabolism is rapidly metabolized by this organism. This is in agreement with previous reports that the biotransformation of RDX and TNT under anaerobic conditions in soil (Funk et al. 1993) and under micro-aerobic conditions in pure culture (Kitts et al. 1994) produces reduced inter-mediates, and yields inorganic nitrogen for utilization by the microorganisms responsible.

The biodegradation of RDX to CO_2 and inorganic ions by *Rhodococcus* sp. strain A thus firmly establishes the aerobic biodegradation potential of RDX, and suggests that aerobic remediation of RDX-contaminated soils may result in a complete reduction of their toxicity. This is in contrast to the biodegradation of RDX under anaerobic conditions that produces mutagenic or carcinogenic hydrazine and dimethylhydrazines (McCormick et al. 1981).

Laboratory-scale bioaugmentation studies using *Rhodococcus* sp. strain A to bioremediate RDX-contaminated soils are currently in progress. An agricultural soil was artificially amended with pure RDX to a final concentration of 50 or 100 mg/kg, and inoculated with an actively degrading culture of strain A. Following incubation for 38 days, higher levels of RDX mineralization were observed in the soils inoculated with strain A compared to uninoculated control soils (Figure 2). However, the ability of this organism to successfully remove RDX from contaminated soil may be influenced by the concentration of RDX,

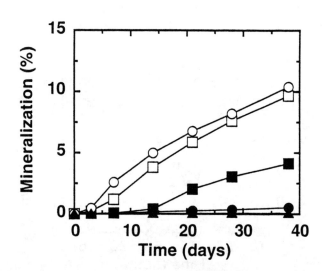

FIGURE 2. Mineralization of [^{14}C]RDX in agricultural soil amended with 50 or 100 mg RDX per kg, and bioaugmented with or without *Rhodococcus* sp. strain A as follows: 50 mg/kg RDX inoculated with (O) or without (●) strain A; 100 mg/kg RDX inoculated with (□) or without (■) strain A. Sterile control soil (▲) also was monitored.

soil/contaminant interactions, and/or nutrient limitation. In this regard, bio-stimulation with various nutrient amendments is currently being evaluated as a means of enhancing the biodegradative ability of *Rhodococcus* sp. strain A in soils contaminated with environmentally relevant concentrations of RDX.

TNT is a more recalcitrant compound, and is the most toxic of the four energetic compounds included in this study. Typically, biotransformations of TNT do not result in ring cleavage, and degradation to CO_2 does not occur (McCormick et al. 1976, Boopathy et al. 1994). To assess TNT mineralization by the indigenous soil microbial populations, microcosms were set up using soils contaminated with TNT over the range of 25 to 12,000 mg/kg. Only soil contaminated with the lowest concentration of TNT (25 mg/kg) exhibited [^{14}C]TNT mineralization activity (Figure 3). Neither glucose nor yeast extract enhanced the rate of mineralization in this soil (Figure 3). This activity was concentration dependent: in soil, as the concentration of TNT increased, minerali-zation decreased and was abrogated at 100 mg/kg. Thus, TNT is an extremely toxic compound with at least limited biodegradation potential.

Despite displaying spontaneous TNT mineralization activity, this soil has as yet yielded no pure cultures able to mineralize TNT to CO_2 under aerobic conditions. However, a *Pseudomonas* sp. was isolated and purified from a soil consortium developed from this soil using TNT (5 mg/L) as the sole nitrogen source. In an axenic liquid culture containing ammonium sulfate (250 mg/L) as a conitrogen source and glucose as a carbon source, this *Pseudomonas* sp. rapidly transformed TNT (5 mg/L) within 26 h to yield 2-amino-4,6-dinitrotoluene

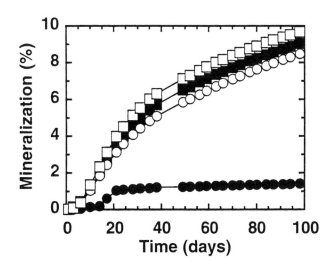

FIGURE 3. Mineralization of [^{14}C]TNT by the indigenous microbial population in soil contaminated with 25 mg/kg TNT. $^{14}CO_2$ evolution was monitored in unamended soil (■), in soil amended with glucose (○) or yeast extract (□), and in sterile soil (●).

(m/z 197) and 4-amino-2,6-dinitrotoluene (m/z 197) as determined by GC/MS. At least three other peaks were detected over the course of the experiment, but these could not be identified due to a lack of standards. In contrast, in the absence of a conitrogen source, this organism did not completely transform TNT even after incubation for 32 days. Despite its extensive biotransformation of TNT, this *Pseudomonas* sp., under the experimental conditions employed to date, exhibits only limited mineralization potential.

Biotransformation of the highly nitrated TNT by *Pseudomonas* spp. under aerobic conditions has been reported previously (McCormick et al. 1976, Schackmann & Müller 1991, Boopathy et al. 1994) but, as observed with the *Pseudomonas* sp. isolated in the present study, TNT appears to be resistant to complete degradation by pseudomonads under aerobic conditions. Studies are currently focused on optimizing culture conditions and characterizing the TNT biotransformation pathway of this *Pseudomonas* sp. In addition, attempts are being made to promote mineralization of the aromatic ring by the addition of various nutrients and enhancers.

GAP is relatively new to the environment and, consequently, it is difficult to identify a reliable source of organisms with significant prior exposure to GAP. Soils collected from a disposal site for energetic compounds displayed moderate levels of [^{14}C]GAP mineralization activity — approximately 14% after incubation for 84 days. An enrichment culture developed from this soil using GAP (100 mg/L) as the sole carbon source yielded a soil consortium with the ability to mineralize GAP. Following incubation of this culture for 40 days, approximately 5.5% of the radiolabel was evolved as CO_2. Due to its use as a carbon source, mass balance studies are required to fully account for the fate of the ^{14}C-labeled substrate in these cultures. No microbial consortia were obtained that could utilize GAP as the sole nitrogen source for growth.

Nitrocellulose with a high degree of substitution is not subject to direct microbial attack, and is generally regarded as persistent in the environment (Walker & Kaplan 1992). Two soils were collected for NC contamination, and initial mineralization studies indicated that both soils possessed indigenous [^{14}C]NC mineralization activity. Following a 25-day lag, $^{14}CO_2$ evolution was observed in both soils, and by 94 days approximately 15 to 17% of the [^{14}C]NC was mineralized to $^{14}CO_2$. That a lag period was observed prior to NC mineralization suggests that a consortium rather than a single microorganism may have been responsible for this activity. Alternately, a cellulase enzyme may have been induced during this lag period that attacked the cellulose component of this polymer, thereby making the molecule more amenable to biodegradation. Enrichment cultures are currently being developed using these soils and NC (100 mg/L) as the sole carbon or nitrogen source. Preliminary indications are that under aerobic conditions, NC is used as a carbon source by the microbial consortia. Future work will focus on isolation and purification of NC-degrading bacteria from these cultures.

Thus, laboratory-scale studies conducted to date firmly establish that bioremediating RDX-contaminated soil is possible under aerobic conditions, and suggest that the potential also exists for aerobic bioremediation of TNT, GAP, and NC.

ACKNOWLEDGMENTS

We thank Suzanne Labelle, Anca Mihoc, Louise Paquet, and Edward Zhou for technical assistance in various aspects of this study.

REFERENCES

Ampleman, G., S. Thiboutot, J. Lavigne, A. Marois, J. Hawari, A. M. Jones, and D. Rho. 1994. "Synthesis of ^{14}C-labelled hexahydro-1,3,5-trinitro-1,3,5-triazine (RDX), 2,4,6-trinitrotoluene (TNT), nitrocellulose (NC) and glycidyl azide polymer (GAP) for use in assessing the biodegradation potential of these energetic compounds." *J. Labelled Cmpd. Radiopharm.* (In press).

Boopathy, R., M. Wilson, C. D. Montemagno, J. F. Manning, Jr., and C. F. Kulpa. 1994. "Biological transformation of 2,4,6-trinitrotoluene (TNT) by soil bacteria isolated from TNT-contaminated soil." *Bioresource Technol.* 47: 19-24.

Dickel, O., and H.-J. Knackmuss. 1991. "Catabolism of 1,3-dinitrobenzene by *Rhodococcus* sp. QT-1." *Arch. Microbiol.* 157: 76-79.

Funk, S. B., D. J. Roberts, D. L. Crawford, and R. L. Crawford. 1993. "Initial-phase optimization for bioremediation of munition compound-contaminated soils." *Appl. Environ. Microbiol.* 59: 2171-2177.

Greer, C. W., J. Hawari, and R. Samson. 1990. "Influence of environmental factors on 2,4-di-chlorophenoxyacetic acid degradation by *Pseudomonas cepacia* isolated from peat." *Arch. Microbiol.* 154: 317-322.

Haigler, B. E., W. H. Wallace, and J. C. Spain. 1994. "Biodegradation of 2-nitrotoluene by *Pseudomonas* sp. strain JS42." *Appl. Environ. Microbiol.* 60: 3466-3469.

Hanson, R. S., and J. A. Phillips. 1981. "Chemical composition." In P. Gerhardt, R. G. E. Murray, R. N. Costilow, E. W. Nester, W. A. Wood, N. R. Krieg, and G. B. Phillips (Eds.), *Manual of Methods for General Bacteriology*, pp. 328-364. American Society for Microbiology, Washington, DC.

Kitts, C. L., D. P. Cunningham, and P. J. Unkefer. 1994. "Isolation of three hexahydro-1,3,5-trinitro-1,3,5-triazine-degrading species of the family *Enterobacteriaceae* from nitramine explosive-contaminated soil." *Appl. Environ. Microbiol.* 60: 4608-4611.

McCormick, N. G., J. H. Cornell, and A. M. Kaplan. 1981. "Biodegradation of hexahydro-1,3,5-trinitro-1,3,5-triazine." *Appl. Environ. Microbiol.* 42: 817-823.

McCormick, N. G., F. E. Feeherry, and J. S. Levinson. 1976. "Microbial transformation of 2,4,6-trinitrotoluene and other nitroaromatic compounds." *Appl. Environ. Microbiol.* 31: 949-958.

Schackmann, A., and R. Müller. 1991. "Reduction of nitroaromatic compounds by different *Pseudomonas* species under aerobic conditions." *Appl. Microbiol. Biotechnol.* 34: 809-813.

Spanggord, R. J., J. C. Spain, S. F. Nishino, and K. E. Mortelmans. 1991. "Biodegradation of 2,4-dinitrotoluene by a *Pseudomonas* sp." *Appl. Environ. Microbiol.* 57: 3200-3205.

U.S. Environmental Protection Agency. 1990. *Nitroaromatics and Nitramines by HPLC*. Second Update SW846, Method 8330.

Walker, J. E., and D. L. Kaplan. 1992. "Biological degradation of explosives and chemical agents." *Biodegradation* 3: 369-385.

Biodegradation of Explosives Using a Known Herbicide Degrader

F. Michael von Fahnestock and Traci L. Forney

ABSTRACT

This research evaluated the application of a pure bacterial culture of M91-3 to degrade 2,4,6-trinitrotoluene (TNT), hexahydro-1,3,5-trinitro-1,3,5-s-triazine (RDX), and octahydro-1,3,5,7-tetranitro-s-tetrazocine (HMX) in liquid medium under aerobic and anaerobic conditions. Tests were conducted in batch cultures and in a continuous plug-flow reactor. In batch cultures, it was established that M91-3 was able to cometabolize TNT as a sole nitrogen source but could not use RDX or HMX under aerobic conditions. Growth using RDX as the sole nitrogen source was demonstrated under anaerobic conditions. Plug-flow reactor tests demonstrated that M91-3 was able to transform TNT in aqueous media from 80 mg/L to below 0.2 mg/L with a 4.3-day residence time.

INTRODUCTION

This research served as an initial feasibility study for the application of a pure, aerobic bacterial culture to degrade TNT, RDX, and HMX in liquid, bench-scale reactor systems. The bacterial species chosen, M91-3, has been proven to mineralize s-triazine herbicides, such as atrazine (Radosevich 1994). Originally, M91-3 was isolated from pesticide-contaminated soil by The Ohio State University. M91-3 is a facultatively anaerobic gram-negative rod.

MATERIALS AND METHODS

Materials

For initial aerobic culture acclimation, Composition B (40% TNT and 60% RDX) and Octol (30% TNT and 70% HMX) were used as sources for TNT, RDX, and HMX. These two explosive compositions were provided by Battelle, Columbus, Ohio. Analytical-grade TNT and RDX (98+% pure with 10 to 20%

water added) were obtained from Chem Services Co., West Chester, Pennsylvania. The analytical-grade TNT was used in all aerobic, continuous-flow reactor and batch growth-curve experiments. Analytical-grade RDX was used in generating anaerobic, batch growth-curve data.

The pure M91-3 culture received from The Ohio State University was incubated at room temperature in basal salts medium (BSM) with 25 mg/L atrazine, then streaked onto a trypticase soy agar (TSA) plate to verify purity. The resulting colonies were medium sized, averaging 2 mm after a 5- to 7-day incubation at room temperature, and a creamy-rose color, which deepened into a more definite pink color after several more days. Gram staining produced small gram-negative rods. A stock culture was maintained on TSA slants. Periodically, the purity of the stock culture was checked either microscopically or by streaking onto a TSA plate.

Methods

The initial culture of M91-3 was grown in BSM containing atrazine as the sole nitrogen source. After turbidity was visible, the culture was transferred to fresh medium in which the nitrogen source was 80% atrazine and 20% Composition B or Octol. This culture was eventually transferred to medium with 50% each atrazine and Composition B or Octol, followed by medium of 20% atrazine and 80% Composition B or Octol and, finally, medium in which the nitrogen source was 100% Composition B or Octol. No visual turbidity was detected in cultures using 100% Composition B or Octol as the nitrogen source. The high-performance liquid chromatography (HPLC) analyses of the batch cultures did not show any degradation of RDX or HMX but did show degradation of the TNT portion of the Octol and Composition B. Sterile controls had no decrease in TNT content. Therefore, aerobic degradation of TNT by M91-3 was pursued in a continuous-flow reactor system, and aerobic RDX and HMX cultures were discontinued.

To measure M91-3 growth using RDX as the sole nitrogen source under anaerobic conditions, 50 mL BSM + RDX (25 mg/L) + glucose (1,600 mg/L) were sterilized and then aseptically added to two sterile serum bottles. The bottles were sealed with a sterile septum. Nitrogen gas was flushed through the bottles to remove any oxygen. One of these bottles was inoculated with 1 mL of M91-3 (5×10^3 CFU/mL) from the first set of aerobic RDX medium cultures described above. The bottle of uninoculated medium was maintained as a control. The culture and control flasks were incubated at room temperature and monitored for growth by measuring the optical density using a spectrophotometer.

The HPLC analyses were based on EPA Method 8330 (U.S. Environmental Protection Agency 1986). The HPLC system consisted of a Waters M-6000 pump, SSI 500 ultraviolet-visual (UV-Vis) detector set at 600 nm, Hewlett-Packard 3390A integrator, and Alltech Alltima C18 reverse-phase column. The column had a 250-mm length and 4.6-mm diameter. The mobile phase was 67% methanol and 33% water with flowrate of 1.0 mL/min. The limit of detection for TNT, RDX, and atrazine was 0.2 mg/L. HMX was also measured but was not easily quantified at concentrations below 3 mg/L, because of an interference with the injection peak. The method was not refined to remove this interference, because studies on HMX were discontinued after the initial culture acclimations. Although the primary emphasis of this research was to measure parent compound concentrations, TNT transformation peaks also were observed. HPLC samples and standards were collected via disposable syringes and filtered into glass screw-top vials through 0.45-μm Gelman Acrodisc GF syringe filters.

Continuous cultures were maintained in an aerated plug-flow reactor (PFR) packed with 2-mm glass beads. Filtered air was fed to the reactor through a sparging stone at approximately 50 mL/min. The net liquid volume of the packed PFR was 370 mL. A Masterflex peristaltic pump metered 60 μL/min of media to the reactor, which resulted in a residence time of 4.3 days. In addition to feed and exit samples, four ports placed at 75-mL volume intervals were used to obtain a cross section of samples from the PFR at different residence times. The reactor system is shown in Figure 1. The

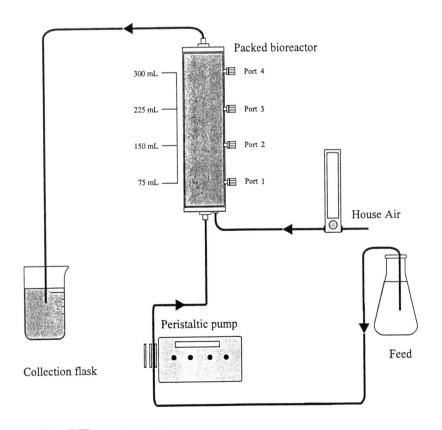

FIGURE 1. PFR reactor system.

sterilized reactor system was inoculated with 10 mL of M91-3 culture that was entering the stationary phase and contained ~5 × 10⁶ CFU/mL. To equilibrate the reactor, the M91-3 culture was allowed to grow for 2 weeks prior to conducting the testing. Throughout the 4-month test period, samples were drawn from the reactor column to check for contamination. Sufficient time was allotted between reaction runs to allow the PFR to equilibrate to the new reactor feed conditions.

RESULTS

RDX Batch Cultures

The aerobic culture maintained in RDX medium produced a constant level of growth, neither increasing nor decreasing in CFU/mL over time. The anaerobic RDX culture increased from 10^2 to 10^6 CFU/mL in 10 days. Figure 2 represents the growth curve of M91-3 using 25 mg/L of RDX as the sole nitrogen source. In the anaerobic RDX batch cultures exhibiting growth, the RDX concentration was reduced by 30% compared to the sterile controls. It was not established whether or not the RDX was partially absorbed by the cell biomass. Further research with M91-3 in an anaerobic RDX-based medium has been planned.

Continuous Cultures

Four PFR trials were conducted using TNT in BSM+glucose. The TNT and glucose concentrations were varied to measure the effects of each on the overall

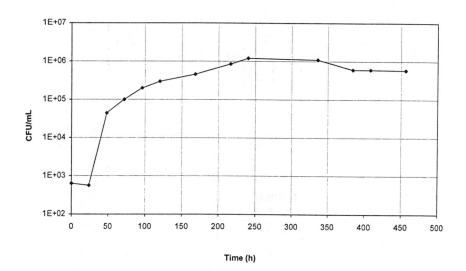

FIGURE 2. **Anaerobic growth curve of M91-3 using 25 mg/L of RDX as the sole nitrogen source.**

disappearance of TNT. Table 1 lists the test matrix used. Runs 1 and 2 used a glucose concentration of 1,600 mg/L and TNT concentrations of 18 and 80 mg/L, respectively. Run 3 utilized a feed with 400 mg/L glucose and 70 mg/L TNT. Run 4 was designed as a duplicate of Run 2. Results from Runs 1 and 2 are shown in Figure 3, where it can be seen that M91-3 achieved >99% degradation

TABLE 1. PFR feed composition matrix to evaluate the degradation of TNT using M91-3.

Run	TNT Conc (mg/L)		Glucose Conc (mg/L)		% Degradation
1	Low	18	High	1,600	99+
2	High	80	High	1,600	99+
3	High	70	Low	400	~50
4	High	80	High	1,600	~40

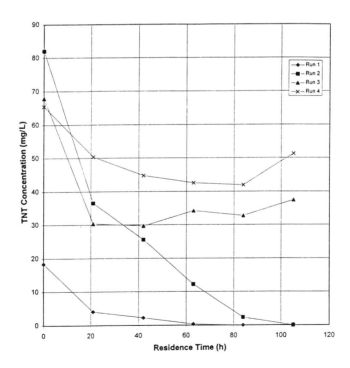

FIGURE 3. TNT degradation by M91-3 in an aerated, glass-bead packed PFR (Run 1—18 mg/L TNT, 1,600 mg/L glucose in feed; Run 2—80 mg/L TNT, 1,600 mg/L glucose; Run 3—70 mg/L TNT, 400 mg/L glucose; Run 4—80 mg/L TNT, 1,600 mg/L glucose).

of TNT in 63 h and 105 h, respectively. In Runs 3 and 4, however, the amount of TNT degradation achieved was around 50%. For Runs 3 and 4, the degradation of TNT proceeded as it did in Run 2 in the first half of the PFR, but the TNT concentration did not continue to decrease in the second half of the PFR. The pH for all four runs changed from 7.0 at the reactor inlet to 5.5 to 6.0 at the reactor exit.

As the TNT degraded, seven degradation compounds were repeatedly detected by HPLC. These peaks, though not identified, were consistent for all four runs. A sample HPLC chromatogram profile of the degradation compounds from Run 2 is shown in Figure 4. With the exception of a peak at 9.2 minutes, the degradation peaks had retention times of 2.9 to 4.5 min. The peak area units of these degradation compounds increased and decreased as a function of reactor residence time. In Figure 4, for example, the degradation peak at 3.93 min appeared as a strong peak in the sample taken from Port 3 (63 h residence time), then disappeared in the Port 4 sample (84 h residence time), and then appeared as a minor peak in the Exit Port sample (105 h residence time).

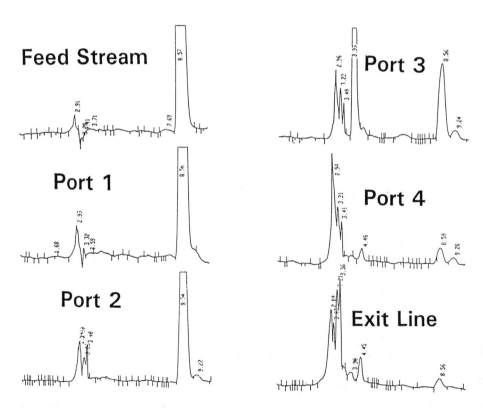

FIGURE 4. PFR concentration cross section as a function of residence time. Residence time increased by 21 h between each sampling port, starting with 0 h at the feed input and ending with 105 h at the exit line. The TNT peak is at ~8.55 min.

However, the overall concentration profile obtained as a function of longitudinal location in the PFR (residence time) remained fairly constant with time for a given run.

DISCUSSION

Batch and continuous test results indicate M91-3 degraded TNT under aerobic conditions. Batch results showed that M91-3 was unable to degrade RDX or HMX aerobically; however, anaerobic batch cultures were able to grow with RDX present as the sole nitrogen source. Based on the initial growth results, further testing is planned to evaluate the ability of M91-3 to degrade RDX and HMX under anaerobic conditions.

Results of PFR Runs 1 and 2 indicate that M91-3 did not experience any toxic effects based on TNT concentration. In Run 3, where the glucose concentration was reduced to 400 mg/L from 1,600 mg/L, the TNT ceased to decrease in the second half of the reactor, which indicated glucose was required by M91-3 to degrade TNT. In Run 4 where the glucose concentration was increased again to 1,600 mg/L, however, TNT degradation continued to be minimal in the second half of the PFR. Although it is unclear why Run 4 did not achieve the level of TNT degradation as in Run 2, one reason could be that a fungal contaminant which was more competitive for the glucose than M91-3 had become established in the PFR. A second possibility is that the population density of M91-3 did not recover from the reduction of glucose during Run 3.

To firm up initial results obtained in this feasibility study, plans have been made to repeat the PFR experiments after sterilizing and reinoculating the PFR. Other future work could include (1) a study of TNT degradation by M91-3 in a soil-based medium, (2) identification of TNT degradation products generated by M91-3, (3) a mineralization study using ^{14}C-ring-labeled TNT, and (4) an evaluation of the ability of M91-3 to degrade other nitroaromatic compounds.

CONCLUSIONS

It was established that M91-3 is able to aerobically use TNT as a sole nitrogen source. A cometabolite, such as glucose, appears to be required as an energy source. In a PFR, it was shown that M91-3 has the potential to transform >99% of TNT in aqueous medium with a 4.3-day residence time. Although several degradation products were observed, the extent of TNT transformation is unclear. Furthermore, inhibition of TNT degradation was observed in the PFR, which could have been caused by a limiting glucose concentration or by competition for glucose by other organisms. M91-3 was not shown to aerobically degrade RDX or HMX, although anaerobic growth using RDX as the sole nitrogen source was observed.

REFERENCES

Radosevich, M. 1994. "Sorptive Constraint on the Biodegradation of Atrazine in Surface and Sub-Surface Environments." Ph.D Dissertation, The Ohio State University, Columbus, OH.

U.S. Environmental Protection Agency. 1986. *Test Methods for Evaluating Solid Waste*. U.S. EPA SW-846, Method 8330, 3rd ed., November.

Degradation of 2,4-DNT, 2,6-DNT, and 2,4,6-TNT by Indigenous Aquifer Microorganisms

Paul M. Bradley, Francis H. Chapelle,
and James E. Landmeyer

ABSTRACT

Microorganisms indigenous to aquifer sediments collected at an explosives-contaminated site removed greater than 99% of added 2,4-dinitrotoluene (2,4-DNT) and 2,4,6-trinitrotoluene (TNT) from aerobic microcosms within 70 days. The initial microcosm substrate concentrations were about 100 μM. Approximately 55% of 2,6-dinitrotoluene (2,6-DNT) was removed within the same period. Decreases in the dissolved concentrations of the DNT and TNT test substrates were accompanied by accumulation of amino-mononitrotoluene or amino-DNT compounds, respectively. The lack of appreciable CO_2 production by aquifer materials in the absence of added test substrate suggests the possibility that a significant fraction of these compounds was completely degraded to CO_2 within 70 days.

INTRODUCTION

The imminent closure of many military facilities throughout the United States has emphasized the need for efficient and economical methods for remediation of explosives-contaminated sites. Various microbiological approaches to remediation of contaminated surface soils have been investigated (Boopathy et al. 1993, Bradley et al. 1994, Carpenter et al. 1978, Fernando et al. 1990, Funk et al. 1993, Hallas and Alexander 1983, Kaplan and Kaplan 1982, McCormick et al. 1976, Osmon and Klausmeier 1972, Parrish 1977, Spanggord et al. 1991, Valli et al. 1992, Won et al. 1974), and several of these may provide viable alternatives to the widespread practice of incineration. At present, however, little attention has been focused on the need for effective and affordable methods for remediating the explosives-contaminated aquifer systems that underlie many of these sites. Most of the existing methods, including incineration, are commonly applied to surface contamination and are not suitable for aquifer remediation. An in situ method based on the activity of indigenous microorganisms may be an effective mechanism for aquifer remediation if these microorganisms are capable of degrading explosives compounds.

MATERIALS AND METHODS

The ability of aquifer microorganisms to degrade 2,4-DNT, 2,6-DNT, and TNT was investigated using sediment collected from a weathered, semiconsolidated, shallow aquifer that underlies a heavily contaminated site at the Weldon Spring Ordnance Works, Weldon Spring, Missouri. Groundwater from the aquifer is characterized by low concentrations (less than 0.1 μM) of 2,4-DNT, 2,6-DNT, and TNT. The organic carbon and carbonate contents of the aquifer sediments were less than the analytical detection limit of 0.01% dry weight. Aerobic microcosms were prepared by placing 20 g (wet weight) of aquifer material in sterile, 40-mL serum vials and adding 20 mL of sterile aqueous solution containing 100 μM 2,4-DNT, 2,6-DNT, or TNT. Experimental microcosms were prepared in triplicate for each test substrate, and triplicate live controls were prepared in the same manner without addition of test substrate. Duplicate abiological controls were prepared for each test substrate by sterilizing the microcosms (5 mM $HgCl_2$; autoclaved at 121°C for 1 h) before adding test solution. Microcosms were sampled regularly for dissolved nitroaromatic compounds and headspace CO_2 over a period of approximately 70 days.

The concentrations of 2,4-DNT, 2,6-DNT, TNT, amino-nitrotoluene compounds, and amino-DNT compounds were determined by high-performance liquid chromatography and ultraviolet (250 nm) detection. Under the conditions used in this study, the detection limits for 2,4-DNT, 2,6-DNT, and TNT were less than 0.5 μM. All compounds were identified by fortification and coelution with known standards. Microcosm CO_2 production was measured by thermal conductivity detection gas chromatography.

RESULTS

The results indicate that the microbial community associated with the aquifer material at Weldon Spring is capable of transforming 2,4-DNT, 2,6-DNT, and TNT. Removal of greater than 99% of 2,4-DNT and TNT from the dissolved phase was achieved in 40 to 60 days (Figure 1). Under the conditions used in this study, the detection limits for 2,4-DNT, 2,6-DNT, and TNT were less than 0.5 μM. The dissolved concentration of 2,6-DNT was reduced by 60%. In all cases, disappearance of the test substrates was attributable to biological activity, because no significant decrease in substrate concentration occurred in sterilized microcosms.

Decreases in the dissolved concentrations of 2,4-DNT, 2,6-DNT, or TNT were accompanied by accumulation of amino-mononitrotoluene or amino-DNT compounds, respectively (Figure 1). The initial and quantitatively most

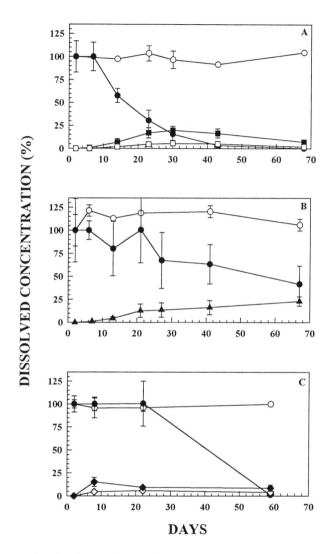

FIGURE 1. Microbially mediated disappearance of nitrotoluene compounds from aquifer microcosms. (A) Dissolved concentrations of 2,4-dinitro-toluene in experimental (●) and killed controls (O) and accumulation and subsequent disappearance of 4-amino-2-nitrotoluene (■) and 2-amino-4-nitrotoluene (□) in the dissolved phase of experimental microcosms. (B) Dissolved concentrations of 2,6-dinitrotoluene in experimental micro-cosms (●) and killed controls (O) and accumulation of 2-amino-6-nitrotoluene (▲) in the dissolved phase of experimental microcosms. (C) Dissolved concentrations of 2,4,6-trinitrotoluene in experimental microcosms (●) and killed controls (O) and accumulation and subsequent decline of 4-amino-2,6-dinitrotoluene (◆) and 2-amino-4,6-dinitrotoluene (◇) in the dissolved phase of experimental microcosms. Data points rep-resent means of triplicates ±SD.

significant intermediate products of TNT transformation were 4-amino-2,6-DNT and 2-amino-4,6-DNT. The fact that the dissolved concentration of 4-amino-2,6-DNT was at least twice that of 2-amino-4,6-DNT indicates that the *para* nitro group was reduced most readily. This pattern is consistent with the findings of previous investigations of microbial TNT transformation (Funk et al. 1993, Higson 1992, McCormick et al. 1978). Likewise, 2,4-DNT was transformed immediately to 4-amino-2-nitrotoluene and, to a lesser extent, 2-amino-4-nitrotoluene as described previously (Liu et al. 1984, McCormick et al. 1976). In contrast, the major intermediate product of 2,4-DNT transformation by the white-rot fungus, *Phanerochaete chrysosporium*, was 2-amino-4-nitrotoluene (Valli et al. 1992). In the present study, 2,6-DNT was transformed to 2-amino-6-nitrotoluene.

Very little CO_2 (approximately 4 ± 2 nmoles of CO_2 per gram dry material per day) was produced by the Weldon Spring aquifer material in the absence of added substrate. In contrast, CO_2 was produced at rates of 110 ± 11, 75 ± 5, and 100 ± 6 nmoles g^{-1} d^{-1} when 2,4-DNT, 2,6-DNT, and TNT, respectively, were present. Although the use of radiolabeled substrates is required to provide conclusive evidence of substrate mineralization, the lack of significant CO_2 production by the aquifer microcosms in the absence of added nitroaromatic compounds suggests the possibility that the aquifer microorganisms are capable of complete degradation of TNT and DNT compounds to CO_2.

DISCUSSION

The results of this study provide direct evidence for degradation of explosives compounds by aquifer microorganisms. These results indicate that an in situ bioremediation approach based on the degradative activity of indigenous microorganisms may be a feasible alternative for the cleanup of explosives-contaminated aquifers.

REFERENCES

Boopathy, R., M. Wilson, and C. F. Kulpa. 1993. "Anaerobic removal of 2,4,6-trinitrotoluene (TNT) under different electron accepting conditions: laboratory study." *Water and Environmental Research* 65:271-275.

Bradley, P. M., F. H. Chapelle, J. E. Landmeyer, and J. G. Schumacher. 1994. "Microbial transformation of nitroaromatics in surface soils and aquifer materials." *Applied and Environmental Microbiology* 60:2170- 2175.

Carpenter, D. F., N. G. McCormick, J. H. Cornell, and A. M. Kaplan. 1978. "Microbial transformation of ^{14}C-labeled 2,4,6-trinitrotoluene in an activated-sludge system." *Applied and Environmental Microbiology* 35:949-954.

Fernando, T., J. A. Bumpus, and S. D. Aust. 1990. "Biodegradation of TNT (2,4,6-trinitrotoluene) by *Phanerochaete chrysosporium.*" *Applied and Environmental Microbiology* 56:1666-1671.

Funk, S. B., D. J. Roberts, D. L. Crawford, and R. L. Crawford. 1993. "Initial phase optimization for the bioremediation of munitions-contaminated soils." *Applied and Environmental Microbiology* 59:2171-2177.

Hallas, L. E., and M. Alexander. 1983. "Microbial transformation of nitroaromatic compounds in sewage effluent." *Applied and Environmental Microbiology* 45:1234-1241.

Higson, F. K. 1992. "Microbial degradation of nitroaromatic compounds." In S. L. Neidleman and A. I. Laskin (Eds.), *Advances in Applied Microbiology*, pp. 1-19. Academic Press, Inc., New York, NY.

Kaplan, D. L., and A. M. Kaplan. 1982. "Thermophilic biotransformations of 2,4,6-trinitrotoluene under simulated composting conditions." *Applied and Environmental Microbiology* 44:757-760.

Liu, D., K. Thomson, and A. C. Anderson. 1984. "Identification nitroso compounds from biotransformation of 2,4-dinitrotoluene." *Applied and Environmental Microbiology* 47:1295-1298.

McCormick, N. G., F. E. Feeherry, and H. S. Levinson. 1976. "Microbial transformation of 2,4,6-trinitrotoluene and other nitroaromatic compounds." *Applied and Environmental Microbiology* 31:949-958.

McCormick, N. G., J. H. Cornell, and A. M. Kaplan. 1978. "Identification of biotransformation products from 2,4-dinitrotoluene." *Applied and Environmental Microbiology* 35:945-948.

Osmon, J. L. and R. E. Klausmeier. 1972. "The microbial degradation of explosives." *Developments in Industrial Microbiology* 14:247-252.

Parrish, F. W. 1977. "Fungal transformation of 2,4-dinitrotoluene and 2,4,6-trinitrotoluene." *Applied and Environmental Microbiology* 34:232-233.

Spanggord, R. J., J. C. Spain, S. F. Nishino, and K. E. Mortelmans. 1991. "Biodegradation of 2,4-dinitrotoluene by a *Pseudomonas* sp." *Applied and Environmental Microbiology* 57:3200-3205.

Valli, K., B. J. Brock, D. K. Joshi, and M. H. Gold. 1992. "Degradation of 2,4-dinitrotoluene by the lignin-degrading fungus *Phanerochaete chrysosporium*." *Applied and Environmental Microbiology* 58:221-228.

Won, W. D., R. J. Heckly, D. J. Glover, and J. C. Hoffsommer. 1974. "Metabolic disposition of 2,4,6-trinitrotoluene." *Applied and Environmental Microbiology* 27:513-516.

Enrichment and Characterization of Anaerobic TNT-Degrading Bacteria

Deborah J. Roberts and Suhasini Pendharkar

ABSTRACT

Three media constitutions were used to enrich for mixed cultures capable of degrading 2,4,6-trinitrotoluene (TNT) under strictly anaerobic conditions. The media were derived from a mineral salts solution buffered to pH 7 with CO_2/bicarbonate and all contained TNT. The cultures were enriched in the TNT mineral salts medium or the TNT mineral salts medium supplemented with glucose, yeast extract, or ammonia in various combinations. Inocula were obtained from a "treated soil," previously contaminated with dinoseb and then treated using anaerobic procedures, or from a bench-top aqueous culture, maintained with an extract from a munitions-contaminated soil for more than 4 years. Several cultures reduced TNT, producing 4-amino-2,6-dinitrotoluene and 2,4-diamino-6-nitrotoluene as the major products. The cultures were unable to effectively remove TNT when cross-transferred to the media they were not enriched on, suggesting that different media had enriched different subcultures from the original inoculum. The "treated soil" provided the most successful inoculum. Two media were chosen for further studies. Medium 1 contained TNT and glucose and produced a culture that might have used TNT as a nitrogen source. Medium 2, containing TNT and yeast extract, enriched cultures that degraded TNT, accumulating small amounts of *p*-cresol during the degradation.

INTRODUCTION

The use of anaerobic treatment technologies is a cost-effective, successful technique for treating munitions-contaminated soils. The procedure used in the field is a simple addition of nutrients, buffer, and an external carbon source to stimulate aerobic activity, which causes oxygen depletion and creates anaerobic conditions (Funk et al. 1993, Kaake et al. 1992, Roberts et al. 1992b). The anaerobic bacteria degrade the munitions compounds, producing fermentation end products from the external carbon source and the munitions. These are then treated with a final aerobic polishing step to produce a fully treated soil (Funk et al. 1994). The process relies on the presence of a competent microbial population in the soil. A munitions-contaminated soil from Umatilla, Oregon, contains high levels of TNT and other munitions compounds (hexahydro-1,3,5-trinitro-1,3,5-triazine (RDX), and octahydro-1,3,5,7-tetranitro-1,3,5,7-tetraazocine (HMX)), but does not have a microbial population capable of degrading TNT (Funk et al. 1994).

Laboratory studies performed to date with this soil have used an external soil inoculum. The soil inoculum was a "treated soil," previously contaminated with dinoseb and treated with the anaerobic procedure (Kaake et al. 1992, Roberts et al. 1992b). This soil inoculum contained the appropriate microbial population both to remove oxygen from the system, lowering the redox potential to below -300 mV, and to degrade the TNT, RDX, and HMX present in the soil (Funk et al. 1993, 1994). It has been characterized microbiologically and contains a wide spectrum of bacteria from a variety of physiological variants (manuscript in preparation). Other sources of inocula that have been used in the past are a bench-top, aqueous-phase reactor culture maintained with an organic extract of the Umatilla soil as its only carbon source for more than 4 years (Roberts & Crawford 1991, Roberts et al. 1992a), and biologically activated carbon, which had been used previously to treat munitions-contaminated water for approximately 3 months (unpublished results).

This research was designed to enrich and characterize bacteria, obtained from different inoculum sources that are capable of anaerobic TNT degradation. The enrichment procedure examined the effects of using glucose or methanol as additional sources of carbon for increased growth and of adding inorganic, organic, or no additional nitrogen source.

EXPERIMENTAL PROCEDURES AND MATERIALS

Media and cultures were prepared using strict anaerobic techniques suitable for cultivating methanogens. The ingredients of the various media were derived from Fedorak and Hrudey (1986) and are listed in Table 1. The headspace of all cultures contained 70% N_2 and 30% CO_2 (both previously scrubbed of oxygen through a heated copper coil). The cultures were incubated at 30°C in 125-mL serum bottles sealed with butyl rubber stoppers. Cultures that had degraded the initial allotment of TNT were used for a 50% (v/v) transfer to fresh medium of the original composition or as cross transfers to the other media. In some cases, the original culture was amended with fresh medium (up to 50% by volume).

Samples of 0.6-mL aliquots of the aqueous phase were removed with a 1-mL syringe, centrifuged to remove fine particulates, and analyzed by high-performance liquid chromatography (HPLC) for the determination of TNT and its degradation products. The HPLC method used a gradient of acetonitrile

TABLE 1. Media compositions for enrichment cultures.

Component	Concentration (mg/L)
Common Components	
TNT	50
NaH_2CO_3	5,700
KH_2PO_4	500
NaCl	500
$CaCl_2 \cdot 2H_2O$	100
$MgCl_2 \cdot 6H_2O$	100
$(NH_4)6Mo_7O_{24} \cdot 4H_2O$	10
$ZnSO_4 \cdot 7H_2O$	0.1
H_3BO_3	0.3
$FeCl_2 \cdot 4H_2O$	1.5
$CoCl_2 \cdot 6H_2O$	10
$MnCl_2 \cdot 4H_2O$	0.03
$NiCl_2 \cdot 6H_2O$	0.03
$AlK(SO_4)_2 \cdot 12H_2O$	0.1
Additional Components	
Medium 1	
Glucose	180
Medium 2	
Yeast Extract	500
Medium 3	
Glucose	180
NH_4Cl	500
Medium 4	no addition, TNT as sole carbon and nitrogen
Medium 5	
Glucose	180
Yeast Extract	500
Medium 6	
NH_4Cl	500
Medium 7	
Methanol	32
Yeast Extract	500
Medium 8	
Methanol	32
Medium 9	
Methanol	32
NH_4Cl	500

and 11 mM phosphate buffer (pH 3.2) as the mobile phase, with detection at 210 nm, as described previously (Ahmad & Roberts 1994).

The cultures were gram-stained before microscopic observation and photography, using an Olympus Model BHT microscope with attached camera. Samples were prepared for scanning electron microscopy by air drying and gold sputtering before observation, using a Cambridge Model Stereoscan 250 MK3 electron microscope.

RESULTS

Effect of Inoculum Source

Of the three inocula tested, the treated soil proved to be the most effective. This inoculum allowed the enrichment of cultures displaying the highest turbidity and the largest number of organisms observed microscopically. The cultures (consortia) were also more efficient at removing both TNT and its degradation intermediates. Cultures, obtained from spent activated carbon used in munitions treatment, were hard to assess due to the absence of any detectable organic compounds in the aqueous phase. Also, these cultures displayed very little growth. Aqueous-phase transfers of these cultures neither grew nor degraded TNT.

Cultures from the aqueous bench-top reactor (previously maintained with aqueous extracts from munitions-contaminated soil) were successful at removing TNT from the medium but accumulated partially reduced TNT intermediates (mono- and diaminonitrotoluene compounds). Because the soil inoculum had been used for the majority of the published research, it was used as the sole source of inoculum for the remaining studies. The following discussion summarizes the effects different media preparations had on the ability of bacterial enrichments from the soil to degrade TNT and its partially reduced intermediates. The removal of TNT intermediates was the most important factor considered in determining the success of a culture or growth medium composition.

Effect of Methanol as Additional Carbon Source

Cultures incubated with no additional carbon source were as efficent as those amended with methanol, indicating that methanol did not enhance the culture. The only methanol-containing medium in which the intermediates did not accumulate to any extent was one that also contained yeast extract (Medium 7 in Table 1). Methanol was abandoned as an exogenous carbon source for TNT-degrading cultures. The culture receiving TNT as the only source of carbon and nitrogen (Medium 4 in Table 1) showed very slow degradation of TNT. This enrichment culture was abandoned due to its inefficiency at removing the intermediates from the culture supernatant.

Effect of Glucose as Additional Carbon Source

The cultures fed glucose as an additional carbon source exhibited good growth as well as efficient TNT degradation compared to the methanol-amended cultures, and the cultures receiving no external carbon source (Medium 4). Analysis of the results indicated that glucose and yeast extract combined (Medium 5 in Table 1) were no more effective than yeast extract or glucose alone as the sole amendments. The original culture (Culture 1.1) from the soil inoculum, after being grown on Medium 1 containing glucose and TNT, was successfully transferred to freshly prepared Medium 1. HPLC analysis of the culture supernatants showed that this culture accumulated low levels of 4-amino-2,6-dinitrotoluene, 2-amino-4,6-dinitrotoluene, and 2,4-diamino-6-nitrotoluene. However, stoichiometric accumulation of the intermediates was not observed and mass balance of the detectable compounds was not complete. The culture did not appear to be using the usual reductive pathway past the initial reductions to the monoamino compounds. Investigations are still under way to determine the pathway used by this culture. Investigations of Culture 1.1 using the electron microscope revealed an almost pure culture of small (0.2 to 0.3 m), irregular-shaped oval bacteria enclosed in large amounts of extracellular material. Observations of gram-stained cells with the light microscope revealed only large aggregates of cells with variable gram reactions.

Effect of Ammonium as Nitrogen Source

When ammonium was present as the only addition to the basic medium (Medium 6 in Table 1), TNT was not degraded nor was any bacterial growth observed. When ammonium was present with glucose as an additional carbon source (Medium 3 in Table 1), the cultures exhibited heavy turbidity (growth). TNT was transformed, and both 4-amino-2,6-dinitrotoluene and 2,4-diamino-6-nitrotoluene accumulated in the medium. The culture enriched on this medium from the soil inoculum was maintained for further experimentation (Culture 3.1). Microscopic observation showed it to be a mixed culture containing both gram-negative and positive organisms. The largest population was of gram-negative rods, although gram-positive cocci and rods also were observed. Small numbers of cell aggregates similar to those in Culture 1.1 also, were observed.

Effect of Yeast Extract as Nitrogen Source

Yeast extract proved to be the most effective amendment for enhancing the degradation of TNT and its reduced intermediates. Culture 2.1 was obtained from the soil inoculum with yeast extract and TNT as the only amendments to the medium (Medium 2 in Table 1). Comparison between Culture 1.1 and Culture 2.1 (Table 2) shows that Culture 2.1 removed TNT at a faster rate and accumulated smaller quantities of the reduced intermediates. This culture has survived many transfers and has shown evidence for the complete removal of nitrogen from the TNT ring by accumulation of *p*-cresol in the culture

TABLE 2. Comparison between Culture 1.1 and Culture 2.1. Rates are for comparative purposes only and do not represent optimized kinetics.

Culture	TNT removal rate (mg/L/day)	Maximum 4-amino-2,6-dinitrotoluene accumulated (mg/L)	Maximum 2,4-diamino-6-nitrotoluene accumulated (mg/L)
1.1	6.5	37.1	20.7
2.1	7.4	8.6	15.5

supernatant. Microscopic analysis has shown the culture to contain large numbers of both gram-negative and gram-positive rods and cocci. The aggregates of small cocci are also present but to a much smaller degree.

Results of Cross Transfer to Different Media

The three most successful or interesting cultures (Cultures 1.1, 2.1, and 3.1) were transferred to the three most successful media formulations (Media 1, 2, and 3) to determine whether each population had increased its TNT-transforming ability after only the first enrichment. The results are summarized in Table 3.

Culture 1.1 was efficient at TNT degradation only when transferred into Medium 1. This suggests that the culture degraded TNT to any extent only when it was forced to use TNT as its nitrogen source. The presence of glucose in Medium 1 has encouraged a population that will degrade TNT as its nitrogen source while growing on glucose. This population will use ammonium or yeast extract when these are present, as evidenced by growth in Media 2 and 3, but will not use TNT when the alternative sources are present. The microbial makeup and metabolic characteristics of this culture make it unique among those obtained to date. Continued viability of the culture away following soil isolation and growth on synthetic media is poor. It is possible that the minimal medium was not supplying the complete nutritional requirements of the culture. Further experiments with the addition of various soil extracts to the cultures will determine if nutritional requirements can be met and stable cultures developed.

Culture 2.1 was capable of degrading TNT in all three media tested, but was most efficient when transferred to Medium 2. The use of yeast extract as an organic source of nitrogen and extra carbon was successful in generating a culture that could degrade TNT at least to *p*-cresol. The accumulation of *p*-cresol by these cultures distinguishes them from other cultures that only accumulate the partially reduced intermediates of TNT metabolism. This culture contains the greatest variety of microorganisms and thus is probably the most representative of the soil population. Again, continued transfers of the

TABLE 3. Summary of enrichment culture and transfer results.

Culture	Media #	Comments
1.1	1	TNT degraded rapidly, very few intermediates observed.
	2	TNT degraded very slowly, increased turbidity observed.
	3	TNT degraded very slowly, increased turbidity observed.
2.1	1	TNT degraded, accumulation of intermediates observed.
	2	TNT degraded rapidly, very few intermediates observed.
	3	TNT degraded, accumulation of intermediates observed.
3.1	1	TNT degraded very slowly, increased turbidity observed.
	2	TNT degraded very slowly, increased turbidity observed.
	3	TNT degraded slowly, accumulation of intermediates observed.

culture away from the soil have diluted the efficiency of the culture to some extent. Further experiments with the addition of various soil extracts to the cultures will determine if nutritional requirements can be met and stable cultures developed.

Culture 3.1 grew well in Media 2 and 3, but was not as efficient for removing TNT and its reduced intermediates as Cultures 1.1 or 2.1. This culture was abandoned so the research could focus on Cultures 1.1 and 2.1.

CONCLUSIONS

The enrichment experiments resulted in two distinct cultures, both capable of degrading TNT while accumulating only small amounts of the common, partially reduced intermediates. Culture 1.1 degrades TNT in the presence of glucose, probably using the TNT as the nitrogen source, whereas Culture 2.1 degrades TNT with yeast extract as the only amendment. These cultures possess different overall populations of bacteria and apparently degrade TNT by different pathways. The cultures are not completely stable when transferred repeatedly using the synthetic media, but are reproducible from the soil inoculum. The enrichment trials revealed that none of the inocula tested was capable of using TNT with methanol as the additional carbon source to minimal medium and that the dinoseb-contaminated soil that had been treated with the anaerobic procedure provided the best inoculum.

ACKNOWLEDGMENTS

This research was supported by a cooperative agreement among the U.S. Environmental Protection Agency, Environmental Research Laboratory in Athens, Georgia, the University of Idaho, and the University of Houston.

REFERENCES

Ahmad, F., and D. J. Roberts. 1994. "The use of narrow bore HPLC-diode array detection for the analysis of intermediates of the biological degradation of 2,4,6-trinitrotoluene." *J. Chromatog. 693*: 167-175.

Fedorak, P. M., and S. E. Hrudey. 1986. "Nutrient requirements for the methanogenic degradation of phenol and *p*-cresol in anaerobic draw and feed cultures." *Water Res. 7*: 929-933.

Funk, S. B., D. L. Crawford, D. J. Roberts, and R. L. Crawford. 1994. "Two-stage bioremediation of TNT contaminated soils." In B. S. Schepart (Ed.), *Bioremediation of Pollutants in Soil and Water. ASTM STP 1235*, American Society for Testing and Materials, Philadelphia, PA, 1994.

Funk, S. B., D. J. Roberts, D. L. Crawford, and R. L. Crawford. 1993. "Initial-phase optimization for bioremediation of munition compound-contaminated soils." *Appl. Environ. Microbiol. 59*: 2171-2177.

Kaake, R. H., D. J. Roberts, T. O. Stevens, R. L. Crawford, and D. L. Crawford. 1992. "Bioremediation of soils contaminated with 2-*sec*-butyl-4,6-dinitrophenol (dinoseb)." *Appl. Environ. Microbiol. 58*: 1683-1689.

Roberts, D. J., and D. L. Crawford. 1991. "Anaerobic degradation of TNT." *Abstracts of the 91st General Meeting of the American Society for Microbiology*, Dallas, TX. Q160.

Roberts, D. J., S. B. Funk, and R. A. Korus. 1992a. "Intermediary metabolism during anaerobic degradation of TNT from munitions-contaminated soil." *American Society for Microbiology 92nd General Meeting*. New Orleans, LA. Q136.

Roberts, D. J., R. H. Kaake, S. B. Funk, D. L. Crawford, and R. L. Crawford. 1992b. "Field scale anaerobic bioremediation of dinoseb-contaminated soils." In M. Gealt, and M. Levin (Eds.), *Biotreatment of Industrial and Hazardous Wastes*, pp. 219-244. McGraw-Hill, New York, NY.

Aerobic Treatment of Explosives-Contaminated Soils Using Two Engineering Approaches

Mark E. Zappi, Douglas Gunnison,
and Herb L. Fredrickson

ABSTRACT

Explosives contamination represents a widespread problem to the U.S. Department of Defense. The potential of two aerobic engineering application approaches—bioslurry and bioagriculture systems—for aerobic biotreatment of explosives-contaminated soils was evaluated using bench-scale reactors. The results were that the addition of a commercially available surfactant dramatically improved the treatment effectiveness of both the bioslurry and bioagriculture systems. Formation and disappearance of aminonitrotoluenes (aminodinitrotoluenes and diaminonitrotoluenes) and nitrobenzenes (di- and trinitrotoluenes) were observed within both systems. The bioslurry systems had much more rapid removal kinetics than the bioagriculture systems. The rationale for this observation is believed to be the superior conditions provided in the bioslurry system over those provided within the bioagriculture system.

INTRODUCTION

The U.S. Department of Defense (DoD) has numerous sites that have become contaminated with explosive compounds such as 2,4,6-trinitrotoluene (TNT). The source of this contamination is from manufacturing and/or assembly activities of munitions. The current technology for soils contaminated with TNT is incineration, which tends to be cost prohibitive (U.S. \$350 to \$1,200 per yd^3 [0.765 m^3] treated) and difficult to implement due to regulatory obstacles and widespread social objections. For these reasons, alternative approaches are under development by the U.S. Army.

Bioslurry is an engineering reconfiguration of other more widely used biotreatment approaches such as landfarming and composting that have been successfully used for soil cleanup. Bioslurry systems are capable of substantially increasing contaminant degradation rates by greatly enhancing process mass-transfer efficiencies. Soils within bioslurry systems are completely mixed within soil-water slurries (soil-to-water ratios ranging from 20% to 50% [w/w]) using mechanical mixers or slurry pumps. The result of these operational features is a biological system that is conducive to optimal biodegradation in terms of rate and completeness of reaction (Zappi et al. 1991). Ross (1990) estimated the cost of applying bioslurry systems ranging from $90 to $200 per yd^3 (0.765 m^3). Montemagno and Irvine (1990) determined that slurry systems provided excellent conditions for removing TNT from contaminated soils using bench-scale units.

Most of the explosives contamination found at DoD facilities appears to be located within the upper few feet (1 to 2 m) of surface soils. Bioagriculture is an engineered approach that treats surface soils in place without requirement of soils excavation/transport. This approach is designed to provide appropriate ecological conditions by merit of scheduled irrigation and possibly aeration via buried sparger lines (bioventing). Bioagriculture is relatively passive compared to bioslurry systems resulting in reduced unit treatment costs. Based on our estimates, these costs range from $20 to $110/yd^3 (0.765 m^3).

Past research on TNT degradation has determined that TNT can be biologically transformed into several by-products, some of which were deemed more toxic than the TNT (Carpenter et al. 1978; McCormick et al. 1976; Kaplan and Kaplan 1982). The primary intermediates of incomplete aerobic biological transformation are aminonitrotoluenes, nitrobenzenes (including tri- and dinitrobenzenes), and azoxy complexes. Additionally, these intermediates are problematic because they tend to bind to organic material in soils and persist (Pennington et al. 1995). Anaerobic biotreatment of soils may minimize the formation of persistent intermediates (Crawford 1994). Alternatively, the problem of persistent intermediates may be circumvented by the initial removal of the nitro groups (Ramos et al. 1994).

The results of recent studies that were performed in support of cleanup efforts of TNT-contaminated soils at a former military facility are summarized in this paper. Two engineering approaches to aerobic biotreatment were evaluated: bioslurry and bioagriculture systems. Discussion is focused on the relative performance of the two approaches. The reason for evaluation of two approaches was that if bioagriculture were to be deemed feasible, then potential cost savings as high as an order of magnitude may be realized over the use of bioslurry systems.

EXPERIMENTAL APPROACH

The soil samples used in this study were passed through a No. 4 sieve prior to loading into either reactor type. Although the samples were collected within close proximity of each other, the initial TNT concentration in the soil used in the bioslurry experiment was 18,000 mg/kg, whereas the soil used in the bioagriculture cells initially contained 11,000 mg/kg of TNT. Major by-products of incomplete degradation via natural processes in the environment initially detected in the soil samples were aminodinitrotoluenes (2-amino-4,6-dinitrotoluene and 4-amino-2,6-dinitrotoluene) and nitrobenzenes (dinitrobenzene and trinitrobenzene). Initial concentrations of these by-products in the bioslurry soils were approximately 15 mg/kg and 10 mg/kg, respectively. The soil used in the bioagriculture experiments had approximate by-product levels of 40 mg/kg and 30 mg/kg, respectively. The total azoxy-toluene level initially detected in the bioagriculture soil samples was 8.35 mg/kg (Note: azoxy-toluene standards supplied by Ron Spanggord, SRI Inc., Menlo Park, California, were not available during the bioslurry experiments).

All soil samples were analyzed for TNT and related by-products using high-performance liquid chromatography (HPLC). Oxygen uptake rates (OURs) were determined by analyzing the change in slurry-phase oxygen concentrations over a 10-minute period using a membrane probe technique. Oxygen levels in the slurries were determined at the initiation of each OUR test which was performed twice weekly.

Prior to selecting the treatment conditions evaluated using the bench-scale reactors, several foreign and native soil microbial consortia were evaluated during this study for their relative ability to degrade and mineralize TNT through addition of a variety of cometabolites using shake-flasks containing 20% explosives-contaminated soils (80% water) and various chemical additives. The native consortia were found to achieve mineralization of TNT aerobically as high as 25% conversion using acetate as a cometabolite. Also evaluated was the use of several commercially available, food-grade surfactants using methods described by Zappi et al. (1991). These efforts determined that Tween 80 at a 3% dose (w/w dry soil) improved the apparent desorption rate of TNT by more than 400% compared to a water control.

The bioslurry experiments were performed using 5-L bioslurry reactors (Zappi et al. 1991) that contained 4.5 L of 30% slurry (w/w dry soil). These experiments generally followed procedures outlined by Zappi et al. (1991). The bioslurry reactors were covered with aluminum foil to prevent photodecomposition. The following treatment conditions were evaluated in the bench-scale bioslurry reactors: (1) acetate amendments; (2) acetate and nutrient amendments; (3) acetate and surfactant amendments; and (4) acetate, surfactant, and nutrient amendments. Acetate was dosed weekly at a 1% (w/w dry soil). Ammonia and phosphate were dosed on an as-needed basis to maintain aqueous-phase concentrations of at least 50 mg/L and 10 mg/L, respectively. Surfactant (Tween 80) was dosed weekly to achieve a 3% (w/w dry soil) dose. Each treatment condition was evaluated using duplicate reactors operated in batch mode for a period of 11 weeks. A poisoned control was established in duplicate reactors using 5,000 mg/kg of mercuric chloride. Enumeration of total heterotrophs and oxygen utilization rate measurements within the poisoned controls indicated no biological activity.

The bioagriculture cells used in this study had a working volume of 250 mL and were constructed of polypropylene (plastic). Moisture sensors were installed in each cell to ensure adequate moisture was present (30% of field capacity). Each cell was charged with a dry soil equivalent dose of 130 g. Water was added to the cells via addition of 30 mL irrigation water every 5 to 7 days. The irrigation water was amended with various amendments based on the targeted treatment condition. Prior to using the plastic cells, no detectable TNT sorptive losses onto the plastic material were observed using Teflon™ as a control. Periodically, air was blown through the cells to maintain aerobic conditions. The gases exiting the cells were monitored for oxygen to determine if the aeration frequency was adequate. All of the cells were covered with black plastic to prevent photodegradation.

Treatment conditions evaluated within the cells were (1) water-only amendments (biotic control); (2) low-nutrient, surfactant, and acetate amendments; (3) high-nutrient, surfactant, and acetate amendments; and (4) poisoned controls. Surfactant and acetate were dosed by amending the irrigation water with 14 g/L of Tween 80 and 67.5 g/L of sodium acetate. Surfactant was added to all of the amended cells because the results of the bioslurry experiments (performed first) indicated that the surfactant dramatically improved the biodegradation rate. Low nutrient levels were maintained by dosing the irrigation water with 126 mg/L NH_4NO_3, 18 mg/L K_2HPO_3, and 18 mg/L $K_2H_2PO_3$. High nutrient levels were established by adding 1,260 mg/L NH_4NO_3, 180 mg/L K_2HPO_3, and 180 mg/l $K_2H_2PO_3$ to the irrigation water. Poisoned controls were established by blending 16.7 g mercuric chloride/kg of dried soil. After that point, the poisoned controls were operated identically to the high-nutrient-, surfactant-, and acetate-amended cells. Enumeration of total heterotrophs within the poisoned controls indicated no biological activity.

Soil samples were collected for explosives analysis from the bioslurry reactors at incubation times of 1, 3, 5, 7, 9, and 11 weeks. The bioslurry poisoned controls were sampled at Weeks 1, 3, 5, and 7. Soil samples from two replicate bioagriculture cells (sacrificial) were collected at test times of 0, 2.5, 4, 6, and 7 months of incubation.

RESULTS

The results of the bioslurry experiments with respect to TNT indicated that the surfactant-amended systems had significantly higher TNT removal rates compared to the reactors not amended with surfactant. For example, by Week 5, the two surfactant-amended bioslurry treatments had TNT levels well below 10 mg/kg, whereas the treatment not involving surfactant addition achieved only 50% removal by Week 5. By Week 11, the treatment conditions not amended with surfactant only reached levels within the 2,000 to 3,000 mg/kg range. The addition of nutrients to the acetate/surfactant-amended system did not appear to enhance the TNT degradation rate.

The bioagriculture experiments indicated an appreciable increase in TNT degradation rate through the use of higher nutrient dosages. This increase differs from the results observed from bioslurry experiments in which the TNT degradation rate appeared independent of nutrient dosing within the nutrient doses evaluated. The high-level-nutrient-dosed bioagriculture system reached over 95% TNT removal within 7 months, while the low-level-nutrient system achieved only 50% TNT removal within the same time frame. The rate of TNT disappearance achieved within the low-level-nutrient-amended cells was slightly greater than the rates observed with both control types. The biotic control achieved only slightly better TNT removal than did the poisoned controls.

The poisoned controls in both the bioslurry and bioagriculture experiments indicated significant TNT transformation (approximately 40% and 60% for the bioslurry and bioagriculture experiments, respectively). These data support recent observations by researchers in our laboratory who evaluated the fate of TNT in soil column experiments. These studies found that much of the TNT was reduced to aminonitrotoluenes during transport through non-aerated columns loaded with soils not previously contaminated with TNT. The mechanism(s) of TNT transformation within these columns are under further investigation; however, abiotic reduction of TNT with simultaneous oxidation of iron has been shown (personal communication with Jim Brannon, U.S. Army Engineers Waterways Experiment Station, Vicksburg, Mississippi, 1994).

All of the reactors in the bioslurry experiments showed conversion of TNT to aminonitrotoluenes (ANTs). The overall yield of these compounds was much greater in the biotic systems compared to the poisoned controls. The surfactant-amended bioreactors had a much higher ANT conversion rate than the systems not amended with surfactant (on the order of 300 mg/kg and 600 mg/kg by Week 7 for the acetate/surfactant/no-nutrient-amended and acetate/surfactant/nutrient-amended systems, respectively). Conversely, in the biotic systems without surfactant addition, ANT production peaked at the 200- to 300-mg/kg level by Week 9. The ANT data for the surfactant-amended systems also indicated a downward trend in ANTs after 9 weeks of incubation, while the systems without surfactant amending did not show a decrease in ANTs by Week 11. The poisoned control indicated a gradual increase of ANTs to a maximum of approximately 60 mg/kg by Week 5 with no ANTs being detected at Week 7.

The bioagriculture experiments generally followed the same ANT trends observed in the bioslurry experiments in that the surfactant-amended systems had appreciable conversion of TNT to ANTs. Within 2.5 months, the surfactant-amended systems resulted in the formation of approximate ANT levels of 250 mg/kg, which were reduced to sub-100 mg/kg levels by Month 7. The two controls indicated little or no change in ANT levels throughout the course of this experiment.

The surfactant-amended bioslurry reactors and controls were the only bioslurry systems to result in an appreciable change in total nitrobenzene concentrations over time. The surfactant-amended bioslurry reactors displayed a steady reduction in concentration over time until reaching less-than-detection-levels by Week 7. The bioslurry-poisoned controls showed a steady increase in trinitrobenzene (TNB) levels over the full course of the experiment (approximately 60 mg/kg by Week 7), possibly due to the abiotic oxidation of the TNT to TNB. The level of total nitrobenzenes in the other biologically active reactors without surfactant amendments generally remained constant over the course of this experiment.

The bioagriculture experiments indicated a high rate of total nitrobenzene conversion within the high-nutrient-dosed system (more than an order of magnitude increase). However, by Month 7, the total nitrobenzene levels within these systems were reduced to levels slightly higher than the initial concentrations. The low-nutrient-level systems indicated little change in total nitrobenzene levels until Month 7, when the levels were reduced by approximately 50%. The total nitrobenzene levels within biotic controls generally remained unchanged over the course of the experiment. The poisoned control indicated an approximate order of magnitude increase in total nitrobenzenes by Month 6; however, these levels appeared to be reduced to levels slightly higher than the levels initially present in the soils by Month 7.

No azoxy-nitrotoluenes were detected within the soil samples collected from any of the biologically active bioagriculture cells throughout this experiment. Final azoxy-nitrotoleune levels detected in the bioagriculture poisoned controls were 830 mg/kg. As stated earlier, the azoxy-nitrotoluene standards were not available during the bioslurry experiments.

Comparison of the performance of the two engineering approaches indicates that the bioagriculture cells appeared to suffer from chemical heterogeneities within the reactors associated with poor mixing. Poor mixing can result in establishment of zones with quite varied ecological conditions ranging from anaerobic to aerobic and unamended to overdosed additive levels. However, it was noted that the bioagriculture system with high-level-nutrient dosing did produce appreciable TNT removal rates (approximately 1,500 mg/kg/month assuming zero-order kinetics). As expected, the bioslurry system clearly represents a process with the potential for achieving significantly higher removal rates than those obtained within the bioagriculture cells. Selection of an approach should be based on available remediation time and the level of concern for potential residual levels of ANTs and nitrobenzenes remaining within the treated soils due to problems likely associated with poor mixing.

CONCLUSIONS

The results of this study indicated that the native microbial consortia had appreciable activity toward the TNT when amended with acetate. The addition of a commercial surfactant, Tween 80, improved the desorption rate of the TNT from the soil by over 400%. This enhancement led to a dramatic improvement in TNT biodegradation within the bioslurry system evaluated. The addition of surfactant also increased the yield and rate of by-product degradation. The results of the bioagriculture experiment indicate that increased levels of nutrients may also enhance TNT removal rate. This dependence was not observed with the bioslurry experiment. The ANT data from both experiments indicated that surfactant addition resulted in systems superior to the other conditions evaluated. The surfactant-amended bioslurry systems achieved TNT removal efficiencies in excess of 95% within a time frame of only 5 weeks compared to 7 months for the high-level-nutrient, surfactant-amended bioagriculture system. The nitrobenzene data for the two systems differed in that no production of nitrobenzene was observed within the biotic bioslurry systems. The high-nutrient, surfactant-amended bioagriculture system and poisoned control both appeared to achieve similar nitrobenzene yields and removal rates. None of the bioagriculture systems tested resulted in a net reduction in nitrobenzene levels.

The poisoned control experiments indicated that TNT is not stable with a large portion being degraded to ANTs, nitrobenzenes, and azoxy-nitrotoluenes. Oxygen utilization rate analysis and microbial enumeration efforts indicated that these controls were indeed abiotic, indicating that the mechanisms associated with TNT loss were likely abiotic.

Based on these results it was decided that bioslurry systems provided better conditions more conducive to rapid explosive and related by-product degradation. Bioagriculture systems have potential for explosive compound removal; however, the rates of contaminant removal were much slower than those obtained within the bioslurry systems. Further research is required to determine if intermediates of TNT biodegradation are also degraded within the bioagriculture systems.

ACKNOWLEDGMENTS

The authors would like to thank the U.S. Army Corps of Engineers, Kansas City District and the Army's Environmental Quality and Technology Program for funding this study. Permission to publish this paper was granted by the Chief of Engineers.

REFERENCES

Carpenter, D. F., N.G. McCormick, J.H. Cornell, and A.M. Kaplan. 1978. "Microbial Transformation of 14C-Labeled 2,4,6-Trinitrotoluene in an Activated Sludge System." *Applied and Environmental Microbiology*, 35: 949-954.

Crawford, R.L. 1994. "Biodegradation of Nitrated Munition Compounds and Herbicides by Obligately Anaerobic Bacteria." In J. Spain (Ed.), *Proceedings of the Symposium on Degradation of Nitroaromatic Compounds*, 22-23 May 1994, Las Vegas, NV. (in press).

Kaplan, D. L. and A.M. Kaplan. 1982. "Thermophilic Biotransformations of 2,4,6-Trinitrotoluene Under Simulated Composting Conditions." *Applied and Environmental Microbiology*, 44: 757-760.

McCormick, N. G., F.E. Feeherry, and H.S. Levinson. 1976. "Microbial Transformation of 2,4,6-Trinitrotoluene and Other Nitroaromatic Compounds." *Applied and Environmental Microbiology, 31*: 949-958.

Montemagno, C.D. and R.L. Irvine. 1990. *Feasibility of Biodegrading TNT-Contaminated Soils in a Slurry Reactor.* Report Number CETHA-TE-CR-90062, U.S. Army Toxic and Hazardous Agency, Aberdeen Proving Grounds, MD.

Pennington, J.C., C.A. Hayes, K.F. Myers, M. Ochman, D. Gunnison, D.R. Felt, and E.F. McCormick. 1995. "Fate of 2,4,6-Trinitrotoluene in a Simulated Compost System." *Chemosphere* (in press).

Ramos, J., A. Haidour, A. Gelgado, E. Duque, M. Fandila, and G. Pinar. 1994. "Potential of Toluene-Degrading Systems for the Construction of Hybrid Pathways for Nitrotoluene Metabolism." In J. Spain (Ed.), *Proceedings of the Symposium on Degradation of Nitroaromatic Compounds,* 22-23 May 1994, Las Vegas, NV. (in press).

Ross, D., 1990, "Slurry-Phase Bioremediation: Case Studies and Cost Comparisons." *Remediation, 1*, Winter 1990/91.

Zappi, M.E., D. Gunnison, C.L. Teeter, and N.R. Francingues. 1991. "A Proposed Laboratory Method for Performance of Bench Scale Bioslurry Studies." Paper presented at the 1991 Superfund Conference, Hazardous Material Control Research Institute, Silver Spring, MD.

Fungal Degradation of Nitrocellulose Under Aerobic Conditions

Anil Sharma, Shanmuga T. Sundaram,
Ying-Zhi Zhang, and Bruce W. Brodman

ABSTRACT

Mycelial fungi were screened alone or in combinations for their ability to degrade nitrocellulose (3 g/L) in liquid medium. All of the fungi tested used nitrocellulose to a varying extent, but a combination of *Sclerotium rolfsii* ATCC 24459 and *Fusarium solani* IFO 31093 was found to be the best because it significantly degraded nitrocellulose. About 38% of the nitrocellulose was degraded by these fungi in a 7-day period when the culture medium was buffered at pH 6.0 with morphilino ethane sulfonic acid.

INTRODUCTION

Nitrocellulose is the main energetic ingredient in gun propellant formulations and is used in a range of other munitions. As a result of its manufacture, use, and destruction, a number of contaminated sites exist. The current remediation method involves the use of a mobile incinerator, requiring soil removal, incineration, and subsequent replacement of the decontaminated soil. This procedure is expensive and time consuming. Biodegradation offers the potential for a much improved remediation method. The purpose of this paper is to describe the biodegradation technology that offers the potential for future accomplishment of this objective.

In the past, a number of studies involving chemical and biological treatments of nitrocellulose have been carried out (Gold and Brodman 1989; Wendt and Kaplan 1976). Kaplan and his coworkers suggested that nitrocellulose is not subject to direct microbial degradation and that a chemical pretreatment is necessary to generate a modified polymer that could be utilized by microorganisms (Wendt and Kaplan 1976). However, Ilinskaya and Leshchinskaya (1988) reported that microorganisms can degrade nitrocellulose. Previously, we reported that partial degradation of nitrocellulose could be achieved by using a combined culture containing *Sclerotium rolfsii* ATCC 24459 and *Fusarium solani* IFO 31093 (Sharma et al. 1995a). We also have reported nitrocellulose degradation by a *Penicillium corylophilum* Dierckx isolated from a double-base propellant (Sharma et al. 1995b). In the present investigation, we have attempted to improve nitrocellulose degradation as determined by the nitrocellulose weight reduction, biomass weight increase, and the presence of cellulose- and nitrate-utilizing enzymes. We do not mean to imply that mineralization has occurred.

EXPERIMENTAL PROCEDURES AND MATERIALS

Growth Medium

Mineral salts medium used in this study contained (g/L): KH_2PO_4, 1.0; $MgSO_4$, 0.5; NaCl, 0.1 and $CaCl_2$, 0.1. A trace metal solution was prepared which contained (g/L): $C_6H_8O_7$ (citric acid), 50.0; $ZnSO_4$, 50.0; $CuSO_4$, 2.5; $MnSO_4$, 2.5; H_3PO_4, 0.5; $Na_2MoO_4 \cdot 2H_2O$, 0.5; and $CoCl_2$, 20.0, and 100 mL of this solution was added to 1 L of the basal salt solution.

Cultural Conditions. The inoculum preparation and the sterilization of nitrocellulose for the biodegradation studies were carried out as described earlier (Sharma et al. 1995b). A known quantity of homogenized mycelia (~ 20 mg dry wt) was transferred to 50 mL of experimental medium (50 mL in a 250-mL Erlenmeyer flask) containing nitrocellulose (3 g/L) as a nitrogen source and starch (1 g/L) as a cosubstrate. Inoculated flasks were incubated for various time intervals (0 and 7 days) in a gyrotary shaker (~ 150 rpm, 28°C). Equal quantities of each homogenized fungal culture (1:1) were used whenever a combination of fungi was used as a source of inoculum. Three different controls were run in parallel to the treatment. One lacked nitrocellulose but contained viable cultures; another contained NaN_3 (1 mM) and $HgCl_2$ (1 mM) along with the cultures (killed control), and the third contained nitrocellulose but lacked the cultures and fungicidal agents. The fungal biomass was harvested at the desired time intervals by centrifugation (8,000 rpm, 4°C, 15 min). The biomass and the supernatant were saved for further analyses.

Analytical Methods. Fungal biomass, residual nitrocellulose, enzyme preparation (Sharma et al. 1995b), total sugars (Dubois et al. 1956), glucose (Sigma glucose estimation kit, 1994), nitrate and nitrite (Nicholas and Nason 1957), carboxymethyl cellulase (Mandels and Weber 1969), nitrate, and nitrite reductases were determined as described elsewhere (MacGreyor 1978; Kakutani et al. 1981).

Organisms. *F. solani* IFO 31093, a denitrifying fungus, was purchased from the Institute for Fermentation, Osaka, Japan. *S. rolfsii* ATCC 24459 was purchased from the American Type Culture Collection, Rockville, Maryland. *P. corylophilum* Dierckx was isolated as described earlier (Sharma et al. 1995b).

Nitrocellulose. Nitrocellulose (smokeless grade) was received as a gift from Hercules Inc., Kenvil, New Jersey. It contained 13.17% nitrogen (~2.33 nitrate ester groups per repeat unit) and 25.07% moisture.

RESULTS

Screening of Fungal Cultures for the Nitrocellulose Degradation

Our earlier work revealed that a combined culture of *S. rolfsii* and *F. solani* could degrade at least 27% of nitrocellulose (3 g/L) in liquid medium under aerobic conditions. In the present study, we made several attempts to improve the rate and amount of nitrocellulose degradation, including the screening of several other fungi. Of the several cultures screened, we observed that the combination of *S. rolfsii* and *F. solani* appears to be the best for the nitrocellulose degradation (Table 1). Therefore, we pursued further studies with these fungi to obtain maximal degradation of nitrocellulose.

Effect of a Nonionic Surfactant on the Degradation of Nitrocellulose

It had been reported previously that the inclusion of a surfactant in the fungal growth medium enhanced the biodegradation of several environmental pollutants. Hence, we have tested the effect of a nonionic surfactant, Makon NF-5, on the degradation of nitrocellulose by *S. rolfsii* and *F. solani*. Results presented in Table 1 indicate that the inclusion of Makon NF-5 did not increase the nitrocellulose degradation. We also have tested the effect of several cationic and anionic surfactants and found that none of them could enhance the degradation of nitrocellulose by these fungi (data not shown).

Effect of a Heavy and a Second Dose of Inoculum

In our previous investigation, we used 6.9 mg (dry weight) of the combined fungal inoculum in the experimental medium. In the present study, we tested the effect of a heavy (30 mg of homogenized mycelia) and a second equivalent dose of inoculum on nitrocellulose degradation. The results (Table 1) revealed no increase in nitrocellulose use by *S. rolfsii* and *F. solani* under these conditions.

Effect of Maintaining a Constant pH

It had been observed previously that, during the growth *of S. rolfsii* and *F. solani* on nitrocellulose, the pH of the culture medium dropped from 6.0 to 2.0. At this acidic pH, certain enzymes involved in the degradation of nitrocellulose might have been inactivated. Therefore, we tested the effect of

TABLE 1. Fungal degradation of nitrocellulose in liquid medium.

Cultural conditions	Residual nitrocellulose (g/L)	Percent nitrocellulose degraded
Control (abiotic)[a]	2.92	—
S. rolfsii	2.42	17.1
F. solani	2.32	20.5
P. corylophilum[b]	2.22	23.9
P. corylophilum+*F. solani*	2.62	10.3
S. rolfsii and *F. solani*	2.14	26.7
S. rolfsii and *F. solani*+Makon NF-5	2.10	28.1
S. rolfsii and *F. solani* (Heavy inoculum)[c]	2.16	26.0
S. rolfsii and *F. solani*+Makon NF-5 (Heavy inoculum)[c]	2.28	21.9
S. rolfsii and *F. solani* (pH maintained at 6.0)	1.82	37.7
S. rolfsii and *F. solani*+Makon NF-5 (pH maintained at 6.0)	1.86	36.3

Starch added at 1 g/L concentration; cultivation time 7 days.
Nitrocellulose was sterilized by using ultraviolet light (254 nm).
Initial weight of nitrocellulose, 3 g/L; Makon NF-5 added at 50 mg/L concentration.
(a) Control lacking cultures and fungicidal agents.
(b) Data taken from Sharma et al. (1955b).
(c) 30 mg of fungal biomass used as an initial inoculum; on the 3rd day, an additional 30 mg of inoculum/flask was added.
No growth was observed in the control lacking nitrocellulose.

maintaining a constant pH on the nitrocellulose degradation. The fungal cultures were cultivated on nitrocellulose and starch dispensed in basal salts medium prepared using 100 mM morphilino ethane sulfonic acid (MES) buffer (pH 6.0). Under these conditions, no drop in pH was measured and the results indicate at least an 11% increase in the nitrocellulose use over the previous experiment. The presence or absence of Makon NF-5, a nonionic surfactant, had very little influence on nitrocellulose degradation in media buffered with MES (Table 1).

Enzymes Involved in the Degradation of Nitrocellulose.

Because *S. rolfsii* and *F. solani* can degrade nitrocellulose, we tested the levels of enzymes likely to be involved. Results indicate that cellulose- and nitrate-utilizing enzymes are present in the growing cultures (Table 2). Depletion of the total sugars along with the nitrate and nitrite ions in the growth medium, compared to the control, indicate that the products of nitrocellulose are used by the fungi. The detection of small amount of sugars, nitrate, and nitrite ions in the control may be due to the fact that nitrocellulose is capable of undergoing some aqueous hydrolysis (Table 1).

TABLE 2. Enzyme levels and the metabolic products released during nitrocellulose degradation by *S. rolfsii* ATCC 24459 and *F. solani* IFO 31093 in
liquid medium.

Culture	Carboxy-methy cellulase (Units/mL)[a]	Nitrate reductase (Units/mL)[b]	Nitrite reductase (Units/mL)[c]	Total sugars (mg/mL)	Glucose (mg/mL)	Nitrate (μg/mL)	Nitrite (μg/mL)
Control	0	0	0	4.0	0.5	11.1	9.4
S. rolfsii and *F. solani*	<0.1	8.4	1.6	0.9	0.05	10.4	2.0

Starch was added at 1 g/L concentration; cultivation time 7 days.
(a) Detected in the presence of Makon NF-5. Unit is defined as the amount of enzyme which generated 1 μmole of glucose equivalent per min.
(b) Unit is defined as the amount of enzyme generating 1 μmole of nitrite per min.
(c) Unit is defined as the amount of enzyme required to reduce 1 nmole of nitrite per min.
No fungal growth was observed in the control lacking nitrocellulose.

DISCUSSION

Our earlier investigation indicated that 27% of the nitrocellulose was
degraded by *S. rolfsii* and *F. solani* in unbuffered media (Sharma et al. 1995a).
In the present study, we have shown about 38% degradation of nitrocellulose
by the combined culture containing *S. rolfsii* and *F. solani* in liquid medium
buffered with MES. As part of our ongoing optimization effort, selected
enzyme and metabolite concentrations were measured. These studies will provide valuable information for optimization of nitrocellulose degradation. The
information on cofactor requirements, rate-limiting steps, pH and temperature
optima, inhibitors, inducers, and other enzyme-related factors controlling the
degradation process can be beneficial in enhancing nitrocellulose degradation.
In the present investigation, we have detected certain nitrate-reducing enzymes that are likely to be involved in the nitrocellulose biodegradation pathway. The cellulose-degrading enzyme, carboxymethyl cellulase, was also
detected, but at lower levels. Attempts will be made to induce the expression
of this important enzyme at higher levels during our optimization effort.
Based on the data obtained, we hypothesize that starch, which was provided
as a cosubstrate, and soluble nitrate and nitrite ions, resulting from the aqueous hydrolysis of nitrocellulose (Brodman and Devine 1981), support initial culture growth. Later, *S. rolfsii* and *F. solani* begin using the nitrate ester group
and, subsequently, the cellulosic part of nitrocellulose.

During the course of this investigation, consideration has been given to the
practical implementation of this technology. For example, the use of starch as
an additional carbon source for the biodegradation of nitrocellulose could be
replaced by cheaper and more available agricultural by-products.

The current data indicate that nitrocellulose utilization can be further optimized. If successful, this technology will form a basis for a cost-effective and environmentally safe soil remediation method.

REFERENCES

Brodman, B.W., and M.P. Devine. 1981. "Microbial attack of nitrocellulose." *J. Applied Polymer Sci.* 26(4):997-1000.

Dubois, M., K.A. Gilles, J.K. Hamilton, P.A. Rebers, and F. Smith. 1956. "Colorimetric method for the determination of sugars and related substances." *Anal. Chem.* 281 (3): 350-356.

Gold, K., and B.W. Brodman. 1989. "Chemical process for the denitration of nitrocellulose." U.S. Patent number 4,814,439.

Ilinskaya, O.N., and I.B. Leshchinskaya. 1988. "Growth of microorganisms on cellulose nitrate." *Biotechnologia* 4(4): 495-500.

Kakutani, T.H., H. Watanabe, K. Arima, and T. Beppu. 1981. "Purification and properties of a copper containing nitrite reductase from a denitrifying bacterium *Alcaligenes faecalis* strain S-6." *J. Biochem.* 89 (2): 453-461.

MacGreyor, C.H. 1978. "Isolation and characterization of nitrate reductase from *E. coli.*" In S.P. Colowick and N.O. Kaplan (Eds.) *Methods in Enzymology, LIII: pp.* 347-355, Academic Press, New York, NY.

Mandels, M., and J.M. Weber. 1969. "Cellulases and their applications" In R.F. Gould (Ed.), *American Chem. Soc. Ser.* 95:391-414. American Chemical Society, Washington, DC.

Nicholas, D., and A. Nason. 1957. "Determination of nitrate and nitrite." In S.P. Colowick and N.O. Kaplan (Eds.) *Methods in Enzymology, 33:* 981-984. Academic Press, New York, NY.

Sharma, A., S.T. Sundaram, Y. Z. Zhang, and B.W. Brodman. 1995a. "Nitrocellulose degradation by *Sclerotium rolfsii* ATCC 24459 and *Fusarium solani* IFO 31093." *J. Ind. Microbiol. 14:* (In Press).

Sharma, A., S.T. Sundaram, Y. Z. Zhang, and B.W. Brodman. 1995b. "Biodegradation of nitrate esters: 2. Degradation of nitrocellulose by a fungus isolated from a double-base propellant." *J. Applied Polymer Sci. 55:* (In Press).

Wendt, T.M., and A.M. Kaplan. 1976. "Chemical and biological treatment process for cellulose nitrate disposal." *J. Water Poll. Control Fed. 48*(4): 660-668.

Bioremediation of a Groundwater Containing Alkylpyridines

James C. Young, Jean-Marc Bollag,
and Cheng-Hsuing Hsu

ABSTRACT

Groundwater at an industrial site has been contaminated with a mixture of alkylpyridines along with other organic materials. Studies were conducted by the authors to evaluate the potential for using bioremediation to clean up this site. Laboratory tests indicated that monosubstituted pyridines were readily biodegradable but 3,5-lutidine was relatively resistant. Both mixed and fixed-film aerobic reactors were effective in degrading the alkylpyridines, but the fixed-film reactor gave better efficiency. Thus, aerobic processes were shown to offer potential for bioremediation of the alkylpyridine-contaminated site.

INTRODUCTION

The contaminated groundwater site was located at a facility of Reilly Industry, Inc. in Indianapolis, Indiana, where alkylpyridines and other chemicals have been produced as industrial products. Contamination of the groundwater beneath the industrial site has been associated with leakage from pipelines and storage tanks, by-product wastes, and leachates from on-site waste disposal lagoons. The contaminants detected at the site included pyridine; monosubstituted pyridines such as 2- and 4-picoline; and disubstituted pyridines such as 2,3-, 2,4-, 2,5-, 2,6-, and 3,5-lutidine (ENSR 1991; Bollag et al. 1992).

Although recent investigations (Rogers et al. 1985; Sims and Sommer 1985) have shown significant potential for alkylpyridine degradation, conventional biological treatment processes may not achieve the permissible concentration of 35 μg/L established in the Record of Decision (ROD) for direct recharge to the aquifer. Such low concentrations generally can be obtained only by maintaining the degradation reaction in a decay environment. Decay occurs in a biological process when operating at an extremely high biomass-to-substrate ratio. This condition also is beneficial in minimizing sludge production. The objective of this study was to compare the biodegradation of alkylpyridines in a decay environment, as exists in the second stage of a two-stage fixed-film process, to that of a mixed reactor.

BACKGROUND

The relationship between substrate utilization, biomass synthesis and decay, and process solids retention time (SRT) in a steady-state reactor can be expressed by the following equation (Young 1993):

$$\frac{1}{SRT} = Y \frac{kS_e}{K_s + S_e} - k_d$$

Where: SRT = solids retention time, days
   S_e = residual substrate concentration, mg substrate/L
   k = maximum substrate conversion rate, h^{-1}
   Y = growth yield coefficient for active biomass, mg biomass/mg substrate removal
   K_s = substrate concentration when conversion rate = 0.5k
   k_d = decay rate, h^{-1}

As SRT becomes very large, the equilibrium substrate concentration (S_e) approaches a minimum (S_{min}) or:

$$S_{min} = K_s \frac{k_d}{Yk - k_d}$$

The above analysis states that operation of a continuously fed biological process at S_{min} can be accomplished only at infinite SRTs, and operation below S_{min} can be accomplished only when there is a net decay environment.

The decay concept was used by Howerton and Young (1987) to develop a two-stage, cyclic process in which the first stage is maintained in a growth phase because of high substrate concentrations while the second stage is operated at low concentrations. Periodic reversal of the stages allows maintenance of a high biomass in the second, or follow, stage. A major feature of the two-stage, cyclic process is that the biological solids in the second stage exist in a negative growth or decay phase. That is, the waste concentration is so low relative to the biomass concentration that the microorganisms are in a starving environment. The substrate concentration is then below S_{min} , and the mass of bacterial solids decreases with time. This condition is known to be beneficial for trace organic matter reduction (Bouwer 1985) and for minimizing sludge production (Young 1993). Periodically reversing the lead reactor to the follow position allows improved removal of organic wastes because the second reactor then contains a high concentration of active microorganisms. Compared to single-stage operation, the cyclic mode has been shown to produce much lower soluble residuals than a single-stage process operating at the same organic loading rate (Howerton and Young 1987).

MATERIALS AND EXPERIMENTAL METHODS

Laboratory-scale tests were conducted using a completely mixed reactor and a two-stage cyclic reactor system (no control reactors of either type were used). The completely mixed reactor consisted of a 2-L Erlenmeyer flask to which alkylpyridine-contaminated groundwater was added daily to give a 16-day hydraulic retention time (HRT) (Bollag et al. 1992: Table 1, column 4). This reactor was seeded with microorganisms obtained from groundwaters and soils at the contaminated site. Oxygen was provided by bubbling air through the culture medium using a small porous stone diffuser.

TABLE 1. Concentrations and removal efficiencies of COD, organic acids, and alkylpyridines in the two-stage, cyclic reactor and the single-stage mixed reactor.

Constituent	Two-Stage, Cyclic [a] Fixed-Film Reactor Concentration, mg/L			Single-Stage[b] Mixed Reactor Concentration, mg/L		
	(1) Inf.	(2) Eff.	(3) % Removal	(4) Inf.	(5) Eff.	(6) % Removal
COD (Day 83)	270.0	BDL[c]	100	—	—	—
Organic Acids (Day 83)						
Acetic	131.1	BDL	100	—	—	—
Propionic	19.3	BDL	100	—	—	—
Butyric	22.1	BDL	100	—	—	—
Alkylpyridines (Day 67)						
2-picoline	2.54	0.003	100	0.494	0.012	98
4-picoline	0.42	0.004	99	0.582	0.073	87
2,3-lutidine	2.11	0.012	99	1.057	0.041	96
2,4-lutidine	2.50	0.016	99	2.285	0.064	97
2,5-lutidine	0.86	0.011	99	0.649	0.014	98
2,6-lutidine	0.43	0.006	99	1.484	0.049	97
3,5-lutidine	9.15	0.197	98	5.276[d]	0.042	99

(a) Loading = 4.32 g COD/L-d and HRT = 3 hours.
(b) Loading = 0.02 g COD/L-d and HRT = 16 days.
(c) BDL = below detection limit.
(d) 3,5-lutidine and 2-methyl-5-ethylpyridine.

The two-stage, cyclic fixed-film reactors were constructed from Corning Pyrex™ glass having medium wall thickness and were 4 in. (100 mm) in diameter and 4 ft (120 cm) in length (Figure 1). The total operating volume of each

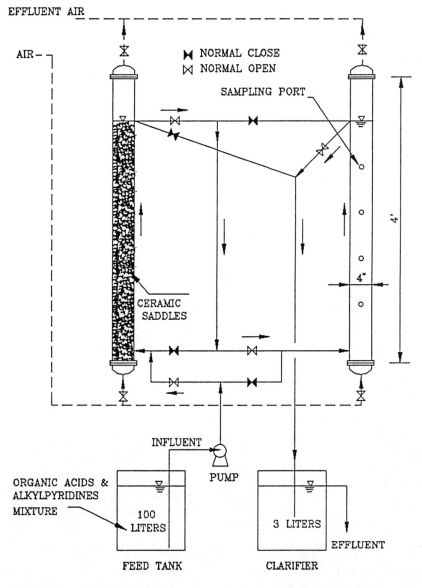

FIGURE 1. Schematic diagram of two-stage fixed-film aerobic reactors.

reactor was 8 L plus a 2-L headspace. The packed section was 1 m in height and the packing medium consisted of 1-in. (25 mm) ceramic Intalox saddles (Norton Chemical Process Products, Akron, Ohio). Aeration and mixing were accomplished by discharging air from a porous stone diffuser at the base of each column. The reactors were seeded with mixed liquor from the Pennsylvania State University wastewater treatment plant and with microorganisms removed from the 2-L mixed reactor.

The feedstock added to the two-stage system contained a mixture of acetic, propionic, and butyric acids, and the alkylpyridines: 2- and 4-picoline and 2,3-, 2,4-, 2,5-, 2,6-, and 3,5-lutidine (Table 1; column 1). This feed was prepared to simulate the composition and strength of the contaminated Reilly groundwater. Nutrients, minerals, and buffers were added to the feedstock solutions for both reactors to support biological growth and to maintain a suitable treatment environment.

The fixed-film reactors initially were operated in parallel at a flowrate of 8 L/day and a 6-h HRT adjusted for 50% porosity, giving an organic loading rate of 1.08 g chemical oxygen demand (COD)/L-day. This condition was maintained to give the microorganisms time to acclimate to the substrate and accumulate in the reactors. The pH was controlled at approximately 7 in the influent to maintain a pH of 7.5 in the effluent. Dissolved oxygen was maintained above 3.0 mg/L in the effluent from each column.

COD concentrations were measured using precharged vials and a DR/2000 spectrophotometer (HACH Company, Loveland, Colorado). Organic acid and alkylpyridine concentrations were measured using standard gas chromatographic procedures.

RESULTS

The mixed reactor was sampled infrequently, but samples collected at the end of the test program showed low levels of alkylpyridine residuals with removal efficiencies ranging from 87 to 98% (Table 1, columns 5 and 6).

After stabilization when operating in parallel, the loading rate to the two-stage, fixed-film reactors was increased in steps from 1.08 to 8.64 g COD/L-d. Analysis of alkylpyridines at various heights throughout the lead and follow reactors on day 67, when operating at a loading of 4.32 g COD/L-day and an HRT of 3 hours, showed that the concentrations decreased gradually with increasing height (Figure 2). The removal of alkylpyridines was essentially completed within 0.4 m of the lead reactor height with overall efficiencies approaching 100% (Table 1, column 3). The 3,5-lutidine removal efficiency was lower than for other compounds, but the removal was 86% in the lead reactor and 98% through both reactors. Residual concentrations are shown in Table 1, column 2, for comparison with the performance of the mixed reactor.

Changing the loading and HRT on Day 70 to 8.64 g COD/L-day and 1.5 h, respectively, caused a significant decrease in alkylpyridine removal in the lead

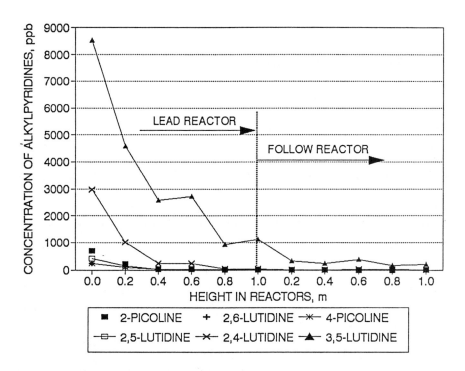

FIGURE 2. Removal of alkylpyridines at various heights throughout the lead and follow reactors when operating at a loading of 4.32 g COD/L-d and an HRT of 3 hours (Day 67).

reactor (data not shown). While the concentration of 3,5-lutidine remained between 6 and 10 mg/L throughout the lead reactor, 2-picoline, 2,6-lutidine, 4-picoline, 2,4-lutidine, 2,5-lutidine, and 3,5-lutidine were removed more effectively. The residuals transferred to the follow reactor were degraded further, except for 3,5-lutidine which showed no significant improvement in removal. The concentration of 3,5-lutidine in the effluent of the follow reactor was measured at 8,595 µg/L, and the degradation of 2,4-lutidine increased from 53 to 94%. Significant improvement of 2-picoline, 4-picoline, 2,5-lutidine, and 2,6-lutidine reduction also was obtained in the follow reactor with removal efficiencies of 87, 85, 95, and 82%, respectively. Because of this low efficiency, an organic loading rate of 8.34 g COD/L-d was considered above the treatment capabilities of the two-stage reactor and it was returned to 4.32 g COD/L-d for subsequent tests (Hsu 1993).

CONCLUSIONS

Tests with mixed and two-stage fixed-film reactors demonstrated satisfactory biodegradation of alkylpyridines in a contaminated groundwater, thereby demonstrating that aerobic treatment would be an effective method for biore-mediation of the contaminated groundwater at the Reilly site. The two-stage cyclic process provided essentially complete removal of alkylpyridines. While 3,5-lutidine was the most resistant to microbial degradation, it was most effectively removed in the two-stage, cyclic process when operating at hydraulic retention times of 3 h or greater. Concentrations of individual alkylpyridines after treatment, except for 3,5-lutidine, were well below the 35 μg/L permissible concentration for direct recharge. However, design optimization is expected to allow reduction of the 3,5-lutidine to concentrations below this limit.

REFERENCES

Bollag, J. M., Z. Ronen, L. Otjen, and G. Gunalan, 1992. *Biorestorative procedures for pyridine-contaminated sites.* Center for Bioremediation and Detoxification. Environmental Resources Research Institute, The Pennsylvania State University, PA.

Bouwer, E. J. 1985. "Secondary utilization of trace halogenated organic compounds in biofilms." *Environmental Progress* 4:43-49.

ENSR Consulting and Engineering, 1991. *Revised focused feasibility study for the groundwater operable Unit Interim Remedial Measure.* Report to Reilly Industries, Indianapolis, IN.

Howerton, D. E. and J. C. Young, 1987. "Two-stage cyclic operation of anaerobic filters." *Journal Water Pollution Control Federation* 59:788-799.

Hsu, C-H. 1993. "Two-Stage, Cyclic Aerobic Treatment of Contaminated Groundwater." Master of Engineering Report, Dept. of Civil and Environmental Engineering. The Pennsylvania State University, University Park, PA.

Record of Decision (ROD). Jan. 5, 1992. *Reilly site declaration for the record of decision: Reilly Tar and Chemical Groundwater Operable Unit.* U.S. Environmental Protection Agency, Region 5.

Rogers, J. E., R. G. Rilley, S. W. Li, M. L. O'Maley, and B. L. Thomas. 1985. "Microbial transformation of alkylpyridines in groundwater." *Water Air and Soil Pollution* 24:443-451.

Sims, G. K. and L. E. Sommer. 1985. "Biodegradation of pyridine derivatives in soil suspensions." *Environmental Toxicology and Chemistry* 5:503-517.

Standard Methods for the Examination of Water and Wastewater. 1985. 17th ed., American Public Health Association, New York, NY.

Young, J. C. Nov. 1993. "The two-stage, cyclic process for wastewater treatment." *Proceedings of the International Conference on Environmental Commerce,* Chattanooga, TN.

In Situ Biodegradation of a Hydrocarbon-Contaminated Landfill

*Settimio Arazzini, Paola Bocchieri, Giorgio Migliorini,
Lucia Rivara, and Giuseppe Tripaldi*

ABSTRACT

The anaerobic and/or low-aeration biodegradation of urban waste, contaminated by polycyclic aromatic hydrocarbon (PAH) compounds and a spill of tar products, is described. Before the industrial plant was designed, laboratory tests were carried out to determine the process feasibility and define the biodegradation rate of the pollutants. Preliminary tests on bacteria growth efficiency in aerobic and anaerobic conditions were carried out in Erlenmeyer flasks and showed interesting results in both cases. Following these tests, four different laboratory reactors were assembled to simulate waste treatment under different operating conditions. During 3 months of continuous treatment, the tar and PAH contents were measured in the waste and in the leachate and the bacteria population growth was registered. Treatment results show pollutant degradation of nearly 90%.

INTRODUCTION

The Sezzadio landfill in Italy is located in an inactive clay quarry and has a total surface area of approximately 8,000 m². About 75% of the quarry's total volume (about 22,000 m³) was filled with urban waste and a quantity of toxic and hazardous wastes. Taking into account possible water lens pollution, a decontamination plan was finalized. The landfill area stratigraphy shows a stratum of covering ground with a mean thickness of approximately 1 m; below this stratum lies a waste mass of up to 5 m in depth. In the lower side the waste borders for the most part on a clay stratum, with the exception of a point where it is in direct contact with the alluvial stratum.

The first survey carried out in 1988 disclosed that the waste was contaminated by a PAH and tar products spill. Because the level of pollution differed according to zone, the first treatment design differentiated the treatment according to pollution level:

- Contamination levels greater than 5,000 mg/kg of tar products—
  wastes to external treatment (special landfill)
- Contamination levels between 200 and 5,000 mg/kg of tar products—in situ bioremediation
- Contamination levels lower than 200 mg/kg of tar products—no
  treatment.

This arrangement is related to Italian national rule 915/82 that defines 50,000 mg/kg as a limit concentration for tar products. Because waste contaminated by tar products can be stored indefinitely in a 2-B type landfill if pollutant is present at 1/100 of the limit concentration, a bioremediation treatment of waste containing between 5,000 and 200 mg/kg was postulated, with a biodegradation efficiency of 80 to 90%.

According to the request to the Authority in 1989, while waiting for final contract agreement definition, the waste landfill was temporarily secured. Initially, to limit pollution diffusion, clearly identifiable hazardous waste was disposed of and the remaining waste was capped to avoid leaching. A drainage system was implemented for surface water disposal.

Remediation began in March 1994, but after capping removal the pollution status was determined to be quite different from the contamination found in 1988. During 5 years the contaminant level had decreased and a diffusion phenomenon had occurred in the waste bulk. Thus, in situ anaerobic and low-aeration biodegradation using indigenous microorganisms was chosen as the optimal method for cleaning up the area. Preliminary tests on bacteria growth efficiency in aerobic and anaerobic conditions were carried out in Erlenmeyer flasks and showed interesting results in both cases. As a second step it was decided to perform in the laboratory a continuous test in pilot reactors to study the degradation phenomena at different operating conditions.

EXPERIMENTAL PROCEDURES AND MATERIALS

Laboratory Reactors

The wastes used in the laboratory tests were collected from the field, transferred to the laboratory, and manually mixed to eliminate larger plastic parts so as to obtain a sufficiently homogeneous substrate. For the experimental tests, four closed reactors with the following characteristics were used: gross volume, 20 L; waste content, 15 L; waste layer thickness, 20 cm; gravel layer thickness, 5 cm; wetting rate, 90 L/h; and air pumping rate, 80 L/h. A support grid was placed on the bottom of each reactor to allow leachate drainage. Each reactor was filled with the gravel and waste. A spray system was provided at the top of each reactor to keep the waste mass wet with recirculated leachate. A wetting rate was adopted at intervals of 4 h for a period of 6 min for each wetting action. Table 1 shows the process parameters adopted for each reactor; Figure 1 is a flowchart for the plant.

All the reactors were provided with recycled leachate; in three reactors aerobic conditions were improved by continually aerating the leachate in an

TABLE 1. Reactor process parameters.

PARAMETERS	Leachate recycle	Leachate aeration	Nutrients	Specific bacteria
Reactor 1 (R1)	X			
Reactor 2 (R2)	X	X		
Reactor 3 (R3)	X	X	X	
Reactor 4 (R4)	X	X	X	X

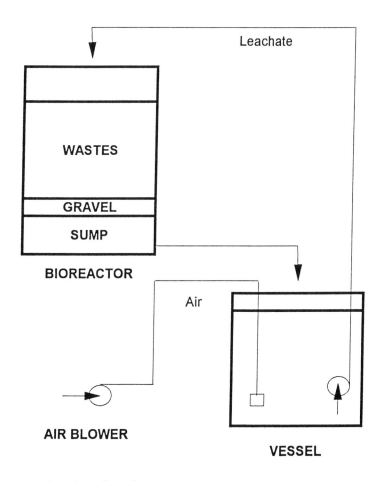

FIGURE 1. Pilot plant flowchart.

external collection vessel by means of submerged ceramic diffusers, making it possible to obtain leachate oxygen saturation. In two reactors, after 45 days of treatment, nutrients were added as NH_4NO_3 to obtain a nitrogen level of 20 mg/L. Specific hydrocarbon-degrading bacteria were added to one reactor only. The tests were carried out at a temperature of 18 to 24°C for approximately 3 months.

The hydrocarbon-degrading bacteria were furnished by a specialized firm and were representative of microorganisms from contaminated areas that are capable of degrading polluted contaminants such as hydrocarbon compounds, PAHs, etc. These bacteria belong to the genera *Pseudomonas, Acinetobacter,* and *Arthrobacter.* Because the indigenous microorganisms had not been taxonomically identified, it was not possible to distinguish them from the laboratory-selected bacteria. The commercial bacteria were added to the one reactor at the beginning of the study and at intervals of 77, 86, and 91 days from startup.

Sampling

The analytical parameters listed in Table 2 were measured to determine the rate of biodegradation. The frequency of sampling for both waste and leachate was weekly up to day 49 and every 15 days after day 49.

Considering that the startup analysis showed low concentrations of anthracene, benzo(a)anthracene, benzo(k)fluoranthene, benzo(ghi)perylene, and indeno(1,2,3-cd)pyrene, only the behavior of the other aromatic compounds, listed in Table 2, was examined.

As far as the sampling method is concerned, the waste was randomly collected from each bioreactor in four subsamples at different depths into the soil using a small corer.

The four subsamples were then carefully mixed together, and a single sample was collected as representative of the bioreactor itself.

Analytical Methods

The analytical methods applied to determine the various parameters are as follows: (1) total hydrocarbon samples are extracted with Freon 113 and the organic phase is purified, concentrated, and analyzed by Fourier-transform infrared (FTIR) analyses between 3200 and 2700 cm^{-1}; (2) PAH samples are extracted with methyl chloride, dehydrated, concentrated, transferred in acetonitrile, and analyzed by high-performance liquid chromatography (HPLC) using an ultraviolet (UV) diode array and fluorimeter; (3) $NO_2, NO_3,$ and PO_4 are analyzed through ionic chromatography; (4) total Kjeldahl N is analyzed through the Kjeldahl method; and (5) organic carbon is oxidized with potassium dichromate and titrated with ferrous iron.

RESULTS

Pollutant content reduction is reported in Figures 2 to 7. For the total hydrocarbon content, the final pollutant reduction in the four reactors was 93%

TABLE 2. Analytical parameters.

Parameters	Waste	Leachate
Total Hydrocarbons	X	X
Naphthalene	X	X
Phenanthrene	X	X
Anthracene	X	X
Fluoranthene	X	X
Benzo(a)anthracene	X	X
Chrysene	X	X
Benzo(k)fluoranthene	X	X
Benzo(a)pyrene	X	X
Benzo(ghi)perylene	X	X
Indeno(1,2,3-cd)pyrene	X	X
Microbial population		X
N	X	
C	X	
P (total)	X	
NO_2		X
NO_3		X
pH		X

(mean value). Taking into account the different reactor process conditions, the first consideration that can be made is that in reactors 2, 3, and 4 the hydrocarbon concentration was reduced to one-half of the initial concentration in 8 days. Specifically, the pollutant contents decrease from 2,297 to 414 mg/kg dry weight for R2, from 1565 to 514 mg/kg dry weight for R3, and from 2263 to 262 mg/kg dry weight for R4. In reactor 1, where the leachate was not aerated,

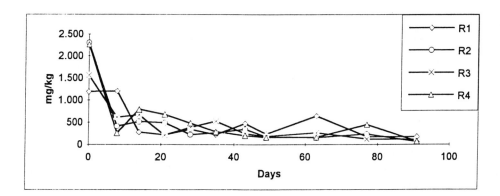

FIGURE 2. Tar products content reduction.

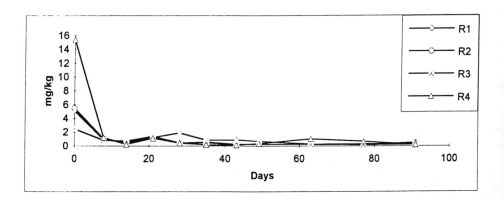

FIGURE 3. Naphthalene content reduction.

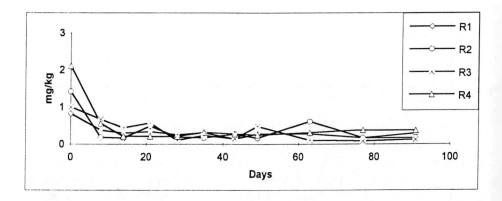

FIGURE 4. Phenanthrene content reduction.

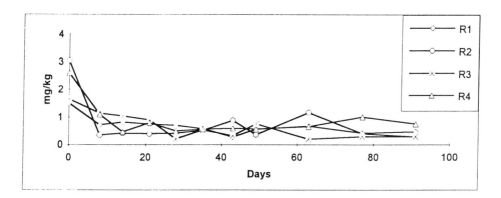

FIGURE 5. Fluoranthene content reduction.

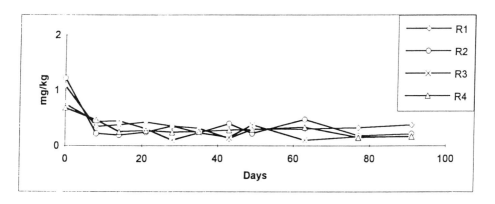

FIGURE 6. Chrysene content reduction.

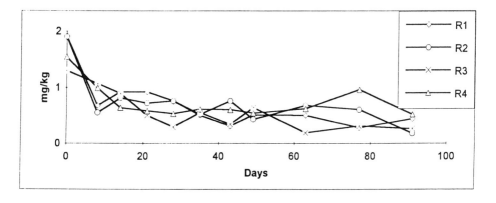

FIGURE 7. Benzopyrene(a) content reduction.

this period increased to 14 days (from 1,187 to 1,200 to 280 mg/kg dry weight). For the PAH concentration, the following percentage reductions (mean value) were registered: naphathalene, 93%; phenanthrene, 81%; fluoranthene, 74%; chrysene, 69%; benzo(a)pyrene, 78%.

The analysis of concentration versus time curves does not show substantial differences among the four reactors; in particular, the nutrients added and the specific bacteria did not improve degradation efficiency, demonstrating the efficiency of the indigenous bacteria. The bacterial activity, tested by plate count, was high during the entire treatment period, with bacteria concentrations changing from 9×10^4 cell/mL to 8×10^7 cell/mL.

The waste and leachate characteristics at the beginning of the experiment and during the treatment are reported in Tables 3 and 4. As shown in Tables 5

TABLE 3. Waste parameters.

Parameters	Reactor 1		Reactor 2		Reactor 3		Reactor 4	
	T_0	T_{28}	T_0	T_{28}	T_0	T_{28}	T_0	T_{28}
pH	7.2	7.0	7.6	7.9	7.6	8.2	7.5	8.4
N Kjeldhal (mg/Kg$_{SS}$)	2311	2100	1988	2048	1790	1954	2028	1981
C organic (% ss)	2.1	2.6	1.7	3.4	1.5	1.2	1.7	2.6
P (PO$_4^{3-}$) (mg/Kg$_{SS}$)	1682	1321	1155	920	1399	666	1606	1158

TABLE 4. Leachate parameters.

Time	NO$_2$ (mg/L)				NO$_3$ (mg/L)				pH			
	P 1	P 2	P 3	P 4	P 1	P 2	P 3	P 4	P 1	P 2	P 3	P 4
T_0	2.07	2.14	2.20	2.02	4.02	4.92	6.29	4.92	7.58	8.27	8.26	8.32
T_8	1.80	2.05	2.80	1.95	4.38	2.54	3.96	2.91	7.67	8.32	8.39	8.39
T_{14}	<0.5	1.25	2.32	2.88	106	136	200	220	7.96	8.39	8.44	8.44
T_{21}	0.86	1.29	1.36	1.34	28.9	41.8	64.0	41.6	8.10	8.50	8.56	8.56
T_{28}	<0,5	<0.5	<0.5	<0.5	29.8	17.4	23.0	31.1	8.41	8.64	8.61	8.61
T_{35}	<0.5	<0.5	<0.5	-	4.77	7.52	11.51	-	8.09	8.54	8.57	-
T_{49}	<0.5	<0.5	136	148	5.15	8.56	477	577	8.00	8.56	8.49	8.53
T_{63}	<0.5	<0.5	<0.5	<0.5	6.14	6.48	148.8	106.7	-	-	-	-
T_{77}	<0.5	<0.5	<0.5	<0.5	4.87	4.77	69.6	104.4	8.32	8.65	8.48	8.48
T_{91}	<0.5	<0.5	<0.5	<0.5	5.32	-	63.8	52.3	-	-	-	-

TABLE 5. Total hydrocarbon (mg/L) and PAHs (μg/L) in the leachate.

		T14	T28	T43	T49	T63	T77	T91
R1	total HC	0.61	0.11	3.50	0.16	<0.10	0.91	<0.10
	naphthalene	0.02	0.166	-	0.06	0.08	0.03	0.02
	phenanthrene	0.05	0.10	-	0.02	0.16	0.01	<0.01
	fluoranthene	<0.01	0.04	-	<0.01	0.04	<0.01	<0.01
	chrysene	<0.01	<0.01	-	<0.01	0.01	<0.01	<0.01
	benzo(a)pyrene	<0.01	0.03	-	<0.01	0.07	<0.01	<0.01
R2	total HC	9.12	<0.10	1.80	0.09	0.11	0.12	<0.10
	naphthalene	-	-	0.05	0.10	0.05	0.05	0.05
	phenanthrene	0.04	0.019	0.016	0.04	0.12	0.02	0.01
	fluoranthene	<0.01	<0.01	<0.01	0.03	0.03	<0.01	0.01
	chrysene	<0.01	<0.01	<0.01	0.01	0.01	<0.01	<0.01
	benzo(a)pyrene	0.01	0.01	0.026	0.04	0.05	<0.01	<0.01
R3	total HC	0.61	<0.10	<0.10	0.31	0.22	0.22	<0.10
	naphthalene	-	-	0.05	0.08	0.04	0.05	0.05
	phenanthrene	0.04	<0.01	0.08	0.03	0.09	0.03	0.02
	fluoranthene	<0.01	0.01	0.02	0.01	0.01	<0.01	0.01
	chrysene	<0.01	<0.01	<0.01	0.03	<0.01	<0.01	<0.01
	benzo(a)pyrene	<0.01	0.013	0.043	0.01	0.01	0.01	0.02
R4	total HC	0.72	0.43	<0.1	0.1	0.27	0.47	<0.1
	naphthalene	0.02	-	0.04	0.08	0.04	0.03	0.05
	phenanthrene	0.04	<0.01	0.07	0.03	0.09	0.01	<0.01
	fluoranthene	<0.01	0.02	0.01	0.02	0.02	<0.01	<0.01
	chrysene	<0.01	<0.01	<0.01	<0.01	<0.01	<0.01	<0.01
	benzo(a)pyrene	<0.01	<0.01	0.014	0.02	<0.01	<0.01	0.01

and 6, there is no evidence of the presence of total hydrocarbons and PAHs in the leachate and in the waste collected from deep in the soil. These results signify that autochthonous microorganisms were able to carry out the biodegrading activity.

TABLE 6. Total hydrocarbon (mg/kg$_{SS}$) and PAHs (mg/kg$_{SS}$) in the waste at T49.

| | R 1 | | R 2 | | R 3 | | R 4 | |
	up	bottom	up	bottom	up	bottom	up	bottom
total HC	231	234	484	171	174	181	153	159
naphthalene	0.29	0.67	0.18	0.23	0.32	0.28	0.27	0.33
phenanthrene	0.25	0.27	0.14	0.16	0.27	0.47	0.08	0.24
fluoranthene	0.54	0.56	0.38	0.36	0.61	0.76	0.19	0.57
chrysene	0.20	0.29	0.18	0.22	0.31	0.37	0.10	0.29
benzo(a)pyrene	0.43	0.51	0.30	0.43	0.60	0.64	0.23	0.54

CONCLUSIONS

The data analysis shows that the indigenous bacteria present in the contaminated soil are capable of degrading organic compounds such as PAHs and total hydrocarbons. This might be due to the evolution of microorganisms that in the past 5 years had adapted to the presence of the pollutants (Galli 1989). Therefore, the right environmental conditions may be the only requirement for microbial degradation.

Our data are comparable to those determined by other researchers. As an example, we cite the experiment carried out by De Kreuk and Annokkèe (1988), who used a bioreactor containing semiliquid soil contaminated by other hydrocarbons and PAH continuously mixed. After 43 days of treatment at ambient temperature, naphthalene and phenanthrene were degraded 99%, fluoranthene 75%, chrysene 82%, and benzo(a)pyrene 77%. In a similar study, Werner (1991) carried out PAH biodegradation experiments using trickling filters. Field bioremediation of oil-contaminated soil also requires several months. Bourquin (1991) reported a 97% decrease of total hydrocarbons in 56 days (from 4,000 mg/kg to 100 mg/kg).

The overall experience in biodegradation of PAH and total hydrocarbons in contaminated waste indicates that biological treatment could be a valid technology for land reclamation. It is often sufficient to improve the environmental conditions to stimulate the biodegradation activity of indigenous microorganisms.

ACKNOWLEDGMENTS

Thanks are due to the staff of the Analytical Laboratory, in particular to Dr. Gabriella Garbarino for assistance in the chemical analysis.

REFERENCES

Bourquin, A. W. 1991. "Risanamento biologico. Una tecnologia per acque sotterranee e suoli contaminati." *Rifiuti solidi V* (4): 297-304.

De Kreuk, J. F. and Annokkèe. 1988. "Applied biotechnology for decontamination of polluted soils. Possibilities and problems." In *Contaminated Soil*. pp. 679-686.

Galli, E. 1989. "Ruolo dei microrganismi nella decontaminazione ambientale: aspetti genetici e biochimici." Paper presented in Atti del Convegno d'Annuale di Genetica Agraria [Annual Convention of Agrarian Genetics]. Alghero, Italy, 23-26 October.

Werner P. 1991. "Trattamento microbiologico dei terreni contaminati." *Rifiuti solidi V* (2): 121-126.

Biodegradation and Bioavailability of *bis*(2-Ethylhexyl)Phthalate in Soil

Sam Fogel, Margaret Findlay,
Chris Scholl, and Michael Warminsky

ABSTRACT

The feasibility of biologically treating process chemicals in soil at a plasticizer manufacturing site containing *bis*(2-ethylhexyl)phthalate, *bis*(2-ethylhexyl)adipate, di-*n*-octylphthalate, and naphthalene was investigated. It was found that direct biological treatment of site soil was not feasible because part of the contamination existed as insoluble crystalline still-bottom material ranging in size from <1 mm to several cm. Bench-scale soil biodegradation studies indicated that spiked phthalates and naphthalene could be biodegraded in site soil, but that only portions of the soil contaminant phthalates and naphthalene were bioavailable.

INTRODUCTION

The process for production of plasticizer at this site involved the oxidation of naphthalene to phthalic anhydride which was further reacted with long-chain alcohols to make phthalic acid esters such as *bis*(2-ethylhexyl)phthalate (BEHP). The soil contained BEHP, di-*n*-octylphthalate (DNOP), *bis*(2-ethyl-hexyl)adipate (BEHA) (a 22-carbon aliphatic ester), and naphthalene (NAP), as well as other process chemicals. During the investigation, it was discovered that the site contamination was heterogeneous, and that in some areas the majority of the contaminant was present in insoluble chunks of still-bottom material as solid phase organics, ranging in size from less than 1 mm to several cm in diameter, which were not viably differentiable from rocks. This paper presents the results of bench and pilot biodegradation tests carried out with site soil that contained solid-phase contaminant, as well as with soil spiked with NAP and BEHP, indicating that the dispersed portion of the contaminant was biodegradable, but not the solid-phase organics.

MATERIALS AND METHODS

Treatment Simulation

Soil samples for bench tests were obtained from area #1 (sandy) and area #2 (gravel/till) of the manufacturing site. Soil was passed through a 2.5-cm screen to break up clumps and remove larger rocks and construction debris, and then mixed and further sieved through a 1.2-cm screen.

Simulations of forced aeration pile treatment were performed using 30-L glass aquaria with airtight lids. The aquaria were fitted with two 30-cm aeration stones in a 5-cm gravel bed and provided with 250 cc/min forced aeration. Soil was amended with 4% to 7% dry weight shredded tree waste as bulking agent to improve permeability, and with 0 to 2% manure to improve moisture retention. A pilot study (Test 5) was set up with site soil crushed and sieved through a 2.5-cm screen, amended with bulking agent and minerals, and provided with forced aeration and recirculation of leachate in a manner similar to that employed in the bench-scale tests.

For Test #3, 24 g each of NAP and BEHP in 250 ml pentane were applied to 24 kg dry weight soil in 10 batches, and mixed to evaporate pentane. For Test #4, approximately 90 g NAP and 115 g BEHP in 600 mL pentane were applied similarly.

A soluble fertilizer containing urea and o-phosphate, as well as 12 additional elements, was applied to the test mixtures. Soil concentrations of ammonia, nitrate, and phosphate, and pH, moisture, and oxygen were measured 3 times per week. The oxygen was maintained at 5% or higher in soil pore space. Water additions were sufficient to generate leachate (which was reapplied), and available nitrogen and phosphate were maintained greater than 10 ppm.

Analytical Methods

Soil samples were Soxhlet extracted with dichloromethane and analyzed by gas chromatography with flame ionization detection (GC/FID) (EPA Method 8100) or by mass spectrometry (EPA Method 8270). Off-gas from one experiment was collected in a SUMMA® canister and analyzed for volatile hydrocarbons by gas chromatography with mass spectrometry (GC/MS). One solid-phase organic crystal was prepared for analysis by dissolution in dichloromethane.

Total bacteria in test samples were enumerated by modifications of EPA Standard Method 9215C (1992 edition). Naphthalene, alkane, and phthalate degraders were determined by plating samples on Noble agar containing only mineral salts, and exposing the plates to vapors of naphthalene, hexadecane, or dimethyl phthalate (DMP).

RESULTS

Test 1

Test 1 was carried out with soil from area #1. The data (Table 1) show the following reductions by 6 weeks: NAP 77%; BEHP 75%; DNOP 61%; and BEHA 89%.

Test 2

For this test, one-third of the treated soil from Test 1 was mixed with two-thirds of the untreated soil from area #2 which contained phthalates but not naphthalene. The data show that NAP introduced from Test 1 was not significantly reduced during the 63 days of Test 2 (Table 2). An off-gas sample taken during the first 6 hours of Day 0 contained NAP equal to 1.2% of the total. A second air sample, taken on Day 5, however, contained no NAP. This result is consistent with the lack of reduction of NAP during 63 days.

The data for BEHP shows no clear trend in reduction of this compound. DNOP and BEHA show reductions of about 86% and 92%, respectively, indicat-

TABLE 1. Test 1. Nonspiked soil from area #1. GC/MS mg/kg dry weight, surrogate corrected.

	NAP	BEHP	DNOP	BEHA
Day 0	340	310	200	560
Day 42	79	78	78	60
Reduction	77%	75%	61%	89%

TABLE 2. Test 2. Nonspiked soil from areas #1 and #2. GC/MS mg/kg dry weight, surrogate corrected.

	NAP	BEHP	DNOP	BEHA
Day 0	35	490	360	350
Day 0 Dup	34	700	340	380
Day 7	46	1,200	590	230
Day 21	13	210	210	40
Day 42	38	230	110	40
Day 63	37	130	60	30
Reduction	none	unclear	83%	92%

ing greater bioavailability in this soil mixture. These three compounds were not detected in the off-gas samples. Weekly bacteria counts of soil samples (Table 3) showed relatively high ($>50 \times 10^6$/g) numbers of total as well as hexadecane and phthalate degraders.

Test 3

Test 1 indicated that about 75% BEHP and NAP were biodegradable in soil from Area #1, but Test 2 indicated that neither the residual naphthalene in Test 1 soil nor the added BEHP in area #2 soil was degradable. Lack of biodegradation could be due to lack of bioavailability, or to presence of toxic substances in the soil. To distinguish between these possibilities, 1,000 mg/kg each of BEHP and NAP were spiked into Test 3 soil. The data (Table 4) indicate that the spiked BEHP biodegraded 91% by Day 70 and that NAP was reduced by 99%. A 200-fold increase in NAP degraders occurred during the first week of treatment. Moderate numbers ($>5 \times 10^6$/g) of total bacteria and NAP degraders were maintained during the test (Table 3).

Test 4

Test 4 was carried out to investigate the biodegradability of spiked NAP and BEHP at higher concentrations. A mixture of soil from areas #1 and #2, which initially contained no naphthalene, but did contain about 300 ppm BEHP and 200 ppm each of DNOP and BEHA, was spiked with about 3,800 ppm of naphthalene and about 4,900 ppm BEHP. The spiked naphthalene decreased about 80% in 14 days, and about 99% in 98 days (Table 5). The BEHP decreased 73% in 14 days, with a total reduction of 92% in 98 days. During the first week of treatment, total bacteria increased about 15-fold, and NAP degraders increased about 200-fold. High numbers ($>60 \times 10^6$/g) of both total and NAP degraders were maintained during the treatment (Table 3).

TABLE 3. Bacteria, mean of weekly counts. Millions per gram wet soil.

	Total Heterotrophs	NAP Degraders	DMP Degraders	HEX Degraders
Test 2 Day 0-48	120 ± 66	—	65 ± 38[a]	69 ± 50
Test 3 Day 0	3	0.02	—	—
Day 7-48	17 ± 20	5 ± 3		
Test 4 Day 0	5	0.3	—	—
Day 7-48	85 ± 35	60 ± 110		
Test 5 Day 0-48	146 ± 63	11 ± 18	3 ± 3	11 ± 6

(a) Week 3-7.

TABLE 4. Test 3. Spiked soil from area #2. GC/MS mg/kg dry sample, surrogate corrected.

	NAP	BEHP	DNOP
Starting Soil	<18	120	90
Spike	1,000	1,000	0
Day 0	640	940	<260
Day 7	190	470	90
Day 21	31	370	50
Day 28	9	230	50
Day 49	3	90	12
Day 70	<20	89	15
Reduction	99%	92%	83%

TABLE 5. Test 4. Spiked soil from areas #1 and #2. GC/MS mg/kg dry weight, not surrogate corrected.

	NAP	BEHP	DNOP	BEHA
Starting Soil	0	300	200	200
Spike	3,800	4,900	0	0
Day 0	3,800	4,300	200	160
Day 11	640	2,500	220	210
Day 14	820	1,400	100	50
Day 21	250	1,100	100	45
Day 35	50	1,300	80	35
Day 56	180	970	110	ND<180
Day 98	30	480	35	30
Reduction	99%	89%	83%	85%

Test 5

Pilot test results (Table 6) showed no significant change in either NAP or BEHP concentrations over 60 days. Both NAP and DMP degraders remained at starting numbers. Numerous solid-phase organic particles, up to 4 cm in

TABLE 6. **Test 5. Nonspiked soil from area #2. GC/FID mg/kg dry soil, surrogate corrected.**

(# of samples)	NAP	BEHP	DNOP	BEHA
Day 0-14 (3)	1,470 ± 350	540 ± 360	82 ± 10	45
Day 56 (5)	1,400 ± 2,000	300 ± 100	44 ± 8	<2

diameter, became evident as a result of leachate recirculation. One of these was analyzed by GC/FID (Figure 1) and was found to contain 22% by weight of GC-analyzable organics, of which 59% was naphthalene. Amounts of other compounds are given in Table 7.

DISCUSSION

Biodegradability of BEHP

Several studies have reported biodegradation of phthalate esters, including BEHP, in liquid culture and slurries (Mathur 1974, Johnson and Lulves 1975, Saeger and Tucker 1976, Taylor et al. 1981). Yu et al. (1993) described biodegradation of DNOP in soil. Our results with forced-aeration soil treatment indicate that BEHP added to soil at 1,000 mg/kg was biodegraded 91% in 70 days, and 4,000 mg/kg was biodegraded 89% in 98 days, by naturally occurring bacteria in soil amended with minerals. DNOP and BEHA also biodegraded during our tests.

BEHP in Site Soils

Test 1 showed that the majority of BEHP in one soil sample from Area #1 was bioavailable. Test 3 and Test 4 showed that *spiked* BEHP would biodegrade in soils from both Area 1 and Area 2, whereas Test 2 and Test 5 showed that unspiked BEHP already present in those samples would not biodegrade under the same conditions. This result is taken to indicate that the disposed BEHP was not bioavailable in some site areas.

NAP in Site Soils

Test 1 showed that most of the NAP was available (biodegraded or volatilized) in one soil sample from Area 1. Tests 3 and 4 indicated that spiked naphthalene was biodegraded—the numbers of naphthalene degraders increased significantly during the first week. The relative portion of biodegradation and volatilization of naphthalene in Tests 3 and 4 is not known because analysis of off-gas was not done. Tests 2 and 5 indicated that NAP was not bioavailable, soluble, or available for volatilization in those soil samples—

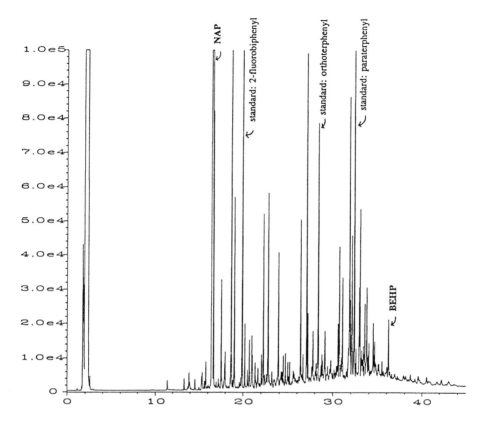

FIGURE 1. GC/FID analysis of solid-phase organic particle. Total GC-analyz-able organics = 220,000 mg/kg.

TABLE 7. Solid-phase organic particle from Test 5. GC/FID, mg/kg, detection limit 150 mg/kg.

Total GC-Analyzable Organics	220,000
Naphthalene	130,000
Dimethylphthalate	770
Butylbenzylphthalate	500
BEHP	375

concentrations of NAP remained unchanged over a 60-day period of forced aeration and leachate re-cycling. In Test 2, off-gas was sampled and found to contain no NAP by Day 5. In Test 5, total heterotrophs were high (146×10^6/g), indicating that conditions were optimized for bacterial growth, but naphthalene degraders were 13-fold less numerous, consistent with the conclusion that NAP was not available. In Test 5, DMP (phthalate) degraders were also low, consistent with the unavailability of BEHP and low concentrations of DNOP. Similar numbers of hexadecane degraders reflect low concentrations of the adipate, BEHA. The unavailability of NAP and BEHP in Test 5 was explained by the persistence of solid-phase organic particles containing high concentrations of those compounds.

Bioavailability

Bioremediation in soil requires that a target chemical be biologically available as well as inherently biodegradable. The inherent biodegradability of a compound is a function of its chemical structure, whereas its availability depends on its physical state (solid, nonaqueous-phase liquid, or dissolved), as well as on its solubility, sorptive properties, and volatility. Because soil bacteria consume chemicals from the dissolved state in the water film on soil particles and in soil pores, it is critical that contaminants be exposed to a water layer. If chemicals are present in physical forms that are not hydrophilic, the mass transfer to the water phase will be very slow, allowing the waste to persist. The treatment strategy for this site therefore requires a preliminary physical separation of solid-phase organics from the soil.

ACKNOWLEDGEMENT

We thank Stan Hatfield for his technical assistance and careful observations.

REFERENCES

Johnson, B. T., and W. Lulves. 1975. "Biodegradation of dibutylphthalate and di-s-ethyl-pthalate in freshwater hydrosoil." *J.Fish. Res. Bd Can.* 32:333.

Mathur, S.P. 1974. "Respirometric evidence of the utilization of di-octyl and di-2-ethylhexyl phthalate plasticizer." *J. Environ. Qual.* 3: 207.

Saeger, V.W., and E.S. Tucker. 1976. "Biodegradation of phthalic acid esters in river water and activated sludge." *Appl. and Environ. Microbiol.* 31: 29.

Taylor, B.F., R.Curry and E. Corcoran. 1981. "Potential for biodegradation of phthalic acid esters in marine regions." *Appl. and Environ. Microbiol.* 42: 590.

Yu, J., M. Diaz-Diaz, C. Kunze, and O. Ward. 1993. "Bioremediation of phthalic esters in contaminated soil." Presented in HMCRI, 14th Annual National Conference (HMC/Superfund 93), Washington DC.

Bioremediation of di(2-Ethylhexyl) Phthalate in Contaminated Soil

Jay J. Yu and Owen P. Ward

ABSTRACT

An efficient process has been developed for bioremediation of di-(2-ethylhexyl) phthalate (DEHP) using soil tilling. The process involves application of a proprietary nutrient formulation and bioaugmentation with a site-specific DEHP-degrading inoculum. Laboratory feasibility studies were conducted to evaluate different factors that affect the process. The effects of moisture content, inoculum, nutrient rate, and initial DEHP concentration on biodegradation were investigated. A novel supplementary system — treatment system 2 — was shown to accelerate DEHP degradation and facilitate remediation of residual persistent contaminant DEHP, which may be tightly bound to soil. In pilot studies, DEHP-contaminated soils (5,000 to 6,000 mg/kg) were remediated to below 100 mg/kg in 70 to 80 days. Use of treatment system 2 resulted in reduction of contaminants to less than 15 mg/kg.

INTRODUCTION

In the past few years, considerable attention has been paid to the biodegradation of persistent hydrophobic compounds such as polychlorinated biphenyls (PCBs) and polycyclic aromatic hydrocarbons (PAHs). Other chemical compounds such as phthalate esters, which are most commonly used as plasticizers in polyvinyl chloride (PVC), have attracted less attention. These chemicals are produced in large quantities, with annual world production of 3 to 4 million tons (Wams 1987). DEHP is a known carcinogen in animals and a suspected teratogen in rodents (Shelton et al. 1984). DEHP is biodegradable, and microorganisms can grow on DEHP and convert the contaminant to CO_2 and H_2O (Kurane et al. 1980). The pathway for degradation of DEHP has been described (Taylor et al. 1981). The first step involves hydrolytic cleavage of the ester bonds catalyzed by extracellular or cell-bound esterase enzymes. All intermediates of phthalate ester metabolism are reported to be more easily degraded than the parent compound (Saeger & Tucker 1976). Aqueous solutions containing low DEHP concentrations were found to be degraded in an activated sludge process (Saeger & Tucker 1976). Extents of degradation of concentrations of 200 mg/kg DEHP in bacterial inoculated aqueous medium in shake flasks were 80% and 99% in 14 and 28 days, respectively.

Fairbancs et al. (1985) spiked New Mexico soils with up to 20 mg DEHP per kg and reported 76 to 93% contaminant degradation in 146 days. Soil spiked with 480 mg/kg DEHP and incubated for 30 days exhibited degradation of the contaminant to a residual level of 40 mg/kg (Shanker et al. 1985). Soil columns augmented with an extremely high concentration of DEHP-degrading bacteria (4×10^{10} per g soil) had the capacity to degrade 1,500 mg DEHP/L/day in percolating aqueous solutions (Kurane et al. 1978).

DEHP, like other hydrophobic contaminants including PAHs and PCBs, manifests low water solubility. These hydrophobic contaminants often bind tightly to soil particles, thereby reducing their bioavailability and rendering them more persistent. As is well established for petroleum hydrocarbon-degrading bacteria, some phthalate ester-degrading species have cell surfaces or produce extracelluar molecules with DEHP-solubilizing properties. A solubilizing agent produced by *Mycobacterium* species enhanced the degradation of DEHP by *Mycobacterium* and *Nocardia* species (Gibbons & Alexander 1989).

Thus, processes for efficient bioremediation of DEHP must address the following key elements:

1. The biological system must involve effective production of the esterase needed for the first degradation step, as well as have the capacity to degrade the breakdown products.
2. The system should be capable of increasing DEHP bioavailability and should degrade DEHP to required criteria.

BIOREM has developed a soil tilling process that takes all these issues into account.

This paper describes studies of DEHP biodegradation by a mixed culture isolated from a contaminated site, with a view to developing the bioremediation process. The study includes laboratory investigations and a demonstration of process effectiveness at the pilot scale.

MATERIALS AND METHODS

Isolation of DEHP-Degrading Mixed Culture

Bacteria were isolated from the plasticizer-contaminated soil, which efficiently degraded phthalate esters using the following general microbiological protocol. A variety of soil samples from the site were used to inoculate Erlenmeyer flasks containing a mineral medium with DEHP as sole carbon source for growth. Cultures were shaken at room temperature on an orbital incubator and subcultured several times using the same medium to select efficient phthalate ester degraders. The mixed culture was subsequently transferred to a cyclone fermenter and maintained during the study.

Laboratory Experiments

Soil for the laboratory experiments was collected from a contaminated site. All soils were sieved prior to use in all experiments to minimize soil sampling errors. After the removal of rocks and bulky organic materials (sticks, leaves, roots, etc.), the soil was thoroughly mixed and homogenized. Distilled water was added to the soil to adjust the moisture levels required for optimal biodegradation activity as necessary. Degradation studies were carried out in soil columns and shake flasks. The soil columns were 6.5 cm in diameter and 50 cm in length. In all cases, soil from a site was packed to the full length of the column. The shake flask experiments were carried out in 250-mL Erlenmeyer flasks containing 100 g of soil and incubated at room temperature on an orbital shaker set at 200 rpm. Experiments were conducted in stepwise fashion in each column and each flask by varying the contaminant concentration or by adding nutrients. Experimental controls were used in all experiments.

On-Site Pilot Experiments

Pilot experiments were conducted at a site in Southern Ontario. Three plots of approximately 16 m^2 were established having different soil depths, i.e., 15, 30, and 45 cm. All parameters monitored in the plots were essentially the same as those monitored in the laboratory experiments. Tilling was carried out daily.

GC Analysis of DEHP

The DEHP was analyzed using a gas chromatography method modified from U.S. Environmental Protection Agency Method 8060. All soil analysis results are expressed on a dry-weight basis.

RESULTS

Fermentation Process

Following isolation of a site-specific mixed culture from the DEHP-contaminated soils, a fermentation process for production of acclimated inocula, capable of efficient DEHP degradation, was developed. Conditions in the fermenter were optimized to give high cell densities. This cell culture in aqueous media had a capacity to degrade 4,000 mg/kg DEHP per day. This culture was used to inoculate DEHP-contaminated soil.

Effect of soil moisture content and amendment type on plasticizer bioremediation was investigated. The contaminated soil was amended with nutrient formula I (containing NH_4^+, NO_3^-, HPO_4^{2-}) or nutrient formula II (containing yeast extract and HPO_4^{2-}), and other constituents, including a site-specific inoculum. The resulting DEHP degradation for each treatment, including different moisture contents, is shown in Figures 1a and 1b. Soil moisture content dramatically affected the rate of degradation. Maximum degradation was observed at a

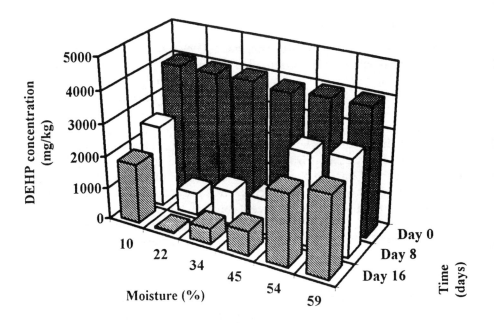

FIGURE 1a. Effect of moisture on DEHP degradation in nutrient formula I.

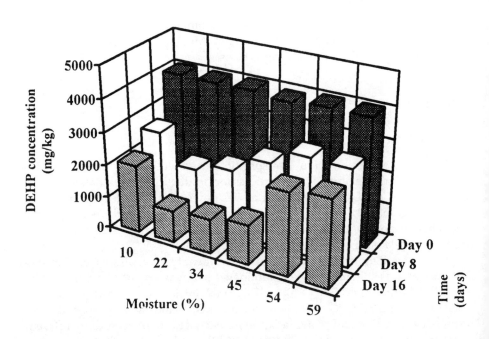

FIGURE 1b. Effect of moisture on DEHP degradation in nutrient formula II.

soil moisture level of 22%. This pattern was found both where soils were amended with organic or inorganic nitrogen. However, nutrient formula I containing inorganic nitrogen promoted a greater extent of DEHP degradation.

As with other soil contaminants, DEHP-contaminated soils may contain indigenous organisms capable of degrading DEHP, which may be stimulated by amendment with nutrients. This is more likely to occur with surface soils than subsurface soils and will depend also on other factors including soil chemistry, contaminant concentration, and the time elapsed since the spill. Bioaugmentation is important where indigenous soil counts are low and with hydrophobic molecules, and can be especially important when contaminant levels are reduced but have not reached criteria i.e., which for DEHP may range from 15 to 100 mg/kg of soil depending on jurisdiction and soil cleanup category. However, even in heavily contaminated soils containing indigenous bacteria, bioaugmentation can accelerate the remediation process. When soils containing 4,000 mg/kg DEHP were supplemented with nutrients and incubated with shaking for 7 days, 40% of the initial DEHP level remained. Augmentation of the soil with the inoculum reduced the residual contaminant level to 25%.

The effect of treatment rate of nutrient formula I was investigated. Soil containing 4,000 mg/kg was amended with different amounts of nutrients (0.6, 1.2, and 3.2 g/kg) and incubated in soil columns for 7 to 14 days. The nutrient treatment rate had a significant effect on DEHP degradation (Figure 2). The effect of DEHP concentration in the range 1,000 to 4,000 mg/kg on degradation of DEHP in soil augmented with nutrients and inocula was investigated (Figure 3). Degradation rates, as would be expected, were slightly higher at the high initial DEHP concentrations. Residual DEHP concentrations after the 7-day

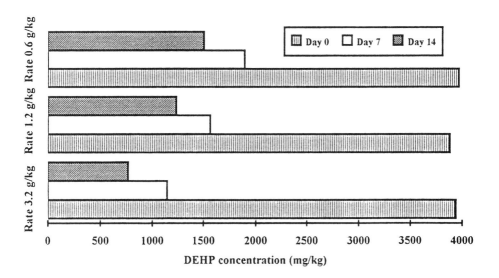

FIGURE 2. Effect of nutrient treatment rate on DEHP degradation.

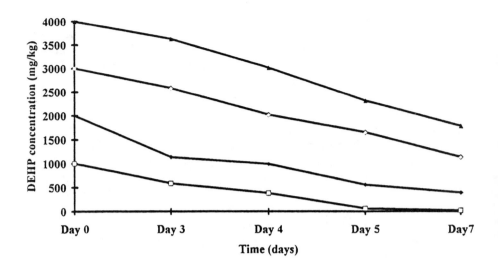

FIGURE 3. **Effect of initial concentration on DEHP degradation.**

incubation were 21, 391, 1,137, and 1,787 mg/kg for soil with initial concentrations of 1,000, 2,000, 3,000, and 4,000 mg/kg, respectively.

Depending on soil type, binding capacity of residual DEHP to soil and the activity of the DEHP-degrading microbial population, the degradation process may slow down and even stop without reaching 15 mg/kg. BIOREM has developed a second system incorporating a biosurfactant, treatment system 2, which promotes DEHP bioremediation of tightly bound DEHP. Soil columns containing 100 to 5,000 mg/kg DEHP with moisture controlled at 20 to 22% were incubated

TABLE 1. **DEHP degradation in treatment system 2.**

	DEHP Concentration (mg/kg)		
		Day 10	
	Day 0	With Treatment 2	Without Treatment 2
Soil Slurry	100	7.3	50.1
	300	26.8	135
	1,000	179	735
	5,000	1,701	4,254
Soil Column	100	27.0	47.5
	300	185	211
	1,000	432	625
	5,000	2,475	3,812

for 10 days with and without treatment system 2. The results are presented in Table 1. The treatment had a dramatic effect on DEHP biodegradation. Similarly, in shake flasks with soil slurry containing 50% moisture, use of treatment system 2 accelerated the DEHP degradation. Residual concentrations of DEHP dropped from 50 to 4,254 mg/kg to 7 to 1,701 mg/kg.

On-Site Bioremediation

Pilot tests using the bioremediation technology DEHP described above were implemented at a contaminated site in Southern Ontario. These plots with soil

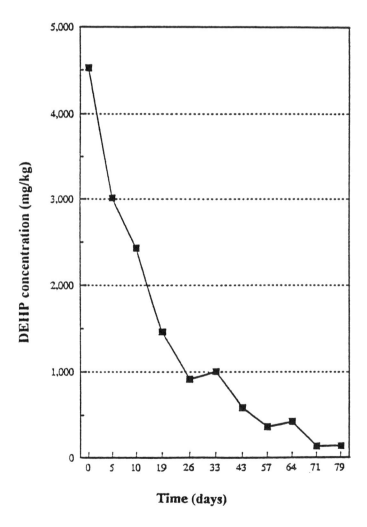

FIGURE 4. Pilot-scale bioremediation of DEHP-contaminated soil using the soil tilling process. (The data indicate DEHP degradation prior to use of treatment system 2.)

depths of 15 to 45 cm were augmented with nutrients and inocula. Soils contained 5,000 to 6,000 mg/kg of DEHP. DEHP degradation in one of the plots over 80 days is presented in Figure 4. During this period, DEHP concentration was remediated to approximately 100 mg/kg. When bioremediation appeared to cease at a DEHP concentration of 20 to 25 mg/kg the soil was amended with treatment system 2. Within 1 week, the DEHP concentration dropped to 10 mg/kg, so that the criteria limit of 15 mg/kg required for soil for residential use was achieved.

CONCLUSION

The study demonstrated both in the laboratory scale, and at the pilot scale that the soil tilling described above could effectively bioremediate phthalate esters in soil. Process effectiveness depended on development of a protocol for selection of the DEHP-degrading site-specific microbial consortium, implementation of a unique fermentation process for production of acclimated inoculum, development of the nutrient formulation, implementation of the tilling and moisture maintenance protocol, and completion of the bioremediation using treatment system 2.

REFERENCES

Fairbancs, B. C., G. A. O'Connor, and S. E. Smith. 1985. "Fate of di-2(ethylhexyl) Phthalate in Three Sludge Amended New Mexico Soils." *J. Environ. Qual.* 14: 479-483.
Gibbons, J. A., and M. Alexander. 1989. "Microbial-Degradation of Sparingly Soluble Organic-Chemicals — Phthalate-Esters." *Environ. Toxicol. Chem.* 8(4): 283-291.
Kurane, R., T. Suzuki, and Y. Takahara. 1978. "Removal of Phthalate Esters in Soil Column Inoculated with Microorganisms." *Agric. Bio. Chem.* 42(8): 1469-1478.
Kurane, R., T. Suzuki, and Y. Takahara. 1980. "Metabolic Pathway of Phthalate Esters by *Nocardia erythropolis*." *Agric. Bio. Chem.* 44(3): 523-527.
Saeger, W., and E. S. Tucker. 1976. "Biodegradation Phthalic Acid Esters in River Water and Activated Sludge." *Appl. Environ. Microbiol.* 31: 29-34.
Shanker, R., C. Ramakrishna, and P. K. Seth. 1985. "Degradation of Some Phthalic Acid Esters in Soil." *Environ. Pollut.* A(39): 1-7.
Shelton, D. R., S. A. Boyd, and J. M. Tiedje. 1984. "Anaerobic Biodegradation of Phthalic Acid Esters in Sludge." *Environ. Sci. Technol.* 18: 93-97.
Taylor, B. F., R. W. Curry, and E. F. Corcoran. 1981. "Potential for Biodegradation of Phthalic Acid Esters in Marine Regions." *Appl. Environ. Microbiol.* 42: 590-595.
Wams, T. J. 1987. "Diethylhexylphthalate as an Environmental Contaminant — A Review." *Sci. Total. Environ.* 66: 1-16.

Application of Variable Nutrient Supplies to Optimize Hydrocarbon Biodegradation

David W. Graham, Val H. Smith, and Kam P. Law

ABSTRACT

The biodegradation in soil of xylenes (BTEX compounds), anthracene and phenanthrene (two polycyclic aromatic hydrocarbons), and n-hexadecane (an n-alkane) was assessed at different nitrogen (N) and phosphorus (P) supply rates and at different N:P supply ratios. The results suggest that the effective biodegradation of each subclass of hydrocarbon requires different nutritional supplies, and that subtle variations in the N and P supply conditions strongly influence their rate of degradation. A simple microcosm screening method is described that approximates a priori the nutrient supply conditions required for effective biodegradation at a given contamination site.

INTRODUCTION

Aerobic biodegradation of hydrocarbon contamination has been demonstrated in both laboratory and field settings (Leahy & Colwell 1990; Prince 1993). Unfortunately, attempts at the bioremediation of hydrocarbon-contaminated soils have not always been successful, and the causes of this variation are not always well understood. Unfavorable oxygen, temperature, and moisture conditions; poor soil permeability; reduced availability of hydrocarbons due to adsorption onto soil surfaces; variable nutrient reserves in different soils; and generally inappropriate nutrient conditions (frequently nitrogen or phosphorus limitation) have all been identified as factors that negatively impact bioremediation success (Bossert & Bartha 1984; Cooney 1984).

Of these factors, the alleviation of nutrient limitation is one of the more easily resolved problems. Bragg et al. (1994) recently demonstrated that nutrient supplementation (primarily as nitrogen and phosphorus amendments) clearly accelerated the rate of hydrocarbon biodegradation at terrestrial sites associated with the *Exxon Valdez* oil spill. However, they also noted that the optimum nutrient supply rate was difficult to predict a priori at a given site, and that the nutrient supplementation was best provided on an incremental basis during the contamination cleanup.

Previous studies of hydrocarbon biodegradation typically have focused either on the influence of contaminant chemistry or on the nutritional conditions (N and P) that are necessary for effective biodegradation (Atlas & Bartha 1972; Atlas & Bartha 1973; Elmendorf et al. 1994; Mahro et al. 1994). Other work has focused on the enumeration, identification, and ecology of microorganisms responsible for hydrocarbon biodegradation (Walker & Colwell 1976; Roubal & Atlas 1978; Song & Bartha 1990; Focht et al. 1990; Lindstrom et al. 1991; Mueller et al. 1994). To our knowledge, however, no systematic study has been made on specific nutrient supply conditions required to optimize the biodegradation of specific hydrocarbons and, in turn, to determine whether the application of hydrocarbon-specific nutrient supply conditions might be used to enhance rates of biodegradation in a given scenario. This key question is the focus of this paper.

EXPERIMENTAL METHODS AND MATERIALS

Experiment Protocols

The mass of CO_2 produced during the biodegradation of xylenes (a combination of the xylene isomers, Fisher Scientific), n-hexadecane, anthracene, and phenanthrene were measured in soil slurries under 12 different supply rates of nitrogen and phosphorus. All experiments employed 165-mL Teflon™ septum-sealed serum vials (microcosms), each containing 60 mL of nitrate salts medium (NSM) (Burrows et al. 1984), 3.0 g of air-dried soil, 0.9 mM of the hydrocarbon source, and supplemental N (as $NaNO_3$) and P (as K_2HPO_4), as required.

Microcosms were prepared with a range of N and P supply conditions for each hydrocarbon source. The N:P molar supply ratios ranged from 5:1 to 40:1, and two levels of N and P (defined as X and 2X) were used for each N:P supply ratio assessed. The absolute nutrient concentrations (per the liquid fraction) added to the microcosms ranged from 1.5 to 10.0 mM N and 0.15 to 0.6 mM P. The background levels (due to natural N and P in the soil) of N and P in the soil slurries were 2.3 mM and 0.15 mM, respectively. The hydrocarbon concentrations in the microcosms were constant at 15.0 mM (per the liquid fraction) for all experiments. Control microcosms, without hydrocarbon amendments, were prepared for each N:P supply ratio to correct for CO_2 evolution resulting from the degradation of the background soil organic carbon.

The soil was collected from the vicinity of a crude oil fueling station near Lawrence, Kansas. Only soil that showed no evidence of direct hydrocarbon contamination was used. The soil was stored under air-dry conditions at 4°C prior to use. The microcosms were agitated on a shaker table maintained at 150 RPM, and all experiments were performed at room temperature (22 to 25°C).

Measurements of the initial soil total nitrogen were made using a Carlo-Erba C/N analyzer. Total phosphorus measurements of the slurry were made at the end of each experiment using standard spectrophotometric methods following acidic persulfate digestion (APHA 1992). Biodegradation activity was monitored by directly analyzing CO_2 accumulation in the headspace of the microcosms using a Carle 311 Analytical Gas Chromatograph with a Poropack Q 80/100 column. To ensure aerobic conditions, the microcosms were flushed with air whenever headspace CO_2 levels exceeded 2% by volume. After flushing, the headspace CO_2 was reanalyzed to normalize the next day's reading. In the case of xylenes, which is volatile, flushing was performed in a systematic manner such that the flushing losses of this hydrocarbon could be estimated. The results reported here are background-corrected CO_2 evolution rates and CO_2 yields (calculated as hydrocarbon-amended CO_2 production minus the CO_2 production from an equivalent no-hydrocarbon control).

Determination of Maximum CO_2 Production Rates and Yields

The model used to estimate the maximum CO_2 production rate was (Krebs 1972):

$$C(t) = K / (1+e^{(a-rt)})$$

where $C(t)$ is the cumulative amount of CO_2 produced over time, r is the maximum rate production of CO_2, t is time (days), and both K and a are constants. The model used to estimate final CO_2 yields was (Grossman & Turner 1974):

$$C(t) = (C(0)Be^{(Bt)}) / (B-C(0)G+C(0)Ge^{(Bt)})$$

where $C(t)$ is the cumulative amount of CO_2 produced over time, $C(0)$ is the CO_2 present at time zero (here we assume $C(0) = 1.0$), the ratio of the curve-fitted constants B and G (B/G) is the asymptotic final CO_2 yield, and t is time (days).

Both models approximate microbial population growth under logistic growth conditions and provide an objective method for estimating the CO_2 evolution rates and final yields. In applying these models, it was necessary to assume that the rate of CO_2 evolution reflected microbial population growth and that the evolved CO_2 was conserved within the microcosm. Although these assumptions are only approximately met, the data fit the models well (the lowest observed r^2 was 0.71 and an r^2 of 0.97 was typical). All modeling was performed using SYSTAT (Wilkinson 1988).

RESULTS

The patterns of CO_2 evolution presented in Figure 1 are typical for each of the four hydrocarbons. Figure 1 describes the cumulative production of CO_2 associated with the biodegradation of each hydrocarbon under 10:1 N:P (2X) nutrient supply conditions.

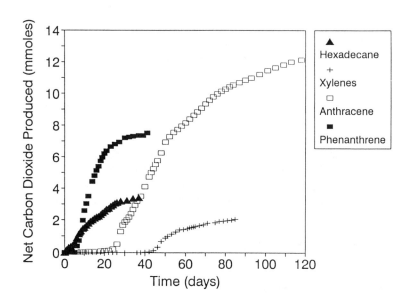

FIGURE 1. **Typical CO₂ production patterns for *n*-hexadecane, xylenes, anthracene, and phenanthrene. Results presented are for the 10:1 N:P-high (2X) absolute supply condition.**

CO$_2$ production from the *n*-hexadecane-amended soils was immediate (with no apparent lag phase), although the final CO$_2$ yields and maximum CO$_2$ production rates were relatively low compared to the other three chemicals. In contrast, CO$_2$ production in the phenanthrene-amended soils was not observed until after 4 to 6 days; after this lag period, CO$_2$ evolution was very rapid. A similar CO$_2$ production pattern was observed in the xylene- and anthracene-amended soils; however, the lags in CO$_2$ production ranged from 8 to 79 days for xylenes, and 5 to 35 days for anthracene. The duration of the lag phase appeared to be influenced by the N and P supply, with longer lags occurring under N and P supplies that were suboptimal.

Figure 2 demonstrates a clear correlation between the maximum rates of CO$_2$ production and final CO$_2$ yields for *n*-hexadecane, phenanthrene, and xylenes in our experiments. Although the individual relationship for each compound subtly differs, the overall relationship between rate and yield for these three contaminants was strong ($r^2 = 0.925$; $p < 0.01$). In contrast, the correlation between these two response variables was not significant for anthracene. For the purpose of the analysis here, maximum CO$_2$ production rates have been used to compare the effects of N and P supplies on biodegradation rate. This is because less time and data are required to produce statistically significant rate approximations.

Our data suggest that the rate of biodegradation of hydrocarbon contaminants in soil microcosms is strongly influenced both by the identity of the hydrocarbon source and by the nutritional supply conditions. In the case of xylenes, a peak in the rate of CO$_2$ evolution was observed at a molar N:P supply ratio of 10:1

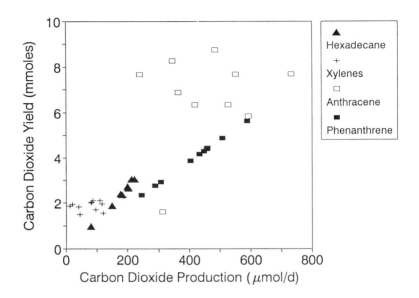

FIGURE 2. Correlation between the estimated maximum CO_2 production rates and the final CO_2 yields.

at the low supply level (Figure 3A). However, when the supply level was doubled to 2X, peak biodegradation of xylenes was observed at a somewhat higher N:P ratio (20:1). In contrast, the biodegradation rate of *n*-hexadecane tended to increase consistently with increases in the molar N:P supply ratio at both absolute levels of supply (Figure 3B).

Phenanthrene and anthracene exhibited very different patterns of CO_2 production. For both compounds, an optimum rate of CO_2 evolution was observed at a 5:1 N:P supply ratio under 2X supply conditions; whereas, at the lower supply level (X), peak CO_2 evolution rates occurred at an N:P supply ratio of 40:1 (Figure 3C and D). Moreover, under the low nutrient supply conditions (X), both compounds exhibited an increased rate of biodegradation with an increase in the N:P supply ratio. However, the exact opposite response was observed when the supply level was doubled (2X).

DISCUSSION

Our results demonstrate that specific nutrient supply levels and ratios can be used to optimize the biodegradation of specific hydrocarbons and that a simple screening experiment can be performed to identify those nutritional supply requirements. This is significant in practical terms because it implies that different nutrient supply conditions should be used for different hydrocarbon contamination sites.

Phenanthrene and Anthracene

For the soil assessed here, phenanthrene and anthracene required only moderate N amendments to maximize biodegradation, and the optimum N:P supply ratio varied depending upon the absolute amounts of N and P added. As the absolute supply levels of N and P were increased by a factor of two, the optimum N:P supply ratio shifted from 40:1 (X) to 5:1 (2X).

Figure 4 presents the maximum CO_2 production rates for the two polycyclic aromatic hydrocarbon (PAH) compounds (data from Figures 3C and D) as a function of the absolute levels of nitrogen provided (irrespective of P supply). A unimodal relationship between the CO_2 production rate and the absolute amount of N supplied is apparent. Reduced CO_2 production rates at the low N supply levels were not unexpected and are likely a result of N-limited growth conditions. However, the negative impact of high N supply levels on CO_2 production rates was not anticipated.

Altered CO_2 production rates at higher nitrogen levels probably resulted from a shift in nutritional limitation. The molar C:N ratios which corresponded to the microcosms described by Figure 4 (see inset) ranged from 1.5:1 (high N) to 10:1 (low N). It is possible that as nitrogen supply was increased, the biodegrading

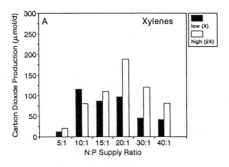
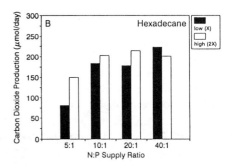
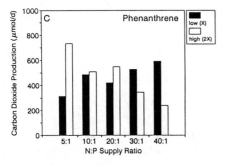
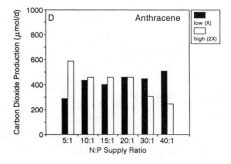

FIGURE 3.   Maximum CO_2 production rates for xylenes, *n*-hexadecane, anthracene, and phenanthrene associated with variable nitrogen and phosphorus supply conditions.

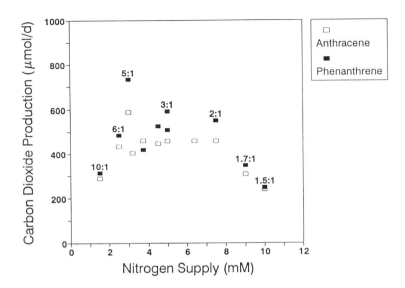

FIGURE 4. Maximum CO_2 production rates for anthracene and phenanthrene as a function of the absolute nitrogen supply. The C:N ratios for each nitrogen supply are inset.

population(s) shifted from N- to C-limitation, or possibly, to carbon/nitrogen colimited conditions. If the populations were strictly C-limited, the rate of CO_2 production would be expected to asymptotically approach a maximum rate determined by C-limitation. As this did not occur, other explanations are required.

For example, Egli (1991) showed that, under C/N colimited conditions, microbial growth rates and cell yields varied dramatically as a function of subtle changes in the composition of the carbon source. We currently are evaluating enrichment populations developed under each C:N:P supply condition to determine whether C/N colimitation could have existed in degradation scenarios assessed here. In conjunction with this further work, we also are assessing whether the reduction in biodegradation rate at elevated N levels was caused by an ecological shift toward less catalytically active microbial strains or was due to a general reduction in metabolic activity of the preexisting strains present.

Xylenes and *n*-Hexadecane

The CO_2 production patterns were very different for xylenes, *n*-hexadecane, and the PAH compounds. The absolute levels of both N and P appeared to be critical to the degradation of xylenes (see Figure 3A), although the optimum CO_2 evolution rates for given absolute N and P supplies consistently occurred at intermediate N:P supply ratios (between 10:1 and 20:1 by moles).

For *n*-hexadecane, however, increases in N:P supply ratio resulted in progressively higher rates of CO_2 evolution (see Figure 3B). The absolute levels of N and P appeared to be less important with *n*-hexadecane biodegradation, as long as the absolute level of N was such that N was not limiting. This strong dependence on the N supply conditions of *n*-alkane biodegradation has been similarly observed in previous work (e.g., Prince 1993; Bragg et al. 1994).

Practical Implications

In interpreting the above results, a series of important practical observations can be made. First, our data show that the biodegradation of specific contaminants is affected by both the absolute quantities and ratios of nitrogen and phosphorus. Previous studies of in situ hydrocarbon biodegradation (e.g., Fedorak & Westlake 1981) found that invariant N and P supply levels resulted in significantly different biodegradation rates for different chemical subcomponents of contaminant spills. Although these types of results can be partially explained by innately different biodegradabilities of different compounds, our results suggest that different degradation rates also can be due to the fact that optimum biodegradation of different chemical contaminants can require different nutritional conditions.

Corroborating evidence for this hypothesis can be found elsewhere. Focht et al. (1990) clearly showed that phenanthrene and hexadecane were degraded by two mutually exclusive guilds of microbes; however, some overlap existed between the microbial guilds responsible for phenanthrene and anthracene biodegradation. The similarity in the nutrient conditions required for the degradation of the two PAH compounds, and the dissimilarity in the nutrient requirements for PAH versus *n*-hexadecane degradation observed here are consistent with the Focht et al. results. It is not improbable that "PAH-degraders" and "non-PAH-degraders" have very different N and P needs, and this is reflected by different degradation responses for different absolute N and P supplies and ratios.

In practical terms, this study suggests that each bioremediation scenario (i.e., a given spill and a given soil) probably requires subtly different N and P supplies and ratios to optimize biodegradation rates. Our results also suggest that suitable nutritional supplements may be approximated based on the chemical composition of contaminants at the site, and case-specific nutrient requirements can be specifically determined using a simple but systematic screening tool such as that presented here.

We emphasize that the microcosm method described here only approximates the nutritional supply conditions required for optimum biodegradation. The effects of other confounding physical and chemical factors must still be overcome in situ to facilitate effective biodegradation (Bossert & Bartha 1984). However, we believe the information provided by a microcosm-based screening method can be used to help guide the spatial and temporal patterns of N and P supplementation needed to optimize contaminant biodegradation at a given site. It should be noted that the methods described here currently are being used to predict optimum nutrient amendments required at two hydrocarbon-contaminated sites in Kansas. Early results are very promising.

REFERENCES

American Public Health Association (APHA), American Water Works Association, and Water Environment Federation. 1992. *Standard Methods for the Examination of Water and Wastewater*, 18th ed. American Public Health Association, Washington, DC.

Atlas, R. M., and R. Bartha. 1972. "Degradation and Mineralization of Petroleum in Sea Water: Limitation by Nitrogen and Phosphorus." *Biotechnol. Bioeng.* 14:196-204.

Atlas, R. M., and R. Bartha. 1973. "Stimulated Biodegradation of Oil Slicks Using Oleophilic Fertilizers." *Environ. Sci. Tech.* 7:538-541.

Bossert, I., and R. Bartha. 1984. "The Fate of Petroleum in Soil Ecosystems." In R. M. Atlas (Ed.), *Petroleum Microbiology*, pp. 435-474. Macmillan Publishing Company, New York, NY.

Bragg, J. R., R. C. Prince, E. J. Harner, and R. M. Atlas. 1994. "Effectiveness of Bioremediation for the *Exxon Valdez* Oil Spill." *Nature 368*:413-418.

Burrows, K. J., A. Cornish, D. Scott, and I. J. Higgins. 1984. "Substrate Specificities of the Soluble and Particulate Methane Mono-oxygenases of *Methylosinus trichosporium* OB3b." *J. Gen. Microbial. 130*:3327-3333.

Cooney, J. J. 1984. "The Fate of Petroleum Pollutants in Freshwater Ecosystems." In R. M. Atlas (Ed.), *Petroleum Microbiology*, pp. 399-434. Macmillan Publishing Company, New York, NY.

Egli, T. 1991. "On Multiple-Nutrient-Limited Growth of Microorganisms, With Special Reference to Dual Limitation by Carbon and Nitrogen Substrates." *Antonie van Leeuwenhoek. 60*:225-234.

Elmendorf, D. L., C. E. Haith, G. S. Douglas, and R. C. Prince. 1994. "Relative Rates of Biodegradation of Substituted Polycyclic Aromatic Compounds." In R. E. Hinchee, A. Leeson, L. Semprini, and S. K. Ong (Eds.), *Bioremediation of Chlorinated and Polycyclic Aromatic Hydrocarbon Compounds*, pp. 188-202. Lewis Publishers, Boca Raton, FL.

Fedorak, P. M., and D.W.S. Westlake. 1981. "Microbial Degradation of Aromatics and Saturates in Prudhoe Bay Crude Oil as Determined by Glass Capillary Gas Chromatography." *Can. J. Microbiol. 27*:432-443.

Focht, J. M., P. M. Fedorak, and D.W.S. Westlake. 1990. "Mineralization of [^{14}C]Hexadecane and [^{14}C]Phenanthrene in Crude Oil: Specificity Among Bacterial Isolates." *Can. J. Microbiol.* 36:169-175.

Grossman, S. I., and J. E. Turner. 1974. *Mathematics for the Biological Sciences*. Macmillan Publishing Company, New York, NY.

Krebs, C. J. 1972. *Ecology: The Experimental Analysis of Distribution and Abundance*. Harper and Row, New York, NY.

Leahy, J. G., and R. R. Colwell. 1990. "Microbial Degradation of Hydrocarbons in the Environment." *Microbiol. Rev.* 54:305-315.

Lindstrom, J. E., R. C. Prince, J. C. Clark, M. J. Grossman, T. R. Yeager, J. F. Braddock, and E. J. Brown. 1991. "Microbial Populations and Hydrocarbon Biodegradation Potentials in Fertilized Shoreline Sediments Affected by the T/V *Exxon Valdez* Oil Spill." *Appl. Environ. Microbiol. 57*:2514-2522.

Mahro, B., G. Schaefer, and M. Kastner. 1994. "Pathways of Microbial Degradation of Polycyclic Aromatic Hydrocarbons in Soil." In R. E. Hinchee, A. Leeson, L. Semprini, and S. K. Ong (Eds.), *Bioremediation of Chlorinated and Polycyclic Aromatic Hydrocarbon Compounds*, pp. 203-217. Lewis Publishers, Boca Raton, FL.

Mueller, J. G., S. E. Lantz, R. Devereux, J. D. Berg, and P. H. Pritchard. 1994. "Studies on the Microbial Ecology of Polycyclic Aromatic Hydrocarbon Biodegradation." In R. E. Hinchee, A. Leeson, L. Semprini, and S. K. Ong (Eds.), *Bioremediation of Chlorinated and Polycyclic Aromatic Hydrocarbon Compounds*, pp. 218-230. Lewis Publishers, Boca Raton, FL.

Prince, R. C. 1993. "Petroleum Spill Bioremediation in Marine Environments." *Crit. Rev. Microbiol. 19*:217-242.

Roubal, G., and R. M. Atlas. 1978. "Distribution of Hydrocarbon-Utilizing Microorganisms and Hydrocarbon Biodegradation Potentials in Alaskan Continental Shelf Areas." *Appl. Environ. Microbiol. 35*:897-905.

Song, H-G., and R. Bartha. 1990. "Effects of Jet Fuel Spills on the Microbial Community of Soil." *Appl. Environ. Microbiol. 56*:646-651.

Walker, J. D., and R. R. Colwell. 1976. "Enumeration of Petroleum-Degrading Microorganisms." *Appl. Environ. Microbiol. 31*:198-207.

Wilkinson, L. 1988. *SYSTAT: The System for Statistics.* SYSTAT Inc., Evanston, IL.

Composting for Remediation of Soil Contaminated with Pharmaceutical Residues

Stuart H. Rhodes and Philip C. Peck

ABSTRACT

Soils excavated from an area used for landfill of waste product at an Australian pharmaceutical manufacturing site contained pharmaceutical chemical residues. Soil composting was investigated as an alternative to incineration. In laboratory trials, a factorial experimental design was used to evaluate organic matter amendment type and concentration, and incubation temperature. Probenecid was reduced from 5,100 mg/kg to less than 10 mg/kg within 20 weeks in mesophilic treatments. An 8 tonne pilot scale treatment confirmed that thermophilic composting was effective under field conditions. In the final stage, 180 tonnes of soil was composted. Initial concentrations of the major contaminants in the compost, probenecid and methaqualone, were 1,160 mg/kg and 210 mg/kg, respectively. Probenecid concentration reached the target level of 100 mg/kg in 6 weeks, and removal of methaqualone to below target levels was achieved after 20 weeks.

INTRODUCTION

Composting has been used for many years for disposal of agricultural, municipal and domestic wastes. Composting has also been investigated for treatment of soils contaminated with nitroaromatic explosives such as TNT (Woodward 1990) and other hazardous wastes (Finstein et al. 1987, Mays et al. 1989). Laboratory and field studies have shown composting can significantly reduce explosive concentrations (Keehan and Sisk 1991). The type of organic amendments used and the ratio of contaminated soil to organic amendment are crucial parameters in making this technique cost competitive with incineration.

In mid-1992, expansion of facilities at a pharmaceutical manufacturing in western Sydney required the excavation of an area where out of date, waste, or off-spec product previously had been landfilled. Excavated soil contaminated with pharmaceutical process wastes was stored in covered waste containers pending disposal. The soil was found to contain pharmaceutical residues including probenecid (4-[(dipropylamino) sulfonyl] benzoic acid) and methaqualone (2-methyl-3-(2-methylphenyl)-4(3H)-quinazolinone; see Figure 1). A quantity of fillers and binders (e.g., lactose) also was present.

The pharmaceutical residues were expected to be relatively nontoxic, but are, by their nature, biologically active compounds. The metabolism of these compounds is well understood, but their fate in the environment has not been previously investigated. The site owner wished to retain the soil on the site for treatment and reuse, for example in landscaping. A feasibility study was undertaken to determine whether concentrations of pharmaceutical residues could be reduced by soil composting.

EXPERIMENTAL METHODS

Sample Selection and Preparation

The excavated material was extremely heterogeneous. Samples were collected for the feasibility study after the material was removed from the waste bins and placed in a single large pile in the warehouse building. These 21 samples were sieved to remove large "non-soil" items such as broken glass, plastic, etc., and thoroughly mixed to prepare a soil composite which was then split into portions for the experimental treatments. This initial composite was found to contain 8,400 mg/kg probenecid and 75 mg/kg methaqualone, with lower concentrations of various organic acids.

Experimental Design

A two-level factorial experimental design was used to investigate the effects of temperature, organic amendment type and concentration. Treatments simulated either a mesophilic process (incubation at 25°C) or a moderately thermophilic process (slowly increasing to 48°C). Soil was mixed with horse manure

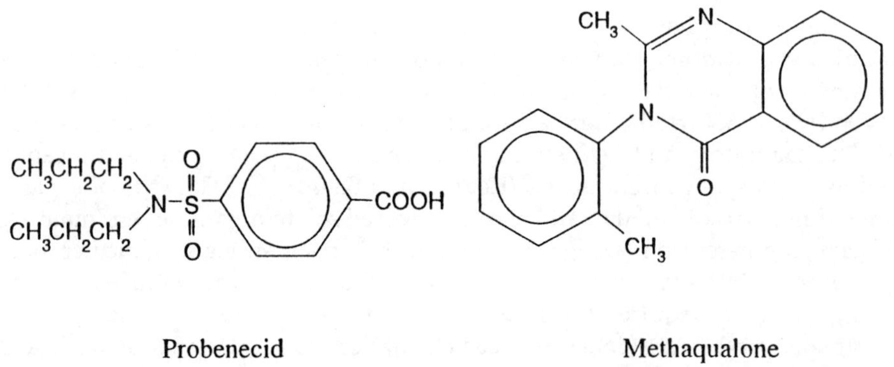

Probenecid   Methaqualone

FIGURE 1. Chemical structures of major pharmaceutical contaminants.

or partially decomposed plant material, added to give either 30% or 60% by weight of the total final wet weight. The treatments were carried out as single 1-kg batches of compost, which were mixed weekly. Mineral nutrients [$(NH_4)_2SO_4$, $MgSO_4 7H_2O$, and K_2HPO_4] were added, and the moisture content was maintained near the initial level for each treatment (in the range 30 to 50%). Two controls included one unamended soil, and one soil poisoned with mercuric chloride.

Analytical Methods

Pharmaceutical Residues. The pharmaceutical residues were analyzed by gas chromatography following extraction by dichloromethane:acetone. For compounds identified by GC-MS, the sample concentration of the corresponding GC-FID peak areas was reported as mg/kg dry matter.

Microbiology. Microbiological populations were determined by standard plate count methods. Total mesophilic, heterotrophic organisms were determined on Tryptone Soy Agar (Oxoid), and thermophilic populations by incubation at 60°C. Fungi, presumptive coliforms, and pseudomonads were enumerated using appropriate selective media (Table 1).

LABORATORY SIMULATIONS

Initial samples from the microcosms contained between 3,500 mg/kg and 7,700 mg/kg probenecid on a dry soil basis (mean 5,100 mg/kg). Significant decreases in probenecid concentration were noted after 42 days, and after 19 weeks two treatments contained no detectable probenecid (Figure 2). The most

TABLE 1. Selected microbial populations in pilot-scale soil compost piles.

Composting Time (days)	Mesophiles (1)	Thermophiles (2)	Yeasts and Fungi (3)	Pseudomonads (4)	Presumptive Coliforms (5)
Soil	2.0×10^8	$< 1.0 \times 10^4$	3.8×10^6	3.2×10^7	7.5×10^5
0	1.1×10^9	3.3×10^7	8.3×10^6	9.0×10^7	2.3×10^7
10	2.2×10^9	2.8×10^7	8.4×10^6	7.8×10^7	1.1×10^6
26	1.4×10^9	2.7×10^7	1.2×10^7	8.4×10^7	7.0×10^5
40	9.3×10^8	2.5×10^7	1.4×10^7	6.4×10^7	1.1×10^5

(1) Tryptone Soy Agar at 28°C.
(2) TSA at 60°C.
(3) Rose-Bengal Chloramphenicol Agar (Oxoid).
(4) Pseudomonas Selective Isolation Agar (Kreuger and Sheikh 1987).
(5) MacConkey Agar (Oxoid).

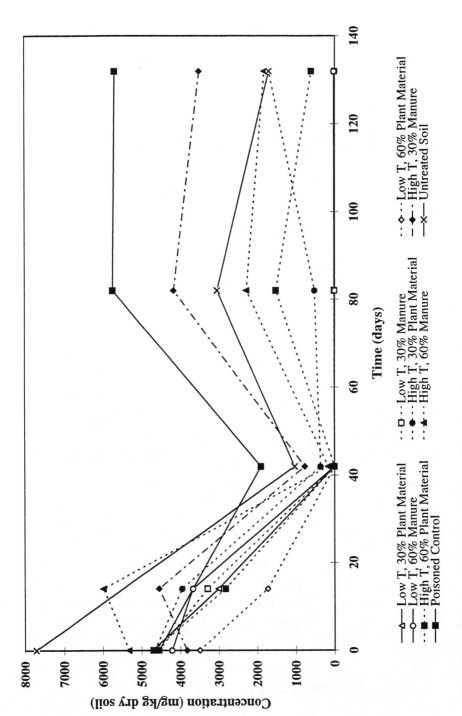

FIGURE 2. Probenecid degradation in soil composting microcosms.

effective treatments were those maintained at 25°C. Probenecid removal in the thermophilic treatments ranged from 30% to 90%. Over the ranges tested, no significant effects were seen from either the type or the concentration of organic amendments.

No decrease was observed in the poisoned control, but the unamended control also showed a substantial reduction in probenecid concentration (70%), probably reflecting the biostimulation effect of mineral nutrients, moisture, and mixing/aeration.

PILOT-SCALE TREATMENT

The contaminated soil was a silty clay, and the pharmaceutical contaminants were present as agglomerates, some extremely large, and fine powder. Preprocessing was required to break up large soil and residue aggregates, and to distribute the contaminants for effective composting.

A pilot-scale treatment was therefore carried out to assess soil processing requirements prior to compost blending; materials handling, bulking, and space requirements; the suitability of the available local organic materials; the rate of contaminant removal in a large-scale operation; and the heat generation characteristics of the compost.

All soil processing and composting operations were conducted within a large warehouse building. Approximately 5 m³ (8 tonnes) of the contaminated soil was mixed with 16 m³ of organic material (commercially available mulch consisting of chipped wood waste and leaves, horse manure, and grass clippings).

The compost pile temperatures rose rapidly after mixing to peak at 57°C after 30 hours. The temperatures then declined slowly as the biodegradable material was decomposed. The piles were regularly mixed to provide aeration using mobile earthmoving equipment.

The initial concentrations of probenecid and methaqualone directly after the blending operations were 1,200 mg/kg and 60 mg/kg, respectively. Probenecid concentrations were reduced to below the target level (100 mg/kg) in 2 to 3 weeks, and to less than 10 mg/kg after 5 weeks. Methaqualone concentrations declined at a slower rate, reaching less than 10 mg/kg at the completion of the pilot trial after 7 weeks.

After amendment of the soil with organic matter the total microbial populations were 10^9 per gram compost, i.e., about 10 times higher than for the contaminated soil. The microbial numbers increased over the first 3 weeks of treatment, then steadily declined as the compost matured. Characterization of significant subpopulations of organisms (thermophiles, yeasts and fungi, pseudomonads and coliforms) showed that addition of organic matter raised the numbers of each type, especially the thermophiles and coliforms (Table 2). This increase most likely reflects the origin and biologically active state of the organic materials as delivered to the site. During soil composting, the thermophilic population declined slightly, and was never more than 3% of the total mesophilic

TABLE 2. Germination and biomass yield in composted soil.

| Species | Soil | Final Germination Survival (%) | Biomass (mg dry weight) | |
			Total	Per Plant
E. eximia	Before treatment	0	0	0
	Composted	58 ± 5	112 ± 6	8 ± 1
	Control topsoil	73 ± 1	134 ± 66	9 ± 2
Oat	Before treatment	27 ± 8	18 ± 6	3 ± 0
	Composted	69 ± 2	326 ± 17	19 ± 0
	Control topsoil	21 ± 2	79 ± 21	14 ± 3
Radish	Composted	86 ± 8	545 ± 42	25 ± 2
	Control topsoil	80 ± 6	249 ± 11	12 ± 0

population. A substantial decrease was measured in the number of coliforms (from more than 10^7 per gram, to about 10^5 per gram).

Because the soil was to be retained on site for use as landscaping material following treatment, samples of contaminated and composted soil were tested for their ability to support early plant growth (OECD, 1984). The results (Table 2) showed that the germination and biomass yield per plant was significantly less in the contaminated soil than in a control topsoil. The composted soil supported germination and early growth of all three test species. Whether the inhibition of plant growth in the untreated soil was the result of chemical toxicity, the physical and nutrient condition of the soil, or a combination of these, it was relieved by composting.

DISCUSSION

The pilot-scale composting resulted in the soil appearance changing from a light grey-brown clay containing obvious white powdery residues to a dark, organic appearance soon after the composting commenced, so that no wastes or residues were visible. No objectionable odors were generated from the process or were noticeable in the treated soil. Moisture additions were managed so that no leachate was produced, and the final product was found to be suitable for reuse on site.

As a result, the remainder of the contaminated soil, approximately 110 m³, was composted with 320 m³ leaf mulch and manure in a pile 15 m by 11 m by 1.7 m high. The initial concentration of probenecid in the compost was similar to that observed previously (1,160 mg/kg). Methaqualone had not been measured previously at concentrations above 100 mg/kg and had not been considered to be a major contaminant, but was found in this case at 210 mg/kg.

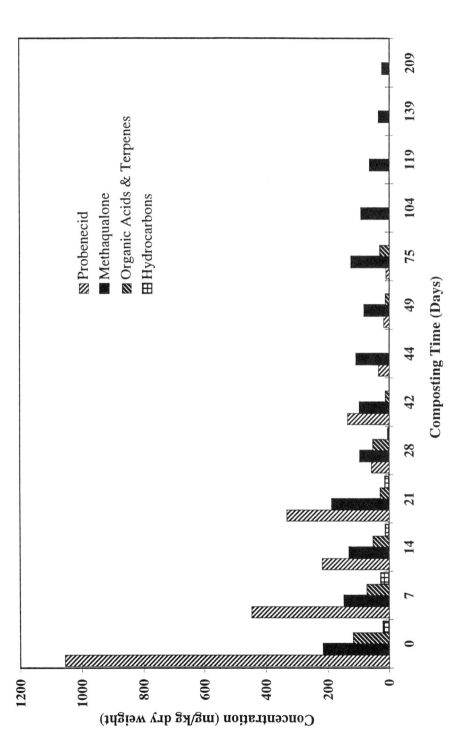

FIGURE 3. Pharmaceutical residues in full-scale soil compost.

The probenecid concentration declined to less than 100 mg/kg in 6 weeks. Extension of the composting period was required to reduce the concentrations of methaqualone to consistently below 100 mg/kg, which occurred after 14 weeks (Figure 3). The compost was allowed to further mature without processing until 30 weeks, when the average methaqualone concentration was 23 mg/kg.

ACKNOWLEDGMENTS

The authors acknowledge the assistance of Dr. I. Swane (Dames & Moore) and Mr. S. Soliman (Merck, Sharp and Dohme Australia Pty Ltd). The phytotoxicity tests were carried out by the Centre for Environmental Toxicology, University of Technology, Sydney.

REFERENCES

Finstein, M. S., J. V. Hunter, J. A. Hogan, and G. R. Toffoli. 1987. *Microbial Decomposition of Hazardous Industrial Compounds Through Composting.* Final report. Hazardous Substances Management Research Center, New Jersey Inst. Tech., Newark, NJ.
Keehan, K. R., and W. E. Sisk. 1991. "Composting of Explosives-Contaminated Soils at a National Priority List Site." Presented at Fourth International IGT Symposium on Gas, Oil and Environmental Biotechnology, Colorado Springs. December.
Kreuger, C. L., and W. Sheikh. 1987. "A New Selective Medium for Isolating *Pseudomonas* spp. from Water." *Appl. Env. Microbiol.,* 53: 895-897.
Mays, M. K., L. J. Sikora, J.W.E. Hatton, and S. M. Lucia. 1989. "Composting as a Method of Hazardous Waste Treatment." *Biocycle,* 30: 298-300.
OECD. 1984. *OECD Guidelines for Testing Chemicals. Terrestrial Plants Growth Test.* OECD Guideline 208. OECD, Paris.
Woodward, R. E. 1990. *Evaluation of Composting Implementation: A Literature Review.* USATHAMA Installation Restoration Program Environmental Technology Development. Report No. TCN 89363.

AUTHOR LIST

Adriaens, Peter
University of Michigan
Dept. of Civil & Environ. Engineering
181 EWRE 1A Building
Ann Arbor, MI 48109-2125 USA

Allen, Bobby
Memphis Environmental Center
2603 Corporate Avenue #100
Memphis, TN 38132 USA

Alvarez, Pedro J.
University of Iowa
Dept. of Civil & Environ. Engineering
1136 Engineering Building
Iowa City, IA 52240-1527 USA

Ampleman, Guy
Defence Research Establishment
    Valcartier
Division des materiaux energetiques
2459 boul. Pie XI Nord, C.P. 8800
Courcelette, Québec G10 1R0
CANADA

Arazzini, Settimio
Castalia S.p.A.
World Trade Center via de Marini 1
Genova 16149
ITALY

Arunachalam, M.
Kansas State University
Civil Engineering, Seaton Hall
Manhattan, KS 66506 USA

Autenrieth, Robin L.
Texas A&M University
Dept. of Civil Engineering
College Station, TX 77843-3136 USA

Bae, Bum-Han
Texas A&M University
Dept. of Civil Engineering
College Station, TX 77843-3126 USA

Banks, M. Katherine
Kansas State University
Civil Engineering Department
Seaton Hall
Manhattan, KS 66506 USA

Barker, James F.
University of Waterloo
Waterloo Centre for Groundwater
    Research
200 University Avenue, West
Waterloo, Ontario N2L 3G1
CANADA

Barkovskii, Andrei
The University of Michigan
Dept. of Civil & Environ. Engineering
181 EWRE Building
Ann Arbor, MI 48109 USA

Beeckman, Marit
University of Ghent
Centre for Environmental Sanitation
Coupure Links 653
B-9000 Ghent
BELGIUM

Bhattacharya, Sanjoy K.
Tulane University
Dept. of Civil & Environ. Engineering
206 Civil Engineering Building
6823 St. Charles Avenue
New Orleans, LA 70118 USA

Bishop, Paul L.
University of Cincinnati
Dept. of Civil & Environ. Engineering
741 Baldwin (ML 71)
Cincinnati, OH 45221-0071 USA

Bocchieri, Paola
Castalia S.p.A.
World Trade Center via de Marini 1
Genova 16149
ITALY

Bollag, Jean-Marc
Pennsylvania State University
129 Land and Water Research Bldg.
University Park, PA 16802-4900 USA

Bonner, James S.
Texas A&M University
Dept. of Civil Engineering
College Station, TX 77843-3136 USA

Bradley, Paul M.
U.S. Geological Survey
Stephenson Center Suite 129
720 Gracern Road
Columbia, SC 29210-7651 USA

Brodman, Bruce W.
U.S. Army
Armaments Research Development
    and Engineering Center
Building 472
Picatinny Arsenal, NJ 07806-5000
USA

Butler, Barbara J.
University of Waterloo
Waterloo Centre for Groundwater
    Research
Department of Biology
200 University Ave., West
Waterloo, Ontario N2L 3G1
CANADA

Cali, Peter R.
U.S. Army Corps of Engineers
Civil & Environmental Engineering
P.O. Box 60267
New Orleans, LA 70160-0267 USA

Chapelle, Francis H.
U.S. Geological Survey, WRD
Stephenson Center Suite 129
720 Gracern Road
Columbia, SC 29210-7651 USA

Chapman, Peter J.
U.S. Environ. Protection Agency
Gulf Breeze Environ. Research Lab
1 Sabine Island Drive
Gulf Breeze, FL 32561-3999 USA

Christian, Barry J.
Earth Technology
5555 Glenwood Hills Parkway, SE
Grand Rapids, MI 49588-0874 USA

Clarke, Bruce H.
Earth Technology
5555 Glenwood Hills Parkway, SE
Grand Rapids, MI 49588-0874 USA

Cole, Michael A.
University of Illinois
Department of Agronomy
1102 South Goodwin Avenue
Urbana, IL 61801-4798 USA

Comeau, Yves
École Polytechnique
Department of Civil Engineering
P.O. Box 6079, Station Centre-ville
Montréal, Québec H3C 3A7
CANADA

Crawford, Don L.
University of Idaho
College of Agriculture
Food Research Center 103
Moscow, ID 83844-1052 USA

D'Haene, Sigfried
Soils N.V.
Haven 1025
Scheldedijk 30
G-2070 Zwijndrecht
BELGIUM

De Saeyer, Nancy
University of Ghent
Centre for Environmental Sanitation
Coupure Links 653
B-9000 Ghent
BELGIUM

Doddema, Hans J.
TNO Inst. of Environmental Sciences
Schoemakerstraat 97
Postbus 6011
2600 JA Delft
THE NETHERLANDS

Donnelly, Paula K.
Sante Fe Junior College
3213 Avenue San Marcos #4
Sante FE, NM 87505 USA

Dooley, Maureen A.
ABB Environmental Services, Inc.
107 Audubon Road, Corporate Place
Wakefield, MA 01880 USA

Dreyer, Günter
BIOPRACT GmbH
Robert-Rössle-Strasse 10
D-13125 Berlin
GERMANY

Felicione, Elise C.
University of Idaho
Food Research Center, Room 105
Moscow, ID 83844-1052 USA

Fermor, Terry R.
Horticulture Research Intl.
Microbial Biotechnology Dept.
Wellesbourne, Warwickshire
     CV35 9EF
UNITED KINGDOM

Findlay, Margaret
Bioremediation Consulting, Inc.
55 Halcyon Road
Newton, MA 02159 USA

Fletcher, John S.
University of Oklahoma
Dept. of Botany & Microbiology
Norman, OK 73019 USA

Fogel, Samuel
Bioremediation Consulting, Inc.
55 Halcyon Road
Newton, MA 02159 USA

Forney, Traci L.
Battelle Columbus
505 King Avenue
Columbus, OH 43201-2693 USA

Fredrickson, Herb L.
U.S. Army Corps of Engineers
USAE Waterways Experiment Station
3909 Halls Ferry Road
Vicksburg, MS 39180 USA

Funk, Stephen B.
University of Idaho
Dept. of Microbiology, Molecular
     Biology, and Biochemistry
Food Research Center 105
Moscow, ID 83844-1052 USA

Graham, David W.
University of Kansas
Dept. of Civil Engineering
2006 Learned Hall
Lawrence, KS 66045 USA

Greer, Charles W.
National Research Council of Canada
Biotechnology Research Institute
6100 Royalmount Avenue
Montréal, Québec H4P 2R2
CANADA

Grey, Gary M.
HydroQual, Inc.
1 Lethbridge Plaza
Mahwah, NJ 07430 USA

Guerin, Turlough F.
Minenco Bioremediation Services
1 Research Avenue
Bundoora, Victoria 3083
AUSTRALIA

Gunnison, Douglas
U.S. Army Corps of Engineers
USAE Waterways Experiment Station
3909 Halls Ferry Road
Vicksburg, MS 39180-6199 USA

Hansen, Susanne Schiøtz
A/S Bioteknisk Jordrens
Maglehøjvej 10
DK-4400 Kalundborg
DENMARK

Hawari, Jalal
National Research Council of Canada
Biotechnology Research Institute
6100 Royalmount Avenue
Montréal, Québec H4P 2R2
CANADA

Hegde, Ramesh S.
University of Oklahoma
Dept. of Botany & Microbiology
Norman, OK 73019 USA

Henrysson, Tomas
Lund University
Biotechnology Dept./Chemical Ctr.
P.O. Box 124
S-22100 Lund
SWEDEN

Hess, Lance
Growth Environmental Services, Inc.
5183 South 300 West
Murray, UT 84107 USA

Hsu, Cheng-Hsuing
The Pennsylvania State University
Dept. of Civil & Environmental Eng.
216 Sackett Building
University Park, PA 16802 USA

Huis in 't Veld, Michel G. A.
TNO Inst. of Environmental Sciences
Schoemakerstraat 97
Postbus 6011
2600 JA Delft
THE NETHERLANDS

Huling, Scott G.
U.S. Environ. Protection Agency
R.S. Kerr Environ. Res. Laboratory
Kerr Lab Drive
P.O. Box 1198
Ada, OK 74820 USA

Jones, Alison M.
National Research Council of Canada
Biotechnology Research Institute
6100 Royalmount Avenue
Montréal, Québec H4P 2R2
CANADA

Karlson, Ulrich
National Environ. Research Inst.
Frederiksborgvej 399
P.O. Box 358
DK-4000 Roskilde
DENMARK

Keith, Jeffrey D.
Brigham Young University
Department of Geology
P.O. Box 24642
Provo, UT 84602-4642 USA

König, Joachim
BIOPRACT GmbH
Robert-Rössle-Strasse 10
D-13125 Berlin
GERMANY

Kupferle, Margaret J.
University of Cincinnati
Dept. of Civ. & Env. Engrg. ML #71
Cincinnati, OH 45221-0071 USA

Landmeyer, James E.
U.S. Geological Survey
Stephenson Center, Suite 129
720 Gracern Road
Columbia, SC 29210 USA

Lavigne, Jacques
Canadian Embassy
Defence Research and Development
501 Pennsylvania Avenue, NW
Washington, DC 20001-2114 USA

Law, Kam P.
University of Kansas
Department of Civil Engineering
2006 Learned Hall
Lawrence, KS 66045 USA

Liu, Xianzhong
University of Illinois at
    Urbana-Champaign
Department of Agronomy
1102 South Goodwin Avenue
Urbana, IL 61801-4798 USA

Liu, Yu-Ting
U.S. Environ. Protection Agency
8644 15th Avenue
Seattle, WA 98115 USA

Maiello, Joy A.
HydroQual
1 Lethbridge Plaza
Mahwah, NJ 07430 USA

Maloney, Stephen W.
University of Cincinnati
Dept. of Civil & Environ. Engineering
741 Baldwin (ML 71)
Cincinnati, OH 45221-0071 USA

Matthews, John E.
JMCO Environmental Consulting, Inc.
P.O. Box 1217
Ada, OK 74821 USA

Mayo, Alan L.
Brigham Young University
Department of Geology
P.O. Box 24642
Provo, UT 84602-4642 USA

McMaster, Michaye L.
University of Waterloo
Waterloo Centre for Groundwater
    Research
200 University Avenue, West
Waterloo, Ontario N2L 3G1
CANADA

Merkley, Brian W.
ESCM and Associates, Inc./
    GeoRem International, Inc.
P.O. Box 7487
394 South Milledge Ave., Suite 258
Athens, GA 30604 USA

Miethling, Rona
National Environ. Research Institute
Fredricksborgvej 399
P.O. Box 358
DK-4000 Roskilde
DENMARK

Migliorini, Giorgio
Castalia S.p.A.
World Trade Center via de Marini 1
Genova 16149
ITALY

Miller, Christopher M.
University of Iowa
122 Engineering Research Facility
Iowa City, IA 52240 USA

Morris, Pamela J.
Medical Univ. of South Carolina
Dept. of Microbiology & Immunology
224 Basic Sciences Building
171 Ashley Avenue
Charleston, SC 29425-2230 USA

Moteleb, Moustafa A.
University of Cincinnati
Dept. of Civil & Environ. Engineering
741 Baldwin (ML 71)
Cincinnati, OH 45221-0071 USA

Omori, Toshio
University of Tokyo
Biotechnology Research Center
Yayoi 1-1-1
Bunkyo-ku, Tokyo 113
JAPAN

Otte, Marie-Paule
University of Québec
6100 Royalmount Avenue
Montréal, Québec H3C 3P8
CANADA

Park, Joon Kyu
University of Southern California
Dept. of Civil & Environ. Engineering
224 A Kaprielian Hall
3620 South Vermont Avenue
Los Angeles, CA 90089-2531 USA

Pasti-Grigsby, Maria B.
University of Idaho
Dept. of Bacteriology & Biochem.
Center for Hazardous Waste Remed.
    Research
Food Research Center 105
Moscow, ID 83843 USA

Peck, Philip C.
Minenco Pty Ltd.
Level 5, 77 Berry Street
N. Sydney, New South Wales 2060
AUSTRALIA

Pendharkar, Suhasini
University of Houston
Dept. of Civil & Environ. Engineering
4800 Calhoun
Houston, TX 77204-4791 USA

Petrova, Krasimira Dekova
Sofia University
Faculty of Biology
8 Dragan Tzankov str.
1421 Sofia
BULGARIA

Pope, Daniel F.
Dynamac Corporation
P.O. Box 1198
Ada, OK 74820 USA

Pott, Britt-Marie
Lund University
Biotechnology Dept./Chemical Ctr.
P.O. Box 124
S-221 00 Lund
SWEDEN

Pugh, Lucy B.
Earth Technology
5555 Glenwood Hills Parkway, SE
P.O. Box 874
Grand Rapids, MI 49588-0874 USA

Rhodes, Stuart H.
Minenco Pty Ltd.
Bioremediation Services
Level 5, 77 Berry Street
N. Sydney, NSW 2060
AUSTRALIA

Ringpfeil, Manfred
BIOPRACT GmbH
Scheiblerstrasse 27
12437 Berlin
GERMANY

Rivara, Lucia
Castalia S.p.A.
World Trade Center via de Marini 1
Genova 16149
ITALY

Roberts, Deborah J.
University of Houston
4800 Calhoun
Dept. of Civil & Environ. Engineering
Houston, TX 77204-4791 USA

Roehl, Marc E.
University of Iowa
Dept. of Civil & Environ. Engineering
125 ERF
Iowa City, IA 52240 USA

Samson, Réjean
École Polytechnique de Montréal
Chemical Engineering Dept.
P.O. Box 6079 Station Centre-Ville
Montréal, Québec H3C 3A7
CANADA

Sayles, Gregory D.
U.S. Environ. Protection Agency
Natl. Risk Mgmt. Research Lab
26 W. Martin Luther King Drive
Cincinnati, OH 45268 USA

Scheible, O. Karl
HydroQual, Inc.
1 Lethbridge Plaza
Mahwah, NJ 07430 USA

Scholl, Christopher
Metcalf & Eddy
30 Harvard Mill Square
Wakefield, MA 01880 USA

Schu, Kirsten
National Environ. Research Institute
399 Fredricksborgvej
P.O. Box 358
DK-4000 Roskilde
DENMARK

Schwab, A. Paul
Kansas State University
Dept. of Agronomy
Throckmorton Hall
Manhattan, KS 66506 USA

Semple, Kirk T.
Horticulture Research Intl.
Wellesbourne, Warwickshire
   CV35 9EF
UNITED KINGDOM

Sharma, Anil
Geo-Centers, Inc.
762 Route 15 South
Lake Hopatcong, NJ 07849 USA

Shaw, Edward A.
ESCM and Associates, Inc.
P.O. Box 7487
394 S. Milledge Ave., Suite 258
Athens, GA 30605 USA

Shelton, Michael E.
University of Minnesota
C/O U.S. EPA-GBERL
1 Sabine Island Drive
Gulf Breeze, FL 32561 USA

Sims, Judith L.
Utah State University
Utah Water Research Laboratory
1600 East Canyon Road
Logan, UT 84322-8200 USA

Sims, Ronald C.
Utah State University
Utah Water Research Laboratory
Logan, UT 84322-8200 USA

Smith, Val H.
University of Kansas
Environmental Studies Program
Haworth Hall
Lawrence, KS 66045 USA

Sorensen, Darwin L.
Utah State University
Utah Water Research Laboratory
Logan, UT 84322-8200 USA

Stacy, Jeff L.
University of Iowa
Dept. of Civil & Environ. Engineering
122 ERF
Iowa City, IA 52240 USA

Stensel, H. David
University of Washington
Dept. of Civil Engineering
309 More Hall FX-10
Seattle, WA 98195 USA

Strand, Stuart E.
University of Washington
College of Forests AR-10
Seattle, WA 98195 USA

Suidan, Makram T.
University of Cincinnati
Dept. of Civil & Environ. Engineering
741 Baldwin (ML 71)
Cincinnati, OH 45221-0071 USA

Sundaram, Shanmuga T.
Geo-Centers, Inc.
762 Route 15 South
Lake Hopatcong, NJ 07849 USA

Taylor, Karen
ABB Environmental Services
Corporate Place 128
107 Audubon Road
Wakefield, MA 01880 USA

Terry, Richard E.
Brigham Young University
Dept. of Agronomy & Horticulture
P.O. Box 24642
Provo, UT 84602-4642 USA

Thiboutot, Sonia
Defence Research Establishment
   Valcartier
Division des materiaux energetiques
2459 boul. Pie XI Nord, C.P. 8800
Courcelette, Québec G10 1R0
CANADA

Tingey, David G.
Brigham Young University
Department of Geology
P.O. Box 24642
Provo, UT 84602-4642 USA

Tittle, Piper C.
Gray and Osborne, Inc.
701 Dexter Avenue North
Seattle, WA 98109 USA

Topalova, Yana Ilieva
Sofia University
Faculty of Biology
8 Dragan Tzankov str.
1421 Sofia
BULGARIA

Towe, Roger D.
Tenneco, Inc.
1010 Milam Street
P.O. Box 2511
Houston, TX 77252-2511 USA

Tripaldi, Giuseppe
Castalia S.p.A.
Via Vitorchiano 151
Roma 00189
ITALY

Uotila, Jussi
University of Helsinki
Department of Applied Chemistry &
   Microbiology
P.O. Box 27
SF-00014 Helsinki
FINLAND

Valentine, Richard L.
University of Iowa
Dept. of Civil & Environ. Engineering
1126 Engineering Building
Iowa City, IA 52240 USA

VanderLoop, Sarah L.
University of Cincinnati
Dept. of Civil & Environ. Engineering
741 Baldwin (ML 71)
Cincinnati, OH 45221-0071 USA

Vanneck, Peter
University of Ghent
Centre for Environmental Sanitation
Coupure Links 653
B-9000 Ghent
BELGIUM

van Veen, Johan
TNO Institute of Environmental &
    Energy Technology
Laan van Westenenk 501
P.O. Box 342
7300 AH Apeldoorn
THE NETHERLANDS

Verstraete, Willy
University of Ghent
Centre for Environmental Sanitation
Coupure Links 653
B-9000 Ghent
BELGIUM

von Fahnestock, F. Michael
Battelle Columbus
505 King Avenue
Columbus, OH 43201 USA

Ward, Owen P.
Biorem Technologies Inc.
450 Phillip Street, Unit #1
Waterloo, Ontario N2L 5J2
CANADA

Warminsky, Michael
Metcalf & Eddy
U.S. Highway, 22 West
Branchburg, NJ 08876 USA

Werners, Joan
TNO Inst. of Environmental Sciences
Schoemakerstraat 97
Postbus 6011
2600 JA Delft
THE NETHERLANDS

Yen, Ten-Fu
University of Southern California
Dept. of Civil & Environ. Engineering
224 A Kaprielian Hall
3620 South Vermont Avenue
Los Angeles, CA 90089-2531 USA

You, Guanrong
University of Cincinnati
c/o U.S. EPA M.S. 420
26 W. Martin Luther King Drive
Cincinnati, OH 45268 USA

Young, James C.
Pennsylvania State University
Center for Bioremed and Detox
Dept. of Civil and Environ. Engrg.
216 Sackett Building
University Park, PA 16802-1408 USA

Yu, Jay J.
Biorem Technologies Inc.
450 Phillip Street, Unit #1
Waterloo, Ontario N2L 5J2
CANADA

Zappi, Mark E.
U.S. Army Corps of Engineers
USAE Waterways Experiment Station
3909 Halls Ferry Road
Vicksburg, MS 39180-5904 USA

Zhang, Liu
University of Illinois at
Urbana-Champaign
Department of Agronomy
1102 S. Goodwin Avenue
Urbana, IL 61801 USA

Zhang, Ying-Zhi
Geo-Centers, Inc
762 Route 15 S.
Lake Hopatcong, NJ 07849 USA

INDEX